ELECTRONICS FOR EVERYONE

Gerald E. Williams, P. E.

SRA
SCIENCE RESEARCH ASSOCIATES, INC.
Chicago, Palo Alto, Toronto, Henley-on-Thames, Sydney, Paris
A Subsidiary of IBM

To Lenore Gibson, my mother-in-law and dear friend

Premise: Nature is consistent. What is true in one area of physics is true in all others. The differences that sometimes seem to exist merely reflect the fact that human knowledge is incomplete.

Acquisition Editor   Alan Lowe
Project Editor   Sara Boyd
Compositor   Bi-Comp
Illustrator   John Foster
Designer   Carol Harris

Cover illustrations are from The Atari Video Computer System, Game Programs: Breakout, Football, Hunt and Score, and Home Run. Used by permission of Atari, Inc.

© 1979 Science Research Associates, Inc. All rights reserved.
Printed in the United States of America

Library of Congress Cataloging in Publication Data

Williams, Gerald Earl, 1931–
  Electronics for everyone.

  Includes bibliographical references and index.
  1. Electronics.   I. Title.
TK7816.W55   621.381   79-15503
ISBN 0-574-21525-5

10 9 8 7 6 5 4 3 2

Preface

Electronics for Everyone offers a broad survey of electrical-electronic theory and concepts at a relatively low technical level. It is primarily intended for use in electronics survey courses. Although the needs and motivations of students who take such courses are varied, most seem to have one characteristic in common—a limited background in science and math. Often these students are liberal arts majors, taking the course as an elective. Or they may take the course because it is a requirement for such relatively nontechnical majors as Radio-TV Broadcasting or Audio-Visual Operations. Some students take the course because they receive science credit for successfully completing it. In many two-year colleges, a survey course also performs a screening function for prospective electronics majors.

Not all survey students are weak in math. In some cases, students majoring in another technical area (civil, mechanical, or chemical engineering, for example) are required to take a basic electronics course.

The incentive for developing this book was the dearth of texts that attempt to meet the needs of such a diverse student population. *Electronics for Everyone* has been carefully designed to serve this student mix as well as possible without overcustomizing for any particular segment.

In the past, instructors who taught this course either had to make the course too general in an effort to avoid the math or had to teach over the heads of many of their students in an effort to cover the material properly—an unsatisfactory outcome in either case. This text offers a solution to the dilemma in the form of a coherent system of analogies which describe the invisible actions that take place in an electrical circuit in terms of the visible actions of wheels, running water, and other everyday happenings.

The analogies are mathematically correct and scientifically proper. More important to the student, however, is that the analogs are expanded as new electronic ideas are introduced. It is the corresponding continuity of the analogs along with the continuity of the electronics concepts that makes the scheme work. The order in which the electronics concepts are presented is quite traditional.

Mathematics of a higher order than simple Ohm's Law is sometimes included for enrichment purposes and is carefully flagged in the text. The text is designed to be complete without this extra math.

Chapter 2 provides a review of some basic math concepts that may be helpful to the student who is weak in the area or to the instructor who wants to include a bit more math in his or her lectures. The approach to these math skills is quite different from the usual "remedial" concept.

An Instructor's Guide is available. It includes, among other things, some suggested laboratory experiments for those courses that include laboratory work.

ACKNOWLEDGMENTS

I would like to thank the following for their invaluable help.

Selby Sharp. "Selby's Simple Sayings" have, over the years, illuminated previously murky corners of my mind. Our many dialogs, which have had an enormous influence on me, are largely responsible for this book.

My wife Patty for her help in so many areas.

My son Kelly for his invaluable help with the text illustrations.

My son Geoffrey for our many computer dialogs.

The professors listed below, all of whom provided many excellent suggestions for improving the manuscript.

W. Murray Black, George Mason University

Morris Brodwin, Northwestern University

Lee W. Churchman, Allan Hancock College

Raymond F. Davidson, Texas State Technical Institute

Erwin C. Hamm, Northern Illinois University

David L. Jelden, University of Northern Colorado

Thomas J. Milleman, College of DuPage

Edward E. Pollock, Cabrillo College

M. B. Rotnem, Kent State University

My editors, Alan Lowe and Sally Boyd, for taking care of all the problems involved in turning a manuscript into a printed book.

William Huddelson, who took most of the text photographs.

Nina Guzman and Patricia Sorrels for their expert typing of the manuscript.

Gerald E. Williams

Contents

	Introduction	xi
Chapter 1	**Electronics: Its Magical Present, Its Past, and Its Science-Fiction Future**	**1**
	1-1 Overview	2
	What Is Electronics?	2
	Computers	4
	1-2 Electronics: Economics and Growth	6
	1-3 The Magical Present	10
	1-4 A Brief Look at the Past	13
	Early Discoveries	16
	Electronics Begins with Edison	20
	1-5 The Science-Fiction Future	22
	Education	23
	Around the Home	24
Chapter 2	**An Introduction to the Queen of the Sciences**	**25**
	2-1 Introduction	26
	2-2 Big and Little Numbers	26
	2-3 Scientific Notation	29
	Fractions	31
	Scientific Notation and Fractions	32
	2-4 How to Use Scientific Notation	36
	Multiplying	36
	Dividing	40
	Adding and Subtracting	40
	2-5 The Metric System	41
	2-6 Graphs	46
	Linear and Nonlinear Graphs	48
	Graph Plotting	48
	Polar and Rectangular Graphs	50
	2-7 Algebra: What Is It and What Good Is It?	51
Chapter 3	**An Introduction to Electricity**	**57**
	3-1 Introductory Comments	58
	3-2 Basic Atomic Theory	58
	Atoms and Electrons	59
	The Structure of the Atom	61
	Electrical Conduction	66
	3-3 Electrical Current	70
	Conventional Current	71
	Basic Electrical Concepts	71
	3-4 Electrical Analogs	84
	3-5 Direct and Alternating Current	87
	Direct Current	87
	Alternating Current	88

Chapter 4 Circuits with Resistors and Voltage Sources — 97

- 4-1 Introduction — 98
- 4-2 Resistors — 98
  - Wattage Rating — 99
  - The Resistor Color Code — 100
  - Resistance in Circuits — 103
- 4-3 Circuits and Analogies — 106
- 4-4 Ohm's Law — 112
- 4-5 Series and Parallel Circuits — 115
  - Series Circuits — 115
  - Examples — 118
  - Parallel Circuits — 123
- 4-6 Short Circuits and Open Circuits — 125
- 4-7 Resistors in Series and Parallel — 126
  - Values Above 100 Ohms — 133
  - Combined Series and Parallel Circuits — 133
- 4-8 Power — 136
  - Power Calculations — 138

Chapter 5 Magnetism and Magnetic Machines — 141

- 5-1 Introduction — 142
- 5-2 Magnets — 143
  - Electrons and Magnetism — 144
  - Magnetism — 145
- 5-3 Electromagnets — 148
  - Solenoids — 151
  - Magnetic Circuits — 153
- 5-4 Magnetic Machines — 158
  - The Solenoid — 159
  - The Electric Motor — 160
  - The Generator — 163
  - Fractional-Horsepower AC Motors — 167
- 5-5 Electrical Meters — 172
  - Voltmeters and Ohmmeters — 174
  - The Watt-Hour Meter — 174
- 5-6 Electrical Power Distribution — 176
  - Delivery of Power to the Home — 179

Chapter 6 Inductance and Capacitance — 184

- 6-1 Introduction — 185
- 6-2 Capacitors — 185
  - Capacitors in Series and Parallel — 188
  - Capacitor Analogs — 189
  - Counter-Voltage in Capacitors — 190
  - Time Constants — 195
- 6-3 The Capacitor and Alternating Current — 197
  - Capacitive Reactance — 200
  - Capacitive Reactance and Resistance in Series — 203
- 6-4 Inductors — 208
  - Inductor Analogs — 211
  - Inductance and Mass — 212
  - The Inductive Time Constant — 214
  - Inductance and Alternating Current — 215

6-5 Resonance	222
Resonant Circuit Applications	226
The Resonant Frequency	229

Chapter 7 Transformers   231

7-1 Introduction	232
7-2 Transformers	234
What Does a Transformer Do?	235
Energy Conservation	237
The Reflected Load Concept	246
Transformer Phasing	251
7-3 The Isolation Transformer	252
7-4 The Autotransformer	252
7-5 Transformer Failures and Testing	252
7-6 Heat Transformers	253
7-7 Antennas	255

Chapter 8 Introduction to Amplifiers   260

8-1 Introduction	261
8-2 What Is Amplification?	261
Analogs	261
8-3 Amplifier Characteristics	262
The Relay as an Amplifier	264
Amplifier Specifications	264
How the Load Affects Amplifier Gain	268
Input Resistance	269
Inverting and Noninverting Amplifiers	269
An Analog Relay Amplifier	269
Gain in Analog Amplifiers	271
Bias	271
8-4 Feedback	274
An Analogy	275
Feedback in Amplifiers	277
Input Resistance	278
Back to the Relay Amplifier	279
8-5 The Operational Amplifier	281
Characteristics of Operational Amplifiers	281
8-6 Semiconductors	285
Silicon Crystals	287
Conduction in Silicon	288
8-7 The PN Junction Diode	291
The Depletion Zone	291
The Reverse-Biased Junction	294
The Forward-Biased Junction	295
Junction Capacitance	296

Chapter 9 Electronic Amplifying Devices and Circuits   298

9-1 Introduction	299
9-2 The Junction-Field-Effect Transistor (J-FET)	299
9-3 The Metal-Oxide Field-Effect Transistor (MOS-FET)	301
The Enhancement-Mode MOS-FET	304
9-4 Vacuum Tubes	304
The Diode Vacuum Tube	305
The Triode Vacuum Tube	307

9-5	The Bipolar (Junction) Transistor	309
	Transistor Construction	311
	Transistor Amplification	312
	Reverse-Polarity Devices	313
	Bipolar Transistor Parameters	313
	Practical Bipolar Transistor Circuits	318
9-6	The Practical J-FET Amplifier	325
9-7	The Differential Amplifier	325
9-8	Switching Amplifiers	326
	Thyristors	328
	Triacs	334
	Unijunction Transistors	334

Chapter 10 Amplifier Systems — 337

10-1	Introduction	338
10-2	Amplifier Classifications	338
	Audio-Frequency Amplifiers	340
	Distortion	342
	Noise	347
10-3	Oscillators	349
	Relaxation Oscillators	350
	Feedback Oscillators	353
	Crystal Oscillators	355
	Phase-Shift Oscillators	358
10-4	Modulation	361
10-5	Amplifier Systems	363
	Heat Sinks	365
	Amplifiers in Radio Receivers	367
10-6	Power Supplies	372
	Regulators	376
	Dual Power Supplies	378
10-7	Switching-Amplifier Circuits	378

Chapter 11 Energy Conversion Devices for Electronic Systems — 382

11-1	Introduction	383
11-2	Lasers and Light Sources	383
	Spontaneous and Simulated Emission	385
	Lasers	386
	Incandescent Lamps	387
	Fluorescent Lamps	389
	The Ruby Laser	392
	Gas Lasers	393
	Semiconductor Lasers and LEDs	393
11-3	Converting Light into Electrical Current	394
	The Silicon Solar Cell	395
	The Photoemissive Cell	397
	Photodiodes and Transistors	397
	Photoresistive Cells	398
11-4	Sound Conversion Devices	398
	The Carbon Microphone	400
	The Crystal Microphone	401
	The Dynamic Microphone	401
	The Loudspeaker	402
	The Velocity or Ribbon Microphone	402

The Capacitance Microphone	402
Phonograph Pickups	405
11-5 Thermoelectric Devices	405
Thermistors	405
Thermocouples	405

Chapter 12 The Recording and Reproduction of Sound — 407

12-1 Introduction	408
12-2 The Decibel	408
12-3 The Great Audio Power Debate	411
12-4 Acoustical Transformers	414
Loudspeaker Enclosures	416
Acoustics	418
Acoustical Feedback	418
12-5 Recording	420
Disc Recording	420
Magnetic-Tape Recording	422
Optical Recording	423
12-6 Interconnecting Audio Equipment	423
Shielded Cable	424
12-7 Electronic Reverberation	425

Chapter 13 Television — 429

13-1 Introduction	430
13-2 How Television Pictures Are Made	430
The Picture Tube	431
The Color Picture Tube	432
The Television Camera	434
Scanning the Picture	437
Synchronization	440
13-3 The Television Receiver	441

Chapter 14 Digital Electronics — 447

14-1 Introduction	448
14-2 What Is Digital Logic?	449
Digital Integrated Circuits	450
Integrated-Circuit Classifications	451
Logic Families	452
The Hierarchy of Digital Systems	453
14-3 Electronic Logic Gates	455
Truth Tables	455
The AND Gate	455
The OR Gate	456
The Inverter	456
The NAND Gate	457
TTL Logic Gates	457
14-4 Boolean Algebra	462
14-5 Numbers for Digital Systems	463
The Binary Number System	463
Hexadecimal	465
Octal	467
Binary-Coded Decimal (BCD)	468
14-6 Making Logic Gates Do Something	468
The Exclusive-OR Gate	469

14-7 Flip-Flops	471
Clocked Flip-Flops	473
The Type T Flip-Flop	473
The J-K and Type D Flip-Flops	474
Flip-Flop Applications	474
Counters	475
Shift Registers	475
Decoders	475
Multiplexers	476
14-8 Memory Systems	478
Magnetic-Core Memory	479
Punched Cards and Tape	479
Magnetic Recording	479
Random-Access Memory (RAM)	480
Read-Only Memory (ROM)	482
Bubble Memory	484
Charge-Coupled Memory (CCD)	484
14-9 Other Digital Subsystems	485
Clocks	485
Programmable Logic	485
The Arithmetic Logic Unit (ALU)	486
Software	486
Firmware	486

Chapter 15 Computers — 487

15-1 Introduction	488
15-2 The Bus	489
15-3 How the Microcomputer Works	490
The Memory	494
The Address Decoder	494
The Arithmetic Logic Unit	495
The Accumulator	495
The General Register	495
The Program Counter	495
The Instruction Register	496
The Instruction Decoder	496
The Clock	496
Peripheral Equipment	496
Peripheral Interface Devices	496
15-4 A Computer Processing Example	497
15-5 Programming Levels	513
Microprogramming	515
Machine-Language Programming	515
Assembly Language	516
Procedure- (or Problem-) Oriented Languages	516
Selecting a Language Level	517
15-6 Mini and Large Computers	518

Answers to Odd-Numbered Problems — **519**

Index — **527**

Introduction

A colleague of mine once defined specialists as "Those who learn more and more about less and less, until they know everything about nothing." The breathtaking march of technology in this century has created an army of specialists and nearly a total void of generalists. In the future though, whether they engage in business or science, people will have to be generalists. We must all live with the complexities of technology and we must all suffer the consequences of its misuse. Just as war is too important to leave to the generals, technology is too important to leave to the experts.

The first industrial revolution reached its peak with the internal combustion engine in all of its forms, and the consequences of that technology are upon us. Now our entire economy is dependent on the internal combustion engine. We find ourselves wondering how long we can continue to fuel it and how long the environment can tolerate the pollution it produces. We have bowed to the "experts," protesting that it was impossible for us to keep up with all facets of technology. Yet there are areas of technology that have such an impact on our lives that we cannot afford to be ignorant of them.

The second industrial revolution is now well along. It is a revolution of electronics—or, more accurately, the communications and the control of our environment that electronics makes possible, a revolution so quick and sweeping that we hardly realize its impact. There are unmistakable signs that television, computers, and other electronics technologies are altering our physical and sociological environment as dramatically as the internal combustion engine once did.

Electronics may help to solve the problems of environmental pollution and excessive energy use caused by the internal combustion engine. We can transport our image, ideas, graphs, pictures, and drawings across the country with a millionth of the energy required to transport physical objects, without generating environmental pollution. The technology exists and at an affordable price.

Can we even guess at the total impact of 10-dollar computers, or robots cheap enough to do everybody's housework! Such ideas used to be science fiction, but they are quickly becoming reality.

This book has been written for generalists. The text is intended to provide enough technical background to replace much of the mystery of electronics with knowledge. One course will not make you an expert, but it can help to make you a generalist and provide the foundation for future study.

We shall make frequent use of analogies in this text. Webster defines analogy as: "In logic, a form of inference in which it is reasoned that if two (or more) things agree with one another in one or more respects, they will (probably) agree in yet other respects." One of the most remarkable things about nature is its consistency. It is reasonable that what is true in mechanics, or acoustics, or heat systems, is also true in electronic systems. Indeed, the evidence is overwhelming that the inference is correct.

Mathematics has been called the *Queen of the Sciences*, but it might more appropriately be called the *Queen of Analogies*. With the Queen of Analogies at their command, scientists rarely use lesser analogies. But those of us who do not have full command of mathematics can find "lesser" analogies as useful for our purposes as mathematics is to the scientist.

One obstacle in learning electronics, particularly without math, is the fact that our senses are nearly useless in aiding comprehension. We cannot see electrons or hear them "whish" through circuits. In short, our senses, which provide us with most of our knowledge about the world, are not much help. Fortunately, through analogy, we can relate electron behavior to things that our senses have been experiencing since birth. We all have seen wheels turn, gears mesh, levers move, and water flow. If we can accurately describe electronic behavior in familiar terms, we have a helpful tool, less powerful and bearing less authority than mathematics, but a tool just the same.

When you slide down a hill, friction warms the seat of your pants and slows your descent. By analogy you already have an intuitive understanding of electrical resistance. If you are also aware that gravity is a force that is pulling you down the hill, then by analogy you have some understanding of the electrical force we call voltage. Further, if you are aware that your body is in motion down the hill, then you also have some knowledge of what electrical current is. This example will be expanded upon later in a more serious way, and enough detail will be added to ensure understanding.

Although we won't attempt to do it, the validity of the analogies we will use in this book can be proved mathematically. We will use careful and accurate analogies whenever they are an appropriate aid in extending the familiar into the realm of the unfamiliar.

Electronics is a Johnny-come-lately science that tends to frighten beginners because of its seeming complexity. In fact it's easier to learn electronic principles than to learn mechanical principles.

This book is designed to provide a greater depth than is normally possible with only a slight involvement in mathematics. The text is written with a light hand—a spoonful of sugar with the medicine, so to speak.

Chapter 1

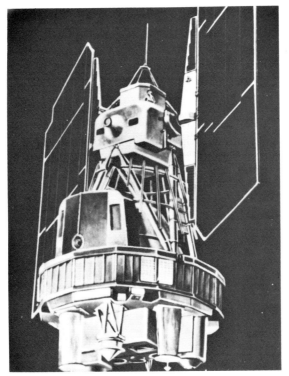

Courtesy of NASA

ELECTRONICS: ITS MAGICAL PRESENT, ITS PAST, AND ITS SCIENCE-FICTION FUTURE

1-1 OVERVIEW

The earliest common applications of electronics were in the field of radio. In 1904 the English engineer John Ambrose Fleming used Edison's discovery that electrons "boil" off a hot wire in constructing his fleming valve, the forerunner of the vacuum tube.

In 1907, Lee DeForest, an American inventor, added a control element to Fleming's valve. This new 3-element device was capable of taking a signal that was too small to be of much use, and producing an exact, but much more powerful replica of it. With the creation of this amplifying device, practical electronics was born.

WHAT IS ELECTRONICS?

Electronics is the technology of controlling electricity—to perform tasks that are difficult, time-consuming, or impossible for people. It is impossible, for example, to transmit or receive radio or television signals without technological assistance. On the other hand, we can do arithmetic with pencil and paper; it is faster and easier to use an electronic calculator.

All of the common electrical devices such as capacitors, coils, motors, generators, switches, lamps, and resistors, are part of electronics. Electronics has become a special technology, beyond the realm of electricity, because of amplifying devices.

Vacuum tubes, transistors, and a variety of special semiconductor devices transform electrical circuits into electronic circuits. These devices not only permit small quantities of electricity to control much larger quantities of electricity, but they also operate much faster than the mechanical or electromechanical devices common in electrical systems.

An ordinary light switch can be used to switch a light on and off at speeds of a few cycles per second. Electromechanical switches can extend that rate to several hundred on/off cycles per second. Electronic amplifier devices can extend the switching rate to thousands of millions of on/off cycles per second. Furthermore, an electronic switch can be actuated by small signals from a distant transmitter, the output voltage from a tiny photoelectric cell, or other minute electrical signals.

Electronics is divided into two broad categories: *digital* and analog (analog is often called *linear*). *Digital* devices are switching devices that switch current on and off. They are used in computers, calculators, and other devices. *Analog* or *linear* devices are continuously variable, like household light dimmers which can vary brightness continuously from very dim to full brightness, as opposed to simply turning lights on or off. Continuously variable devices are called *analog* or *linear*. At one time there was a sharp distinction between these categories, because a satisfactory digital simulation of a continuously variable (analog) device using electronic switches was extremely complex and expensive. Now it is possible to *integrate* a number of digital switches into a circuit that simulates an analog circuit. These integrated circuits once were expensive; now they're dirt cheap.* Many tasks that once required analog circuits are now being performed by integrated digital circuits.

A good example is, again, the solid-state lamp dimmer. Lamp dimmers using analog (linear) devices have been too expensive, but inexpensive semiconductor switching devices have been developed in recent years. Such devices provide a good simulation of continuous control and are far cheaper than analog devices for household applications. Using them, it only *seems* that we are causing a continuous variation of lamp brightness when we turn the knob on a lamp dimmer. The electronic device inside the box is actually turning the light on and off at a rate of 60 to 120 times per second. The ratio of *on* time and *off* time controls the apparent brightness of the lamp. Our eyes do not respond fast enough to see the light go on and off, so we see only a variation in brightness. Motion pictures take advantage of this same characteristic of the human eye.

Speech and music systems have long been the special province of linear amplifiers; but the digital simulation of analog high-fidelity music systems and public-address systems is already a reality. They will probably soon become less expensive than traditional analog (linear) high-fidelity and public-address systems. Digital synthesis of music and even human speech is now possible.

Any kind of information can be transmitted and received digitally, if there are a sufficient number of small pieces involved. Photographs in a book or magazine are essentially digital. We see what seems to be a continuous spectrum of colors and brightness levels, but we are actually looking at many evenly spaced dots, arranged in such a way that our eyes are deceived.

Information can be used by both people and machines equally well in either digital or analog form. Until recently analog electronics was relatively inexpensive; the same information in digital form required prohibitively expensive electronics.

In recent years the pendulum has swung to the opposite extreme. Digital information transmission has become far cheaper and far

---

*Dirt cheap is a good expression here, because modern integrated circuits are made of silicon, the principal element in the sands of all of the deserts and beaches of the world.

more reliable than analog. The U.S. space program used digital electronics to transmit pictures (and other data) from outer space. Furthermore, telephone company research in verbal communications has demonstrated that the digitizing of telephone communication lines is more efficient.

COMPUTERS

To most people, the word *computer* conjures up either the idea of large-scale data-processing systems or scientifically oriented machines spitting out numbers and equations. Only a few years ago this was a valid view of computers. They were expensive and affordable only for large-scale business data handling or well-funded government projects. Now integrated-circuit technology has made computer power available at small cost for such trivial applications as electronic games played on a home television screen.

The first industrial revolution began when steam and gasoline engines took over much of the work formerly done by humans and horses. Norbert Weiner and others prophesied a second industrial revolution as automation began to take over many of the human chores involved in operating industrial machines. To labor, this trend presented the frightening prospect of massive unemployment. This fear still exists, but automation has moved slowly enough to allow workers to shift into jobs that build the machines and keep them going.

The relatively slow march of automation is closely tied to computer technology. Most of industry's robots are "dumb" robots—machines with almost no brain power. Until the last few years, computers were so expensive that only elaborate, high-priority automatic machines would be computer-controlled. Inexpensive microcomputers are rapidly changing that picture. Microcomputers can now be used economically to control traffic signals, automotive equipment, and many other systems where only a very cheap computer can be justified. A $30,000 computer to control the fuel-air mixture and ignition in a $5,000 automobile is out of the question; a $50.00 computer, though, is a practical and economical means of increasing gas mileage and reducing harmful emissions.

The impact of integrated circuitry on the communications industry may change our lives significantly. Prediction at this point is risky, because social conditions change much less rapidly than technology. However, increased communication capacity may well decentralize industry; instead of transporting people from one place to another we may simply transport images, speech, charts, and other information. The need to conserve limited energy supplies may force our

ELECTRONICS: ITS PRESENT, PAST, AND FUTURE 5

FIGURE 1-1. A modern personal microcomputer, the Radio Shack TRS-80. (Courtesy Radio Shack)

FIGURE 1-2. Computer terminal with graphics capability. (Courtesy Tektronix, Inc.)

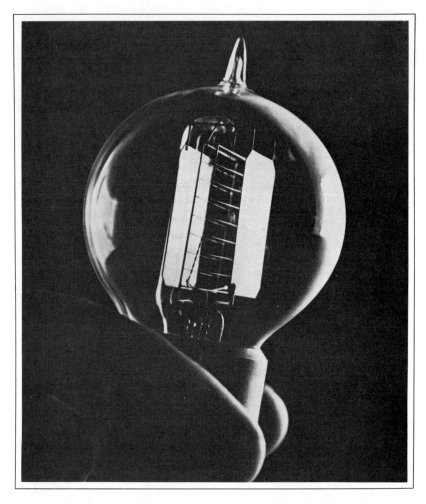

FIGURE 1-3. Western Electric's first commercially successful vacuum tube was used in 1915 in the amplifiers that made possible the first transcentinental telephone call. (Reproduced with permission of A.T.& T. Co.)

society to make some changes. It takes several hundred thousand watts of power to transport a person from California to New York, but only a few thousandths of a watt are needed to transport a person's ideas, image, and information over the same distance.

1-2 ELECTRONICS: ECONOMICS AND GROWTH

With the help of digital or analog amplifying devices a small electrical signal can control the spots of light on a TV screen that form moving pictures, control the course of a space vehicle, or the speed of a 1000 horsepower industrial motor. The applications are endless. The control of power by electrical signals—amplification—is what electronics is all about. Amplifying devices, vacuum tubes, transistors, and many other devices have expanded our use of electricity beyond the wildest dreams of early electrical pioneers.

Electronics was still moving at a comparative crawl in 1948 when Doctors William Shockley, John Bardeen, and Walter Brattain invented the transistor at the Bell Telephone Laboratories. They received a Nobel prize for the invention. The transistor (the solid-state equivalent of the vacuum tube) is more efficient, smaller, and cheaper to make than a vacuum tube. Transistors 2 to 3 thousandths of an inch in diameter by 20 millionths of an inch thick are commonplace in modern electronics. Such small size permits circuit densities millions of times greater than had been possible with vacuum tubes.

The U.S. space program provided the incentive to put thousands of transistors on a small wafer of silicon ($1/8$ inch to $1/4$ inch square) and interconnect them to form miniature electronic systems. These integrated circuits have revolutionized every aspect of electronics. Pocket calculators, computers in a package the size of a large postage stamp, and color television receivers with most of the circuitry in 3 or 4 integrated-circuit packages are becoming commonplace. Personal computer systems are creating a mushrooming hobby in computer use, as well as radically changing the way in which more serious computing is handled.

The integrated-circuit revolution is largely an economic one. Many electronic systems that once were extremely expensive have become quite cheap. Twenty years ago who would have foreseen the personal electronic calculator as a consumer item? A machine with that kind of computing power would have cost thousands of dollars, and would not have been pocket size.

The little slice of silicon that forms the heart of an integrated circuit is called a *chip*. At the time of this writing a computer central-processing unit on a chip costs about 10 dollars. Hardware with the same computing power in 1960 would have cost at least $30,000. Figure 1-5 is a graph showing the cost pattern of this chip between 1975 and 1980, where the cost levels off to about $5.00 per unit.

The economic impact of integrated circuits has been more dramatic in computers and calculators than it has in other areas of electronics because these systems are very complex, and the cost savings were obvious from the beginning. Developing the fabrication processes for these chips was expensive; the cost was more easily justified for such complex systems. Now that these processes have

FIGURE 1-4a. Large-scale integrated circuit on a printed circuitboard. (Courtesy Telesensory Systems, Inc.)

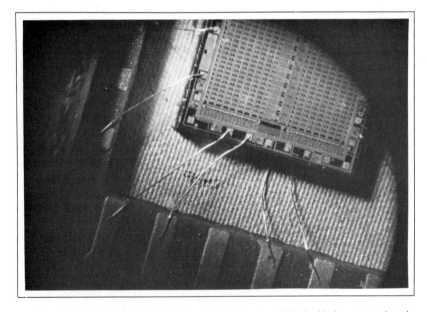

FIGURE 1-4b. A photomicrograph showing how the integrated-circuit chip is connected to the outside world.

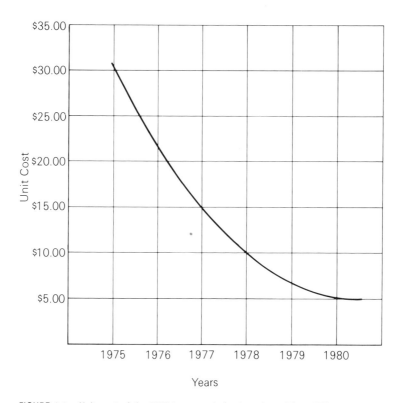

FIGURE 1-5. Unit cost of the 8080 integrated circuit projected into 1980.

become routine, other areas of electronics can have an almost free ride in converting their less complex hardware to integrated-circuit form.

Figure 1-6 summarizes the current electronics economic picture, with a total dollar value of more than 50 billion, and rising. Electronics sales exceed General Motors sales by 6 billion dollars. The computer industry alone accounts for over 6 billion dollars, and that figure can be expected to rise dramatically as computers become even more established as a consumer item.

Other technological advances such as the laser and fiber optics will increase the demand for electronics. With laser light sources and optical fibers instead of telephone wires it is possible to transmit hundreds of thousands of telephone conversations over an optical fiber no larger than a conventional telephone wire. Fiber optics can also carry television pictures, which can now be carried only over complex and very expensive coaxial cables (one or two channels at a time) or over microwave relay systems. Because they are designed for

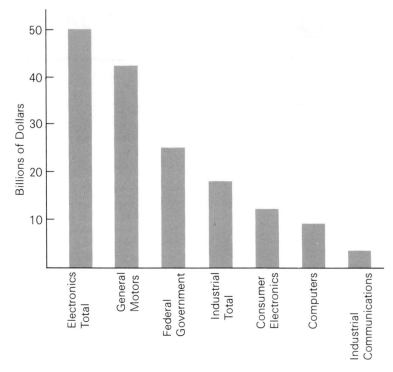

FIGURE 1-6. Electronics dollar volume. General Motors is included for comparison. (U.S. Government statistics, 1976–77.)

transmitting human speech, telephone lines cannot be used for television pictures. Computer data can be transmitted over telephone lines, but only at a very slow rate. Laser fiber-optic systems can handle both TV pictures and high-speed data transmissions.

The use of optical fibers could increase the Bell system's information-transmission capabilities a million-fold in a few years. This increased communication capability will require an enormous amount of electronics, both as part of the system and at the telephone user's location. With modern integrated circuits, customer electronic equipment will be easy to afford for both business and personal use.

1-3   THE MAGICAL PRESENT (A SAMPLING)

Electronics is so much a part of life that we tend to pay little attention to these wonders of technology. Color television, tape recorders, stereo systems, and the like are old hat. That wonder of mathematical wizardry, the pocket calculator, already seems to have been with us

ELECTRONICS: ITS PRESENT, PAST, AND FUTURE 11

FIGURE 1-7. The laser, one of the most revolutionary developments in modern physics. (Reproduced with permission of A.T.& T. Co.)

forever. Pocket calculators are so inexpensive they have become disposable items. Yet, not very many years ago such a powerful computing machine would have been far too expensive for any but the wealthy.

Suppose we leave the more common devices for the moment and take a look at some of the far-reaching recent developments in electronics technology.

The computer is more than a data processor. Car manufacturers have already ordered microcomputer systems for on-board control of ignition timing and fuel mixtures, anti-lock braking systems, and nearly every other automotive system. The automotive computer can also be used to diagnose potential troubles, and predict failures before they are dramatically called to your attention some dark night on the freeway. Anti-collision and automatic steering systems are also possible; the necessary computer power is already here.

Microcomputers are used in everything from home appliances to sophisticated laboratory, medical, and industrial instruments. With

FIGURE 1-8. A talking calculator for the blind. (Courtesy Telesensory Systems, Inc.)

the aid of microcomputers, formerly "dumb" machines are exhibiting a surprising degree of "intelligence." In medicine, electronically controlled pacemakers, diagnostic equipment, and continuous biological monitoring systems are becoming commonplace.

Microcomputers are waiting in the wings to control the electrical and chemical processes for artificial organs as soon as the mechanical and biological problems are solved. Talking calculators and devices that can read a printed page aloud or translate it into braille are available for the blind. Artificial voice boxes are available for those who cannot speak.

Paralyzed patients can type, or control any device that is electrically actuated, with nothing more than eye movement. And computer-based devices exist that can translate human speech into a printed display on a TV screen, but so far they are too expensive and too bulky for individual use.

Satellite communications have put us in instant touch with people and events everywhere. Electronics helped make it possible to explore the moon and Mars. Electronic typesetters speed the publishing of the enormous volume of printed information demanded by our high-technology society.

FIGURE 1-9. Electronic technicians at work checking out a communication satellite. (Reproduced with permission of A.T.& T. Co.)

1-4 A BRIEF LOOK AT THE PAST

Figure 1-11 summarizes the history of electronics. Basic electrical knowledge was developed in the first 90 years, culminating in the telegraph. The telegraph was chosen to mark the end of the era because it was the first electrical communication device. The history of electronics is largely the history of communications. If we take a

FIGURE 1-10. Telegraph sounder.

broader view of communications that includes communications between machines or between parts of machines, electronics and communications are nearly synonymous terms.

The most significant information in Figure 1-11 is the time required to develop technology from basic discoveries. The first 90 years yielded no more sophisticated technology than the telegraph, electric motors, and generators. Telephone technology, which began as early as 1820 with Oersted's experiments with electromagnets, came to fruition 56 years later in 1876. In the case of radio, progressing from discovery to technology required only 15 years. Transistors went from the laboratory to industry in about 3 years. Integrated microelectronic circuits took a bit longer—6 years. Their development involved not only very sophisticated electronics, but also some of the most difficult metallurgical and automated process technology ever attempted. Less than 3 years were required to reduce a full-size computer to a package the size of a pack of cigarettes. The microprocessor on a chip (or on a few chips, depending on the system) emerged from a number of years of computer technology and a very few years of experience with integrated microcircuits.

Integrated circuits are made almost entirely by automated processes, at low labor cost. Although highly purified silicon is fairly expensive, very little is needed to make an integrated circuit. Many well established and proven integrated circuits can be bought in quantity for pennies each, including packaging. The major cost is a one-time design and art-work cost. Even this task is now being relegated to computers.

There is also a spectacular spin-off factor in modern technology. Good examples are the development of light-emitting diodes, common in calculator displays, and semiconductor laser diodes. Both of these

ELECTRONICS: ITS PRESENT, PAST, AND FUTURE 15

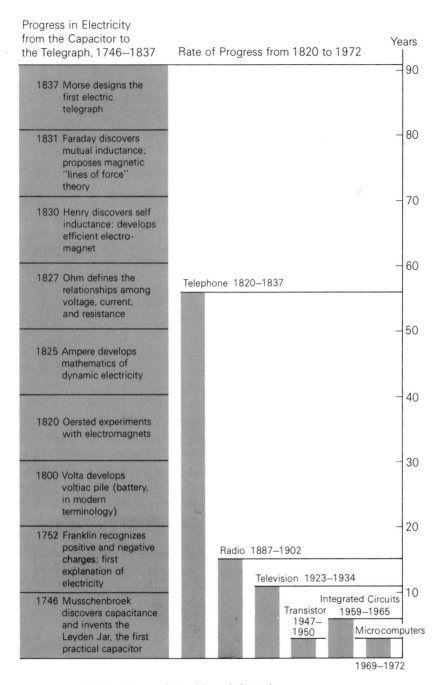

FIGURE 1-11. Graphical history of electricity and electronics.

show great promise for use with optical-fiber communication lines, and are a direct result of knowledge gained in the manufacturing of transistors and integrated circuits. Solar power cells are also a silicon technology spin-off.

As fast as electronics is growing now, we can expect even more rapid growth in years to come. In this brief summary of electronics history, let us examine the major advances in the first years of experiments with electricity, and some of the pioneers responsible for them.

EARLY DISCOVERIES

The history of electrical current properly begins in about 1715 with Pieter Van Musschenbroek and his invention of the Leyden jar, in Leyden, Holland. The Leyden jar was the first device that could store a significant quantity of electricity and deliver a useful current for any length of time. All earlier electrical experiments had involved static electricity with currents too small to measure or to do any real work. The Leyden jar permitted experiments with dynamic electricity, the kind of electricity that relieves humankind of so much of its drudgery. The Leyden jar was the precursor of the modern capacitor. (Units of capacitance were later named for Joseph Faraday, perhaps because it was too difficult to ask for a 10 micromusschenbroek capacitor.)

In Philadelphia, Benjamin Franklin (1706–1790) proved that a lightning discharge is electricity. His famous, but potentially fatal,* kite-and-key experiments led to the invention of the lightning rod, which was quickly adopted for the protection of munitions stores. In addition to the rebel Franklin's other disagreements with King George III, the two men were drawn into a quarrel over whether lightning rods should be pointed or tipped with knobs. King George went so far as to try to convince the President of the Royal Society, Sir John Pringle, to make *knobs* the society's formal position on the matter. Sir John stood firm with the reply, "Sire, I cannot reverse the laws of nature." Franklin, who was then in France seeking French help in breaking the American colonies away from the King's rule, expressed the hope that the King would bring the thunder of heaven upon himself by dispensing with lightning rods altogether.

Franklin's most important contribution to the knowledge of electrical current was his concept of electrical fluid. His was the first satisfactory, experiment-based explanation of electricity. He origi-

---

*Many experimenters were killed trying this experiment, including some experienced researchers.

nally theorized that there were two kinds of moving charges: one positive and one negative. His theory has proved to be essentially correct. We know now that only the negative charges are usually free to move. The positive charges are captive within the nucleus of the atoms and move only if the entire atom moves. His theory led to practical experiments that otherwise would not have been made so early.

In Italy, Luigi Galvani (1737–1798) was experimenting with the biological effects of electricity on muscles in frog legs. Galvani attributed the twitching muscles to a special form of animal electricity.

Meanwhile the Italian physicist, Alessandro Volta (1745–1827), interpreted Galvani's experiments as a simple chemical reaction. He demonstrated that the source of electricity was the contact between two different metals in a solution. His investigations, without benefit of the frog, led to the development of what we now know as the battery. Volta's device provided, for the first time, a *continuous* source of electrical current, which was essential to further electrical research. The practical unit of potential difference, the volt, was named for Volta.

Another development was even more important to the growth of the electrical industry. In 1820 the Danish physicist, Hans Oersted (1777–1851), discovered that a magnetic field surrounded a current-carrying conductor. Almost immediately afterward, Andre Ampere (1775–1836) determined the relationship between electrical-current intensity and the strength of the magnetic field. The practical unit of electrical current was named for Ampere.

In 1831, eleven years after Oersted's discovery, the English chemist Michael Faraday (1791–1869) first used magnetic fields to produce mechanical motion. Faraday's research laid the foundation for the development of generators, motors, and most of the other electromechanical devices we know today. Faraday caused a wire carrying an electric current to circle around a fixed magnet. Though not practical, it was the first electric motor. Practical versions soon appeared. Also, because almost any electric motor will generate electricity when the shaft is rotated, Faraday had also pointed the way toward practical generation and distribution of electric power. The practical unit of capacitance, the farad, was named for Faraday.

In 1830, a year before Faraday's "lines-of-force" theory of magnetism and his discovery of the relationship between moving magnetic fields and electrical current, Joseph Henry (1797–1878) had developed a highly efficient electromagnet. In 1831 Henry devised the first telegraph and demonstrated its effectiveness at a distance of over a mile. Henry took out no patent on his invention nor did he try to commercialize it. His discovery of self-inductance, though it was not immediately exploitable, is of vital importance to electronics and the electrical power industry. The practical unit of self-inductance, the henry, was named for Henry.

The beginning of the communications industry was initiated by Samuel F. B. Morse (1791–1872), artist and promotor. He met Henry by accident and, using Henry's freely provided ideas, designed and patented the telegraph and developed his well-known telegraph code. He performed the minor miracle of persuading the reluctant U.S. Congress to invest 30 thousand dollars in a 40 mile experimental telegraph system between Washington, D.C., and Baltimore, Maryland. The later invention of the telephone proceeded directly from the telegraph receiver or sounder, and Bell's personal discussions with Henry, the telegraph's original inventor.

Now it is necessary to go back 10 years before Morse to examine George S. Ohm's contribution. Ohm (1787–1854) was a German physicist at Cologne University when he formulated the mathematical relationships among voltage, current, and resistance in an electrical circuit. Ohm's law is one of the first things a student of electronics learns. Nevertheless, Ohm's "law" was so ridiculed by his colleagues at Cologne that he resigned his position and lived in relative obscurity from 1827 to 1833. His work then began to be recognized as valid and important. Part of Ohm's problem with his colleagues rested on the fact that his theory was developed largely by analogy from Fourier's research on heat conduction. Ohm saw an analogy between temperature difference and potential difference, equated heat flow with current flow, and electrical conduction with heat conduction. Ohm's colleagues failed to see the significance of analogies in natural phenomena.

Ohm also made a number of advances in the field of acoustics, again leaning heavily on analogies with better-understood physical systems. Acoustical engineers continued to rely on analogies in solving acoustical problems until recently, when powerful computers became easily available. The practical unit of resistance is named for Ohm.

Alexander Graham Bell (1847–1922) is generally given credit for the invention of the telephone. However, interest in the idea was keen at the time, and many of Bell's contemporaries were at work on similar devices. Professor Elisha Gray filed with the U.S. Patent Office only a few hours after Bell's patent application was submitted, and some 600 lawsuits were filed contesting Bell's patent rights.

Bell's interest in sound dated back to his early childhood in Edinburgh, Scotland. His father was a pioneer in teaching speech to the deaf, a task that Alexander also took up.

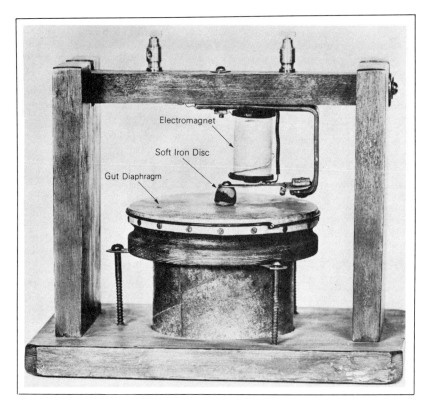

FIGURE 1-12. Bell's first telephone, 1875. (Reproduced with permission of A.T.& T. Co.)

Tuberculosis took the lives of Alexander's two brothers and ultimately drove the family to Canada. In 1871 Bell came to the United States, where he became a professor of vocal physiology at Boston University.

Bell's telephone receiver was an improvement on the electromagnetic telegraph sounder. It had a flexible diaphragm in place of the rigid sounding bar. The modern telephone receiver is a much improved version of Bell's original. The transmitter in Bell's telephone had a diaphragm that moved a metal cup inside a larger cup that held an acid solution. In this way, speaking into the diaphragm affected the cup of acid, which varied the current flow to the receiver. In modern telephones acid has been replaced by a pile of carbon granules. The carbon microphone was invented by Hughes in 1878 and later improved by Edison. Telephone technology was used in early radio microphones and earphones. Loudspeakers were made by attaching a metal horn to a telephone-type receiver.

FIGURE 1-13. William Shockley, Walter Brattain, and John Bardeen (left to right), inventors of the transistor. (Reproduced with permission of A.T.& T. Co.)

ELECTRONICS BEGINS WITH EDISON

An experiment performed by Thomas Edison (1847–1931) planted the seed that began the electronics industry. In 1883, in an effort to end the problem of vaporized metal coating the inside of his incandescent lamp, Edison introduced an extra electrode into the lamp. He observed that a sensitive current meter connected between the filament and the extra electrode indicated a current flow. Edison had no explanation for the phenomenon, nor could he see any application for it. Being practical, he patented it anyway.

Sir John Ambrose Fleming (1849–1945) became acquainted with the "Edison Effect" while working as a consultant to the Edison and Swan Lighting Company in London. He was able to obtain some of the Edison lamps with the extra electrode. A few years later, while working with Marconi and the newly developed wireless telegraph, he concluded that the Edison Effect could be used as a detector for the wireless. What was needed was a sensitive one-way electronic "valve." The "cat whisker" galena crystal and "coherer" detectors

FIGURE 1-14. Lee De Forest. (Reproduced with permission of A.T.& T. Co.)

of the time required fairly strong signals. A more sensitive detector would extend the practical distance of the wireless. Fleming built a modified version of the two-element Edison lamp. This first vacuum tube was called the *fleming valve;* in England, vacuum tubes are still called *valves*.

The story of the broadcast industry begins with Lee De Forest (1873–1961) and his Audion, the first practical amplifier. The Bell Telephone Company had previously developed a magnetic amplifier, but it was large, heavy, costly, and not very effective. De Forest's Audion produced 50 times the amplification of magnetic amplifiers, and it was compact and cheap. De Forest added a third element, called a *grid*, to Fleming's valve. This tube was capable of controlling not only the direction but also the quantity of electron flow by using a much smaller control signal on the grid. De Forest went into the broadcast business and later applied his Audion, now known as the triode, to the development of talking motion pictures. The vacuum tube remained the backbone of electronics for almost 50 years.

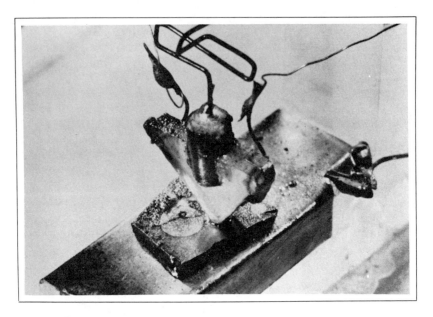

FIGURE 1-15. The first transistor. (Reproduced with permission of A.T.& T. Co.)

In 1948 Bell Laboratories announced that Doctors John Bardeen, William Shockley, and Walter Brattain had invented the transistor. This miniature replacement for the vacuum tube began another technological revolution. By 1965 it was possible to place hundreds of transistors in a package no larger than a single transistor, and another industrial revolution exploded into being. Within 10 years the integrated circuit had revolutionized the field of computers and communications and had invaded every other area of electronics. Integrated circuits containing over 100,000 interconnected transistors in a package the size of a stamp are now common. The power requirement for an integrated circuit with 100,000 transistors is just a little over 1 watt. The same system built around vacuum tubes would require nearly half a million watts of power to operate.

1-5 THE SCIENCE-FICTION FUTURE (SOME SPECULATION)

Before long, many of us may work at home or in neighborhood offices shared by a number of firms. Each site will have visual, audio, printed, and graphic communications with other offices. We may be able to live in California or Nebraska and work in New York without leaving our home town.

Much manufacturing work may be done by semi-intelligent robots under the remote supervision of human directors. Not all manual jobs would be eliminated, and people displaced by automation could program and direct robots. Programming and so on would not necessarily be the highly skilled task it is now. There would be many levels of technology direction, ranging from the remote driving of a semi-intelligent forklift, for example, to the design and implementation of new manufacturing processes. Verbal commands to computers and robots will probably be commonplace.

Eventually, perhaps 50 percent of the available jobs will cease to depend on geographic location. We could continue to live in our home town and find work anywhere in the nation. Businesses and manufacturers would have greater freedom to choose locations. They would no longer be dependent on local labor and educational resources. We could also free-lance, working at one location for firms scattered wherever the work happens to be.

EDUCATION

Individual education for the lower grades may be available at home. Youngsters would spend less time at the local schools, easing overcrowded classrooms and making local schools less dependent on local resources. College students could get their general education at local colleges while taking advantage of the best available courses in specialty areas. Handicapped children and adults would have access to the best special programs no matter where they live.

AROUND THE HOME

Robots could take over much of the drudgery of home maintenance. Each Tuesday, for example, a semi-intelligent robot would come out and, weather permitting, mow the lawn. If the robot encountered an obstruction, it would simply work elsewhere on the assumption that the obstruction might be human or animal and would be likely to move, given a little time. When it came back later, it could perhaps make a decision to either move the obstruction out of the way if it is still there, or call it to human attention for further instructions. Robots might also take care of household chores. The robots would have on-board computers, but could also communicate with people or the central household computer for help with problems beyond its capabilities. Robots could take care of many repetitive chores under the direction of the central household computer.

Video and audio recorders will have no moving parts. Bubble memories or charge-coupled memories (or something new) will store

TV programs and music in little blocks of plastic. You will also be able to store "classroom" notes, correspondence, documents, graphs, pictures, and so on for future playback.

These speculations are based on existing technology. Nothing we have mentioned requires any technological breakthroughs. The mobile household robot is the most speculative, but it is at least crudely possible with existing technology.

Whether these things will come to pass depends more on economic and social factors that on electronic technology. Questions about artificial intelligence and its impact on the study of human psychology, and the social implications of future technology need to be given serious thought. Speculations about the future of technology and its human impact should not be left solely to science fiction. That future will be here very soon indeed.

Chapter 2

AN INTRODUCTION TO THE QUEEN OF THE SCIENCES

2-1 INTRODUCTION

This chapter is intended to provide you with some extra background to enrich your study of electronics. This book has been designed so that you can skip this chapter and still understand the rest of the text. There is some mathematics in the following pages, but don't be alarmed if you are not particularly good at math. Most of it is a review of things you learned long ago in grade school. There they approached it in a slightly different way. Those things you are not already acquainted with are very simple.

In studying this chapter you may find that much of it is familiar to you. Scan the familiar material just to see if there is a detail or two you've forgotten. If you find something new, put a little effort into learning it. It will not be difficult and it might prove valuable. If you encounter unfamiliar words and ideas, continue reading. They will be explained later on.

2-2 BIG AND LITTLE NUMBERS

The Bare Facts

1. When numbers exceed the range of an inexpensive calculator, some special ways of writing those numbers are in order.

2. Scientific calculators show numbers in this special form, but you must know how to read them.

3. Any number can be written in scientific notation, to bring very large or very small numbers into the range of a simple calculator.

4. *Scientific notation* sounds impressive, but it is actually a simplified (short-cut) technique for dealing with very large and very small numbers.

5. The number of times a number is multiplied by itself is called the *exponent* of the number:

$$10^3 = 10 \times 10 \times 10 = 1{,}000$$

The number taken as an example here is 10. Its exponent is 3, meaning it is multiplied by itself 3 times. The exponent is written above the number itself. The 10 is called the *base*, and the number 10 is the common base for exponents.

6. 10^3 is read as *10 to the third power*.

AN INTRODUCTION TO THE QUEEN OF THE SCIENCES 27

Examples:

$10^1 = 10$
$10^2 = 10 \times 10 = 100$
$10^3 = 10 \times 10 \times 10 = 1{,}000$
$10^4 = 10 \times 10 \times 10 \times 10 = 10{,}000$
$10^5 = 10 \times 10 \times 10 \times 10 \times 10 = 100{,}000$
$10^6 = 10 \times 10 \times 10 \times 10 \times 10 \times 10 = 1{,}000{,}000$

7. By convention, a number written without an exponent actually has the exponent 1. Thus 10 and 10^1 are the same. In a number without a decimal point, the decimal is assumed to be located after the last digit. In the number 204 the decimal point is after the digit 4 (204.).

8. For the case of decimal fractions (smaller than 1) the exponent is negative.

Examples:

$10^{-1} = 0.1 = 1/10$
$10^{-2} = 0.01 = 1/100$
$10^{-3} = 0.001 = 1/1{,}000$
$10^{-4} = 0.0001 = 1/10{,}000$
$10^{-5} = 0.00001 = 1/100{,}000$
$10^{-6} = 0.000001 = 1/1{,}000{,}000$

9. When a number is not exactly in tens, hundreds, thousands, and so forth, it can be converted into two parts: a units part and a power of ten. The base 10 is used with an exponent.

Examples:

$10 = 1 \times 10 = 1 \times 10^1$
$20 = 2 \times 10 = 2 \times 10^1$
$75 = 7.5 \times 10 = 7.5 \times 10^1$
$99 = 9.9 \times 10 = 9.9 \times 10^1$
$100 = 1 \times 100 = 1 \times 10^2$
$750 = 7.5 \times 100 = 7.5 \times 10^2$
$950 = 9.5 \times 100 = 9.5 \times 10^2$
$1000 = 1 \times 1{,}000 = 1 \times 10^3$
$7500 = 7.5 \times 1{,}000 = 7.5 \times 10^3$
$9000 = 9 \times 1{,}000 = 9 \times 10^3$
$9530 = 9.53 \times 1{,}000 = 9.53 \times 10^3$

Note: The exponent is the number of places the decimal point moves to the left.

10. For fractions (smaller than 1), the negative exponent is the number of places the decimal point is moved to the right.

   Examples:
   $$0.01 = 1 \times 0.01 = 1 \times 10^{-2}$$
   $$0.02 = 2 \times 0.01 = 2 \times 10^{-2}$$
   $$0.05 = 5 \times 0.01 = 5 \times 10^{-2}$$
   $$0.08 = 8 \times 0.01 = 8 \times 10^{-2}$$
   $$0.092 = 9.2 \times 0.01 = 9.2 \times 10^{-2}$$

11. Numbers greater than 1 have positive exponents. An exponent with no sign is understood to be positive.

12. Numbers less than 1 (fractions) have an exponent with a negative (−) sign. The minus sign is *always* written for the exponent of a fraction.

13. Scientific notation is convenient for multiplying or dividing numbers that are very large or small because the process is reduced to adding or subtracting exponents.

   Example:
   $$(1 \times 10^4) \times (1 \times 10^2) = 1 \times 10^6$$
   $$10{,}000 \times 100 = 1{,}000{,}000$$

14. Only the exponents of base 10 are added. The unit parts are still multiplied.

   Example:
   $$(2 \times 10^4) \times (3 \times 10^2) = 6 \times 10^6$$
   $$20{,}000 \times 300 = 6{,}000{,}000$$

15. When multiplying with negative exponents, exponents add to produce a larger negative exponent.

   Example:
   $$(1 \times 10^{-2}) \times (1 \times 10^{-1}) = 1 \times 10^{-3}$$
   $$0.01 \times 0.1 = 0.001$$

16. For multiplying with both negative and positive exponents, take the difference between the two and give it the sign of the larger exponent.

   Example:
   $$(1 \times 10^5) \times (1 \times 10^{-3}) = 1 \times 10^2$$
   $$(1 \times 10^{-5}) \times (1 \times 10^3) = 1 \times 10^{-2}$$

17. For division, divide the unit parts and subtract the exponents.

Examples:
$$(1 \times 10^6) \div (1 \times 10^3) = 1 \times 10^3$$
$$1{,}000{,}000 \div 1{,}000 = 1{,}000$$
$$6 \times 10^6 \div 3 \times 10^3 = 2 \times 10^3$$

18. When the denominator has a negative exponent, change its sign and add.

    Example:
    $$(1 \times 10^5) \div (1 \times 10^{-3}) = 1 \times 10^8$$
    $$(1 \times 10^{-5}) \div (1 \times 10^{-3}) = 1 \times 10^{-2}$$

19. A special case of division occurs when numerator and denominator are equal. Then the quotient must be 1.

    Example:
    $$1 \times 10^4 \div 1 \times 10^4 = 1 \times 10^0 = 1 \times 1 = 1$$

    The example shows that for any base with a zero exponent, the number is 1.

20. To add or subtract numbers expressed in powers of 10 (10^2, 10^3, etc.), they must have the same exponent. Then add or subtract the units part but keep the same exponent.

    Example:
    $$(5 \times 10^3) + (2 \times 10^3) = [5 + 2] \times (10^3) = 7 \times 10^3$$

    If the numbers do not have the same exponents, they must be changed to this form before you can add or subtract. Any exponent can be used, but they must all be the same.

21. Another special case is called the *reciprocal*. The reciprocal always has a 1 in the numerator.

    Examples:
    $$\frac{1}{2}, \frac{1}{10}, \frac{1}{.001}, \frac{1}{1.5}$$

2-3 SCIENTIFIC NOTATION

In almost any science-fiction film you will hear something like this:

"But professor, the mysterious object is only two times ten to the twelfth power light years away."

"Two-point-five times ten to the twelfth would be more accurate."

"Perhaps, perhaps, but the point is that it is rapidly approaching, and with our mass only one point five times ten to the minus four, we haven't a chance if it gets much closer . . ."

All we get from this is that some object is bearing down on the actors and that catastrophe is coming if it gets any closer. All of this jargon, "one point five times ten to the twelfth power" (1.5×10^{12}) and so on, is designed to impress us with how scientific it all is.

This jargon is impressive only when we do not understand it. Fortunately for the motion picture and television people, not many do understand it. If too many people understood it, movie and TV directors would have to either abandon it or make some effort to get it right. It is called *scientific notation* and is quite valuable when used properly. It is about as difficult to learn as the game of monopoly, or how to take out the garbage without spilling too much. In short, it's a cinch to learn. Most people feel they have no real need to learn it, so they don't. However, it is worth learning (since it is so easy) if only to get a laugh or two out of an otherwise boring science fiction program. However, if you get at all involved with it, scientific notation can become a very useful tool.

When we multiply a string of equal numbers together such as $10 \times 10 \times 10$, we often use a shorthand notation. The notation 10^3 tells us to write down 3 tens, and multiply them together. The raised 3 is called an *exponent*, and it tells us how many 10s to write down to multiply together. The 10 is called the *base*, and simply specifies what number is to be multiplied. The base in ordinary scientific notation is always 10, but the notation is also useful for other bases.

Examples:

$10^4 = 10 \times 10 \times 10 \times 10$ (multiply four 10s)
$2^3 = 2 \times 2 \times 2$ (multiply three 2s)
$3^2 = 3 \times 3$ (multiply two 3s)

For small numbers the shortcut notation is of little value:

10×10, or 100, is as easy to write as 10^2.

But for big numbers:

10^{20} is much easier to write than $10 \times 10 \times 10 \times 10 \times 10 \times 10 \times 10 \times 10 \times 10 \times 10 \times 10 \times 10 \times 10 \times 10 \times 10 \times 10 \times 10 \times 10 \times 10 \times 10$.

Now, of course, we could multiply the tens out, get a number that says exactly the same thing, eliminate the ($\times$) sign, and get:

100,000,000,000,000,000,000

That does simplify it somewhat. But suppose you read the number to a friend over the telephone. The most reasonable way is to tell your friend to write a 1 followed by 20 zeros. You could also say "ten, exponent twenty," or "ten to the twentieth power," as it is most commonly expressed. 10^{20} is simply an instruction to either write a 1 followed by 20 zeros or write 10^{20}. One big advantage of scientific notation is that no one has to count zeros, and this reduces errors. Keep in mind that the most disastrous error in mathematics is an error in zeros. If you make $1,000 per month and your paycheck for the month comes printed out at either $100 or $10,000, you have a problem! If it comes out $100 you won't eat, and if it comes out $10,000 you may not be able to cash it, and still go hungry until it is straightened out.

We do have names for hundreds, thousands, millions, and so on, but how many of us know for certain how many zeros to write for a trillion? Some of us have trouble with a billion, and the U.S. billion is different than the English billion. In the U.S. it is 1,000,000,000 and in England it is 1,000,000,000,000—a mere 1000 million dollars difference in money terms. A billion in scientific notation is 10^9 in the U.S. and 10^{12} in England. Scientific notation is understood anywhere in the world, even over the telephone.

FRACTIONS

The first arithmetic we learned in school was addition and subtraction. For the most part, scientific notation is more trouble than it is worth for addition and subtraction. Next we learned multiplication and division. At this point we should have learned scientific notation, the easy way to avoid our mathematical frustrations. We should have learned scientific notation most of all to get us through the horrors of fractions.

Suppose we take a look at how scientific notation can make mathematics far less painful and a lot simpler. We have been taught that ½ is somehow more acceptable than 0.5.* Perhaps no one ever specifically told us that ½ was more acceptable, but the so-called common fraction still gets the lion's share of emphasis in most grammar schools. The opinion that common fractions are superior is encouraged by fractions such as ⅓, which in decimal fractions becomes 0.333333 . . . with no end at all.

However, there is another, less favorable view of common fractions that puts them in the second-rate category (where some of us firmly believe they belong). The notation ⁴/₅ is an instruction to divide 4 by 5. If we carry out that instruction we get a decimal fraction of 0.8, a very convenient number. If we view common fractions as no more than a division instruction, they lose much of their mystery. The $\sqrt{4}$

---

*The zero ahead of the decimal point is included mostly because of the fear of errors in typesetting. The zero says, "this is not a mistake, no digit 1, 2, 3 . . . was intended."

means nothing to us until the instruction to extract the square root of 4 is carried out. We then have a simple digit, 2. If we view common fractions in much the same way, as an operation not completed, we rid them of mystery.

What we were taught about arithmetic using common fractions was important primarily because of our peculiar English measuring system. The adoption of the metric system will make it virtually unnecessary to perform arithmetic with common fractions.

Adding Fractions

We can add fractions in the following way:

$$½ + ¼ = ?$$
Divide: 1 by 2 = 0.5
Divide: 1 by 4 = 0.25
Add: 0.5 + 0.25 = 0.75

The two fractions in the above example are special division instructions. When the numerator is a 1, the fraction is called a *reciprocal*. This special kind of fraction occurs frequently in electronics (and science in general).

SCIENTIFIC NOTATION AND FRACTIONS

If $10^2 = 100$, how do we write $1/100$ in scientific notation? We write it as 10^{-2}. The negative sign tells us that the number is a fraction. It has nothing to do with negative numbers.

The expression 10^{-2} tells us that the number occupies two places to the right of the decimal point and is written as 0.01.

↑
2 digits

$10^{-3} = 0.001$ $10^{-4} = 0.0001$
3 digits 4 digits

Rules

1. With whole numbers the exponent tells how many zeros there are between the decimal point and the left-most digit.

2. With fractions the exponent tells how many digits there are to the right of the decimal.

3. A positive sign for the exponent (no sign at all is the same as a plus sign) tells us that the number goes on the left of the decimal point.

4. A negative sign for the exponent tells us that we are dealing with a fraction (less than 1) and that the number goes to the right of the decimal.

Examples:

The following are illustrations of Rules 1 and 2 above:

$1000 = 10^3$ $10,000 = 10^4$

| 3 zeros |

| 4 zeros |

$100,000 = 10^5$ $1,000,000 = 10^6$

| 5 zeros |

| 6 zeros |

$0.001 = 10^{-3}$ $0.0001 = 10^{-4}$

| 3 digits |

| 4 digits |

$0.00001 = 10^{-5}$

| 5 digits |

Here are some other cases:

If $10^2 = 100$, then $2 \times 100 = 2 \times 10^2 = 200$.
Thus,
$$300 = 3 \times 100 \text{ or } 3 \times 10^2$$
$$400 = 4 \times 100 \text{ or } 4 \times 10^2$$

If $1/100 = 0.01$ and $0.01 = 10^{-2}$, then $1/200 = 1/(2 \times 10^2) = 1/2 \times 1/10^2 = 0.5 \times 10^{-2}$ (or 5×10^{-3}).

Study Problems

Write the following in scientific notation:

1. 500   2. 600   3. 20,000   4. 50,000
5. 0.03   6. 0.002   7. 0.0004   8. 0.000,005
9. 0.000,000,6

Standard Form

Suppose we have the number 4500, and we wish to write it in scientific notation. We know that

$$4500 = 45 \times 100$$

Since $100 = 10^2$, we can write it as 45×10^2. This is perfectly correct.

We can also write it as:

$$4.5 \times 10^3$$

We obtain this from:

$$4000 + 500 = 4{,}500$$

$$\begin{array}{r} 4 \times 10^3 \quad (4000) \\ + .5 \times 10^3 \quad (0.5 \times 1000) \\ \hline 4.5 \times 10^3 \end{array}$$

Units — Power of 10

Rule

When adding, the exponents must be alike and the sum uses the common exponent. More about this later.

The form in which the units part is always between 1 and 10 is called *standard form*. Standard form is a big help in avoiding errors in arithmetic.

Writing a number in standard form is easy. Let's take the example 4500. In any whole number a decimal point is assumed to exist at the right-hand end, even if it is not shown. 4500 is actually 4500.00. To write 4500. in standard form, place the decimal to make the number between 1 and 10 and affix the proper power of 10.

Example:

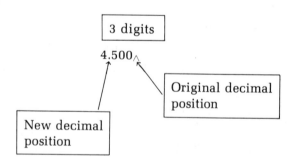

The decimal has jumped 3 digits, so the exponent is 3. The number in standard form is: 4.5×10^3.

More examples:

The following are further illustrations of standard-form scientific notation:

$$3900. \quad ; \quad 3{.}900_\wedge = 3.9 \times 10^3$$

AN INTRODUCTION TO THE QUEEN OF THE SCIENCES 35

$$270. \quad ; \quad 2.70_\wedge = 2.7 \times 10^2 \quad \boxed{\text{2 digits}}$$

$$68{,}000{,}000 \quad ; \quad 6.8000000_\wedge = 6.8 \times 10^7 \quad \boxed{\text{7 digits}}$$

$$4700. \quad 4.700_\wedge \quad 4700 = 4.7 \times 10^3 \quad \boxed{\text{3 digits}}$$

$$39{,}000. \quad 3.9000_\wedge \quad 39{,}000. = 3.9 \times 10^4 \quad \boxed{\text{4 digits}}$$

$$560{,}000. \quad 5.60000_\wedge \quad 560{,}000 = 5.6 \times 10^5 \quad \boxed{\text{5 digits}}$$

When there are more nonzero digits, as in the number 637,000, the process of writing the number in standard-form scientific notation is the same.

Examples:

$$637{,}000 \quad ; \quad 6.37000_\wedge = 6.37 \times 10^5 \quad \boxed{\text{5 digits}}$$

$$42{,}300 \quad ; \quad 4.2300_\wedge = 4.23 \times 10^4 \quad \boxed{\text{4 digits}}$$

$$3{,}330{,}000 \quad ; \quad 3.330000_\wedge = 3.33 \times 10^6 \quad \boxed{\text{6 digits}}$$

Study Problems

Write the following in standard-form scientific notation:

1. 49,000
2. 3,300
3. 64,000,000
4. 69,000,000,000
5. 490
6. 67,300
7. 43,300,000
8. 121,000
9. 521,300,000

Standard Form for Small Numbers

The procedure for writing fractions (values less than 1) in standard-form scientific notation is basically the same except that the decimal moves to the right, and the exponent gets a minus sign.

Examples:

$$0.00015 \quad ; \quad 0.0001.5 = \underset{\text{Units}}{1.5} \times \underset{\text{Powers}}{10^{-4}} \quad \boxed{4 \text{ digits}}$$

(Notice that the units value is selected to be between 1 and 10 just as it was for whole numbers. The leading zeros are normally dropped.)

$$0.000033 \quad ; \quad 0.00003.3 = 3.3 \times 10^{-5} \quad \boxed{5 \text{ digits}}$$

$$0.0054 \quad ; \quad 0.005.4 = 5.4 \times 10^{-3} \quad \boxed{3 \text{ digits}}$$

$$0.00273 \quad ; \quad 0.002.73 = 2.73 \times 10^{-3} \quad \boxed{3 \text{ digits}}$$

Study Problems

Write the following in standard-form scientific notation:

1. 0.00011   2. 0.000032   3. 0.0046   4. 0.00253

2-4 HOW TO USE SCIENTIFIC NOTATION

MULTIPLYING

If you wish to multiply 22,000 × 21,300,000 you will have to do it by hand, the hard way, unless you have a calculator that works in scientific notation. Ordinary pocket calculators can't handle an answer this big. The hard way is:

$$\begin{array}{r} 21,300,000 \\ \times \quad\quad 22,000 \\ \hline 42,600,000,000 \\ 42,600,000 \quad\;\; \\ \hline 468,600,000,000 \end{array}$$

AN INTRODUCTION TO THE QUEEN OF THE SCIENCES 37

To do this with scientific notation, first write the numbers to be multiplied in standard form:

$$2.13 \times 10^7$$
$$\times \ 2.2 \times 10^4$$

Multiply by hand or with any calculator:

$$2.13$$
$$\times \ 2.2$$
$$\overline{4.686} \quad \text{(Answer for the units part)}$$

Now let's write the original problem and include the answer for the units part:

$$2.13 \times 10^7$$
$$\times \ 2.2 \times 10^4$$
$$4.686 \times 10^{11}$$

The units part ⟵ Adding the exponents

To finish it off, we have simply added the exponents $7 + 4 = 11$. The final result is:

$$4.686 \times 10^{11}$$

Doing it the hard way, we got: 468,600,000,000.
Converting this into standard form, we get:

| 11 digits |

4.68600000000 ⟵

$$4.686 \times 10^{11}$$

Both methods give the same answer.
Let's try a problem without the long-winded explanation.

Example:
  Multiply:

$$462{,}000{,}000{,}000$$
$$\times \quad 32{,}000{,}000$$

You can try this the hard way if you like, but here is the easy way:

$$4.62 \times 10^{11}$$
$$\times \ 3.2 \times 10^7$$
$$\overline{14.784 \times 10^{18}}$$

Oh, oh! The units part is correct, but it is not in standard form. What do we do now?

Write the units part in standard form and add the exponents:

$$\underset{\leftarrow}{1.4}784 = 1.4784 \times 10^1 \quad \boxed{1 \text{ digit}}$$

If we include the 10^{18}, we get $1.4784 \times 10^1 \times 10^{18} = 1.4784 \times 10^{19}$.

Rule

To multiply powers of 10, always *add* the exponents.

$$10 \times 100 = 1000$$
$$10^1 \times 10^2 = 10^3$$
$$10 \times 10 \times 10 = 1000$$

Multiplying Fractions

The following example would be even more prone to longhand arithmetic error than the previous examples. It is also beyond the range of the ordinary calculator.

Multiply:

$$\begin{array}{r} 0.000025 \\ \times\ 0.0000045 \\ \hline \end{array}$$

If we write the numbers in scientific notation (standard form) we get:

$$\begin{array}{r} 2.5 \times 10^{-5} \\ \times\ \ 4.5 \times 10^{-6} \\ \hline 11.25 \times 10^{-11} \end{array}$$

When the exponents are negative, we again add the exponents and affix the negative sign to the sum. The answer to the units part came out larger than 10, so the entire result is not in standard form. If we wish to put it in standard form, we proceed as follows. Write the units part in standard form:

$$11.25 = 1.125 \times 10^1$$

Including the powers part of the result, we have:

$$1.125 \times 10^1 \times 10^{-11}$$

Rule

To multiply powers of 10 when the signs of the exponents are different, subtract the smaller from the larger exponent and use the sign of the larger exponent. The final answer in standard form is given by:

$$1.125 \times 10^1 \times 10^{-11} = 1.125 \times 10^{-10}$$

Multiplying a Fraction by a Whole Number

Example:
Multiply:
$$0.00015 \times 1{,}200{,}000$$

Write the numbers in standard form:

$$\begin{array}{r} 1.5 \times 10^{-4} \\ \times\ 1.2 \times 10^{6} \\ \hline 1.8 \times 10^{2} \end{array}$$

The previous rule governing exponents of opposite sign applies here also.

If you find it difficult to think in scientific notation, it can easily be converted back into ordinary numbers after the arithmetic is done. In the previous example we got an answer of 1.8×10^{2}. To convert back into ordinary numbers, write the 1.8 with some zeros added. The exponent 2 tells you to make the decimal jump two digits.

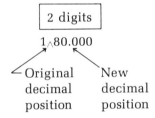

Now throw away any extra zeros to the right of the new decimal and you have it:

$$1.8 \times 10^{2} = 180$$

But suppose you have a number such as 1.8×10^{-2} to convert into ordinary numbers. The minus sign before the exponent tells you that the number is going to be less than 1. The only way that can happen is if the decimal moves to the left. The exponent 2 tells you to make the decimal jump 2 digits. As before, add some zeros and make the original decimal jump 2 digits to the left.

```
            2 digits
           000.01.8
          ↗      ↖
       New       Original
      decimal    decimal
      position   position
```

The result in ordinary numbers is 0.018.

DIVIDING

Rule

To divide using scientific notation:

1. Write the number in standard form.
2. Divide the units part as you would divide ordinary numbers.
3. *Subtract* the exponents.

Example:
   Divide:

$$250{,}000 \div 1500$$

In standard form:

$$\frac{2.5 \times 10^5}{1.5 \times 10^3} = 1.66 \times 10^{5-3} = 1.66 \times 10^2$$

One of the many mathematical shortcuts built into scientific notation is one that makes those special fractions called *reciprocals* a cinch. Reciprocals always have a 1 for the numerator.

Example:

$$\frac{1}{10^2} = 10^{-2} \quad \text{and} \quad \frac{1}{10^{-2}} = 10^2$$

One simply does away with the fraction and changes the sign of the exponent. Thus, in conventional form:

$$\frac{1}{100} = 0.01 \quad \text{and} \quad \frac{1}{0.01} = 100$$

In scientific notation, an answer can be obtained by inspection, no matter what the size of the number.

ADDING AND SUBTRACTING

In some cases scientific notation is more trouble than it is worth for addition and subtraction. The typical scientist would perhaps argue with that statement because scientific notation is his native language. But for the beginner, the statement is valid. The difficulty is this:

Rule

The power of 10 *must* be the same for both the number to be subtracted and the one it is to be subtracted from (and for 2 numbers to be added).

In most problems this is not the case, and decimals must be shifted and powers of 10 adjusted to satisfy the rule. However, once scientific notation has become as automatic as shifting gears in your car, such adjustments are no problem.

Example:
  Add:

$$250 + 125$$

Written in scientific notation this is:

$$\begin{array}{r} 2.50 \times 10^2 \\ + \ 1.25 \times 10^2 \\ \hline 3.75 \times 10^2 \end{array}$$

This example is a trivial one, but so are most of them that don't require considerable adjustment.

2-5 THE METRIC SYSTEM

The Bare Facts

1. The metric system is based on multiples of 10.
2. If you can count money, the metric system is simple.
3. Conversions between the English and metric systems can never be exact.
4. Conversions can be as accurate as needed, but never perfect.
5. Often, approximate conversions between values in the two systems are adequate.
6. The metric system has been a legal measuring system in the United States since 1866.

The controversy about metric measurement in the United States began in 1799. It has always been an emotional controversy. Scientists throughout the world, including the United Kingdom and the U.S., have used metrics since about 1866 when the metric system was made legal in the U.S. Oddly enough, our English system has never been formally legalized in this country. The National Bureau of Standards in the U.S. has no standard yard or pound weight, but rather a standard meter and kilogram. English measurements are derived mathematically from those metric standards. John Adams and Benjamin Franklin both strongly recommended the adoption of the metric system in this country. However, because of political events Congress was not in a mood to deal with the problem at the time. Today, over 100 years after the metric system became a legal measurement system, it is slowly coming into general use.

Those who have used metrics understand its advantages, and are

Basic Units
Meter: a little longer than a yard (about 1.1 yard)
Liter: a little larger than a quart (about 1.06 quart)
Gram: about the weight of a paper clip (comparative sizes are shown)

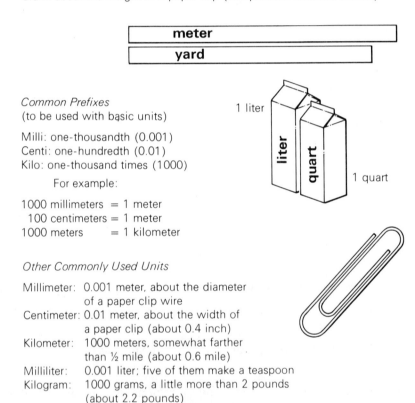

Common Prefixes
(to be used with basic units)

Milli: one-thousandth (0.001)
Centi: one-hundredth (0.01)
Kilo: one-thousand times (1000)

    For example:

1000 millimeters = 1 meter
 100 centimeters = 1 meter
1000 meters   = 1 kilometer

Other Commonly Used Units

Millimeter: 0.001 meter, about the diameter of a paper clip wire
Centimeter: 0.01 meter, about the width of a paper clip (about 0.4 inch)
Kilometer: 1000 meters, somewhat farther than ½ mile (about 0.6 mile)
Milliliter: 0.001 liter; five of them make a teaspoon
Kilogram: 1000 grams, a little more than 2 pounds (about 2.2 pounds)

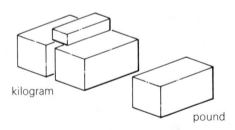

FIGURE 2-1. Common metric and English units compared.

TABLE 2-1 METRIC PREFIXES AND VALUES

	Value	Prefix	Symbol	Example*
Numbers Larger than 1	1 000 000 000 000 = 10^{12} 1 000 000 000 = 10^{9} 1 000 000 = 10^{6} 1 000 = 10^{3} 100 = 10^{2} 10 = 10	tera giga mega kilo hecto deka	T G M k h da	THz = 10^{12} Hz GHz = 10^{9} Hz MHz = 10^{6} Hz kV = 10^{3} V hm = 10^{2} m dam = 10 m
Numbers Smaller than 1	0.1 = 10^{-1} 0.01 = 10^{-2} 0.001 = 10^{-3} 0.000 001 = 10^{-6} 0.000 000 001 = 10^{-9} 0.000 000 000 001 = 10^{-12}	deci centi milli micro nano pico	d c m μ n p	dm = 10^{-1} m cm = 10^{-2} m mA = 10^{-3} A μs = 10^{-6} s ns = 10^{-9} s pF = 10^{-12} F

*Hz stands for hertz (cycles per second)   A stands for amperes
 V stands for volts   s stands for seconds
 m stands for meters   F stands for farads

in favor of the change. The arithmetic is easier. In the following section we will examine the metric system. The metric system is far simpler than the English system. Conversions between metric and English measurements have generally been presented as being very complicated and confusing, especially if they must be extremely accurate.

Recently both metric and English measurements have been used on speed limit signs and so on, in the belief that people would absorb the conversions in this way. Any educator can tell you that this teaching method is unreliable. Like the multiplication tables, English–metric conversions must be memorized. However, they need not be difficult to learn if some small inaccuracies are acceptable. In this book we shall assume that such small conversion inaccuracies are acceptable, and that more accurate values can easily be found in published tables when necessary.

The metric system is based on our normal decimal counting system. New units occur in multiples of 10. Figure 2-1 shows the most common metric units and compares them to English units. Table 2-1 shows the entire range of international metric units. The relationships among metric units are the same as those in our money system. If you can count money, you can think metric.

Converting Fractions of an Inch to Millimeters

1. Divide the numerator of the fraction by its denominator.

2. Multiply by 25 for a close approximation, or by 25.4 for a more exact answer.

TABLE 2-2 APPROXIMATE METRIC-ENGLISH CONVERSIONS

a. Lengths greater than an inch.

Metric	English	The English System Has This Much Extra
1 meter	1 yard	3 inches
1/3 meter	1 foot	1 inch
2 1/2 centimeters	1 inch	Less than 1/10 inch
5 centimeters	2 inches	Less than 1/10 inch
10 centimeters	4 inches	Less than 1/10 inch
1 kilometer	1/2 mile ⟶ (actually about 0.6 mile)	

b. Lengths less than an inch.

Metric	English
1.5 millimeters	1/16 inch
3 millimeters	1/8 inch
6 millimeters	1/4 inch
12 millimeters	1/2 inch
18 millimeters	3/4 inch
25 millimeters	1 inch

c. Weights.

Metric	English
1 kilogram (1000 gm)	2 pounds
1/2 kilogram (500 gm)	1 pound
30 grams	1 ounce
60 grams	2 ounces
90 grams	3 ounces
120 grams	4 ounces
150 grams	5 ounces

d. Volumes.

Metric	English
4 liters	1 gallon
1 liter	1 quart
1 liter	4 cups
1/2 liter (250 ml)*	1 pint
1/2 liter (250 ml)*	2 cups
1/4 liter (125 ml)*	1 cup
30 milliliters	1 ounce (liquid)
15 milliliters	1 tablespoon
5 milliliters	1 teaspoon

*ml means milliliter.

e. Temperatures.

Metric (Celsius)	Example	English (Fahrenheit)
0°	Water freezes	32°
25°	Room temperature	70°
37°	Body temperature	98.6°
100°	Water boils	212°

TABLE 2-3 MORE ACCURATE CONVERSION VALUES

Length and Volume

1 inch (in.) = 2.54 centimeters (cm)
1 foot (ft) = 30.48 centimeters
1 mile = 1.609 kilometers (km)
1 pound (lb) = 0.4536 kilogram (kg)
1 quart (qt) = 946 cubic centimeters (cm^3)
1 centimeter = 0.3937 inch
1 meter (m) = 1.094 yards = 39.37 inches
1 kilometer = 0.6214 mile
1 kilogram = 2.2046 pounds
1 liter (l) = 1,000 cm^3 = 1.057 quart
1 micron (μm) = 10^{-6} meter = 0.0001 centimeter
1 angstrom (Å) = 10^{-10} meter = 0.00000001 centimeter
1 gallon (gal) = 231 cubic inches = 3.785 liters
            = volume of 8.337 pounds of water

Examples:

To convert $1/2$:

a. Divide 1 by 2: $1 \div 2 = 0.5$

b. Multiply by 25: $0.5 \times 25 = 12.5$ millimeters, or multiply by 25.4 for a more exact result: $0.5 \times 25.4 = 12.7$ millimeters

To convert $19/64$:

a. Divide 19 by 64: $19 \div 64 = 0.296$

b. Multiply by 25: $0.296 \times 25 = 7.4$ millimeters, or, more exactly, $0.296 \times 25.4 = 7.5184$ millimeters.

Table 2-2 provides approximate conversions for metric and English units. Table 2-3 provides some more exact conversion factors.

It is important to understand that, with one exception, there is no *exact* conversion betweeen the two systems.* Conversions can be as *precise* as required, but they will never be exact.

Converting Fahrenheit Temperatures to Celsius

For temperatures above about 100° Fahrenheit, you can get a fair approximation of the temperature in Celsius by simply dividing the Fahrenheit temperature by 2.

Example:

$$400°F = 200°C \text{ (approximately)}$$

---

* The exception is: 1 inch = 2.54 cm, exactly.

2-6 GRAPHS

The Bare Facts

1. A graph presents numerical values in a visual form.

2. Graphs normally compare two related quantities; for example, earnings compared to years (time).

3. In electronics, *time* is the most common factor to which other things are compared.

4. A linear graph is one that relates a constant rate of change with time (or some other factor). For example, a constant rate of pay (by the hour) plotted against the time worked yields a linear curve. The line on a graph is called a *curve* even if it is a straight line.

5. A nonlinear curve is one that relates something that is not constant with time or some other quantity. For example, an hourly rate that increases after 40 hours to time-and-a-half would produce a curved line on the graph. A curved graph is nonlinear.

People are visually oriented creatures; they find pictures much easier to interpret than columns of figures. Much of our raw information in science, technology, and business is first available as neat arrays of numerals and notations. We can put raw data into visual form whenever something varies with time or when two kinds of data are somehow related.

The most common kind of graph represents things changing with time. Other relationships are also common. A medical investigator trying to wipe out the plague would find a graph of incidents of the disease versus the rat population in the area very instructive. The doctor could probably examine the figures and find a correlation without making a graph of it, but if he must make an argument before the city fathers to convince them of the importance of rat abatement, he would be wise to make a graph. In addition to their efficiency in visual communications, graphs have another useful and interesting application. Science finds that certain patterns of behavior recur often in nature. When such a pattern is made into a graph, it is often immediately recognizable as one we have seen before. If mathematical methods of evaluating a particular natural pair of relationships have been developed, they need not be developed again if that new pattern is a familiar one. For example, a graph of how bacteria multiply in a culture, a graph of human population growth, and a graph of how an electrical capacitor charges, all look alike. The shape of the curve is the same in each case.

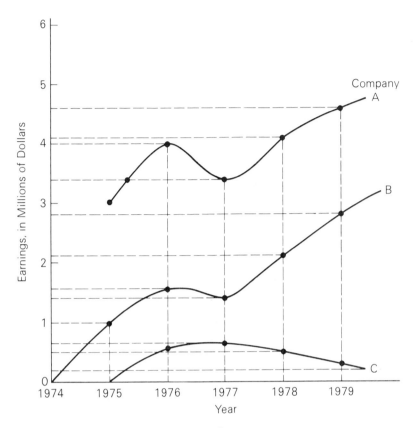

FIGURE 2-2. A family of curves: earnings vs. time.

A graph, theoretically, contains an infinite number of plot points. In real situations the plots are based on a finite number of measurements. Often, once the graph is drawn, it becomes obvious that no measurement was taken at some important point along the graph. If there is a certainty that the measurements are following one of the natural curve shapes, it is easy to add essential plot points without having to go through the time-consuming measurement process.

In many cases a graph consists of several related curves to allow visual comparison of several factors. Figure 2-2 compares 3 curves depicting the earnings of 3 different companies over a period of years. Hypothetical Company A has shown good growth from 1975 to 1979 and shows a trend of rapid growth. Data about Company A are unavailable for 1974 to 1975.

Company B shows a growth curve very close to that of Company A. If we had enough curves for different companies and they all (except Company C) had the same general shape, we might conclude that

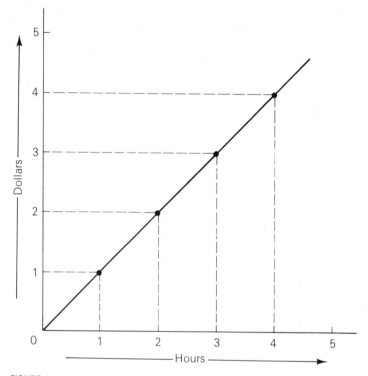

FIGURE 2-3. A linear graph: total earnings at the constant rate of $1.00 per hour.

Companies A and B have followed the country's economic fluctuations. We are likely to have some reservations about investing in Company C because it has not followed the normal growth pattern for other companies.

LINEAR AND NONLINEAR GRAPHS

A linear graph simply shows an increase or decrease that occurs at a constant rate. Figure 2-3 is an example.

Nonlinear graphs describe things that increase or decrease at varying rates. Figure 2-4 shows a nonlinear graph about wages and overtime. The actual dollar values are selected for convenience and do not represent realistic wages.

GRAPH PLOTTING

When a graph is to be plotted from laboratory measurements it is good practice to scan the range of measurements first, if possible. The

Assume that from over 8 hours, but less than 12 hours, the rate is $1.50 per hour; for over 12 hours, the rate is $2.00 per hour.

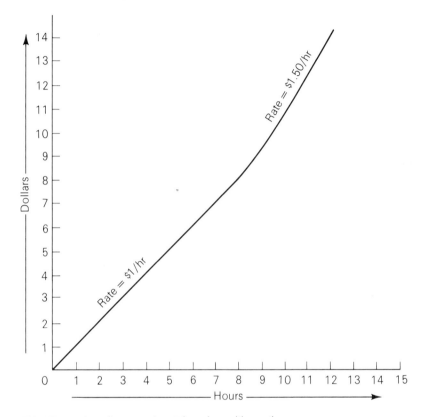

FIGURE 2-4. A nonlinear graph: total earnings with overtime.

investigator can then make a note of where significant changes occur in the slope of the curve. In most cases, the most significant part of a graph is where it changes direction or slope. If you are plotting a graph from columns of figures, you can often spot and flag all sudden changes in the pattern before you begin to plot the curve.

A basic assumption in curve plotting is that the curve is most probably a smooth one. An attempt is usually made to draw a smooth, flowing curve rather than straight line segments connecting individual plot points.

Study Problems

1. In the graph in Figure 2-5, pick out the most important plot points. What are the least important plot points?

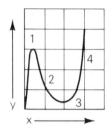

FIGURE 2-5. Graph for study problem.

TABLE 2-4 GRAPH STUDY PROBLEM

x	y
0	0
30	0.50
60	0.87
90	1.00
120	0.87
150	0.50
180	0

2. Using the data in Table 2-4, plot a graph. (Use any size grid that is handy. A grid of ¼ inch squares is a good choice.)

POLAR AND RECTANGULAR GRAPHS

The graphs we have examined so far use a rectangular grid on which to locate the plot points. Graphs that use rectangular grids are called *rectangular-coordinate* graphs. In applications where angles and distances form the data to be plotted, polar-coordinate graphs are often more useful. Figure 2-6 compares the arrangement of polar and rectangular grids.

The position of a particular plot point on a rectangular-coordinate graph is located by counting squares along the X axis and then moving along the Y axis the proper number of squares. The intersection of lines in the two directions locates the plot point. (See Figure 2-6b.) The location of a point on a polar-coordinate graph is defined by the angle of rotation counterclockwise from the zero-degree point and the distance away from center.

If you will notice, the rectangular-coordinate graph accommodates negative and positive values on both axes. Negative values can be used for such things as temperatures below zero or to indicate which direction current is flowing in an electrical circuit. In the case of the polar graph, a negative value often rotates the vector in a clockwise direction. A vector can be defined as a line of specific length, with a specific angle. In Figure 2-6a, a line drawn from the center to the plot point would be the vector of that particular plot point. Vectors are most handy for solving certain alternating-current problems without getting too involved in mathematics. You will get a chance to play with vectors in Chapter 6.

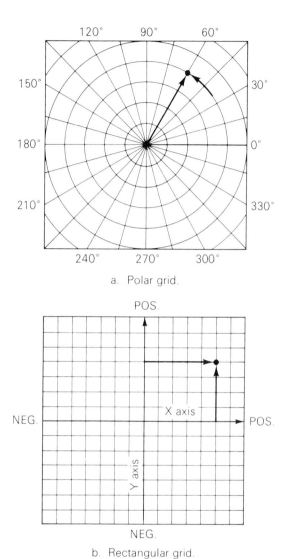

FIGURE 2-6. Comparison of polar and rectangular grids.

2-7 ALGEBRA: WHAT IS IT AND WHAT GOOD IS IT?

The Bare Facts

1. Algebra is a tool that lets you solve mathematical problems before all the numerical values are known.

2. In an algebraic equation, letters are substituted for numbers that are, at least temporarily, unknown.

3. Equations must always be balanced.

4. Quantities can be added to or subtracted from equations.

5. The same amount must be added to or subtracted from *both* sides of the equation.

6. An equation can be altered by multiplying *both* sides of the equation by the same quantity.

7. An equation can be altered by dividing *both* sides of the equation by the same quantity.

8. When all but one of the numerical values in an equation are known, the unknown value can be found.

Algebra is a tool, and a language. Unfortunately, the traditional method of teaching algebra gives us little information about its practical use. You can't help but question the value of any tool that seems to have no meaningful purpose. It is a bit like learning all about lawnmowers, but never being told about lawns. When you leave school and discover that lawns have been around for some time, your understanding of lawnmowers suddenly gains meaning.

Algebra is basically very simple, but it is so flexible that if you get very deeply into it there are a lot of variations and combinations to learn. It takes a great deal of time and effort to learn them all. However, most people never have use for much more than a few simple algebraic ideas. Those simple concepts, once learned (and it doesn't take much effort), can prove to be quite useful.

The scale in Figure 2-7 illustrates the balanced nature of an equation. An equation is an equality statement. In its simplest form it might say $5 = 5$ or $3 + 2 = 5$. The value of equations, however, lies in the fact that *one* unknown value can be discovered when the rest of the values are known. For example: $X + 2 = 5$. In this case a letter is used to represent a number because we don't know what number should occupy X's spot in the equation. Of course it is simple here to examine the equation and figure out the numerical value of X. When the answer to the problem is not obvious, a couple of basic rules of algebra can allow you to get the answer.

Rule

An equation must always be balanced. The numerical value on both sides of the equal sign *must* be equal. When the numerical value of the unknown represented by some letter is to be found, the procedure

FIGURE 2-7. A balanced equation.

is to rearrange the equation so that the unknown symbol stands *alone* on one side (or the other) of the equal sign. The prime directive is: *keep the equation balanced.*

If you will examine the scale in Figure 2-7 you will see that the block labeled 5 is also labeled X. Let us assume that X is the only notation on the block and we need to find its numerical value. Suppose we remove some clutter from the scale. Our objective is to get X in a pan by itself. So let's get rid of the 3 block, leaving the X block by itself. We can't stop at that, however, because we will no longer have a balanced condition. We must remove the 3 block from the right-hand pan to restore the balance. We now have $X = 4 + 1$. We get $X = 5$. We can perform any arithmetic operations that will reduce the equation to the final form of: $X =$ a numerical value. We can do whatever is necessary, as long as we do the same thing to both sides of the equation—that is, as long as we don't upset the *balance*. Let's look at the previous example without benefit of scales and weights.

$$X + 3 = 4 + 3 + 1$$
Subtract 3 from both sides
$$X = 4 + 1 \text{ (adding)}$$
$$X = 5$$

An equation can have only a single unknown value if it is to be solved. We often encounter formulas such as:

$$V = IR$$

where

V is the symbol for voltage (volts) (The symbol E is also used.)
I is the symbol for current (amperes)
R is the symbol for resistance (ohms)

The formula sssumes that you already know the numerical value of two of the quantities. If we are solving for V, and know the values for I and R, we simply substitute the numerical values for the symbols and multiply.

Example:

$$V = 2 \text{ amperes} \times 10 \text{ ohms}$$
$$V = 20 \text{ volts}$$

The real power of algebra lies in the ability to manipulate formulas and equations before any numerical values are plugged in. For example, suppose you have the previous formula V = IR, but you already know the numerical values for V and I. R is what you are interested in finding. If we have the numbers we can plug them in and then work on the solution, but we can do it just as well without the numbers. Let's look at it both ways. First, *with numbers* (the same numbers we used in the previous example):

The formula: V = IR
We know: V = 20 and I = 2

If we substitute the numbers for the symbols, we get:

$$20 = 2R$$

The objective is to get R by itself. This means getting rid of the 2 on the right-hand side of the equation. The only arithmetic operation that will work is to divide the 2 by 2. If we divide the right-hand side by 2, we must do the same with the left-hand side to maintain the balance. So we get:

$$20/2 = 2/2\,R$$

or:

$$10 = 1R, \text{ or } 10 = R$$

Now let's try it *with letters only*. The object is still to get R by itself on one side of the equal sign.

$$V = IR \text{ (the original formula)}$$

We can still divide I by I, because any number divided by the same value is equal to 1. To keep the balance we must also divide the left-hand side by I.

It goes like this:

$$\text{The original: } V = IR$$

$$\text{Dividing by I: } \frac{V}{I} = \frac{I}{I}R$$

Or,

$$\frac{V}{I} = 1R$$

Or,

$$\frac{V}{I} = R$$

If we plug in the numbers:

$$20/2 = 10, \quad R = 10$$

This has been a very short introduction to algebra, but it contains some useful ideas. The little bit that is here is more than adequate to help you get the most out of this text.

Study Problems

Write the following numbers in scientific notation:
1. 450,000   2. 695,000,000   3. 3000   4. 4900
5. 49,000   6. 490,000   7. 49,000,000   8. 0.005
9. 0.01   10. 0.00025   11. 0.000,000,39
12. 0.000,001,25   13. 0.00143

Study Problems (continued)

Multiply the following:
1. $10^3 \times 10^4$
2. $(5 \times 10^3) \times (4 \times 10^2)$
3. $(2 \times 10^9) \times (6 \times 10^4)$
4. $(2 \times 10^{-5}) \times (2 \times 10^{-3})$
5. $(2 \times 10^{-2}) \times (5 \times 10^{-2})$
6. $(6 \times 10^{-5}) \times (5 \times 10^{-3})$
7. $(2 \times 10^{-2}) \times (2 \times 10^2)$
8. $(2 \times 10^{-3}) \times (5 \times 10^3)$

Divide the following:
1. $(10 \times 10^5) \div (5 \times 10^2)$
2. $(6 \times 10^8) \div (3 \times 10^4)$
3. $(1.5 \times 10^6) \div (3 \times 10^5)$
4. $(1.6 \times 10^{15}) \div (8 \times 10^7)$
5. $(8 \times 10^{-2}) \div (4 \times 10^{-4})$
6. $(4 \times 10^{-11}) \div (4 \times 10^{-12})$
7. $(4 \times 10^{-11}) \div (2 \times 10^{11})$
8. $(4 \times 10^{-5}) \div (2 \times 10^2)$

Find the value of the *unknown* in each of the following equations:
1. $2V = 8$
2. $5I = 15$
3. $250R = 500$
4. $15 = 7V$
5. $R/2 = 6$
6. $1/R = 5$
7. $R + 3 = 12$
8. $2R + 5 = 20$
9. $V - 5 = 10$
10. $3V - 2 = 17$

Chapter 3

Painting by Kelly Williams

AN INTRODUCTION TO ELECTRICITY

3-1 INTRODUCTORY COMMENTS FOR THE SCIENTIFICALLY ORIENTED

In this chapter some liberties have been taken with the precision of certain scientific definitions. Some closely related terms have been used as synonyms even though they are not quite synonymous. These liberties have been taken to make it easier to understand the ideas involved. Here are some working definitions:

Force—the ability to impart motion to or to cause a change in speed and/or direction of a moving body.
Work—force applied to a body, moving it over a distance.
Energy—the ability to do work.

In many of the analogies we have equated electrical current and velocity. This is sound enough, but the situation is actually a bit more complex. Electrical current is a product of volume and velocity, but the velocity is constant, and only the volume (number of electrons moving per second) varies. The motion of a solid can also be considered a volume-velocity product, but in this case the volume is fixed and only the velocity can change. Mechanical fluids such as air or water move as a volume-velocity product, but in this case both volume and velocity are variable.

3-2 BASIC ATOMIC THEORY

The Bare Facts

1. All matter, including gases, liquids and solids, is made up of atoms.

2. There are about 100 different kinds of atoms.

3. All matter is made up of these few kinds or combinations of these few kinds of atoms.

4. If two atoms of hydrogen and one atom of oxygen are combined chemically, we have a compound called *water*, H_2O.

5. Hydrogen and oxygen are both gases (and both elements).

6. If hydrogen and oxygen are brought together (with a little heat), there will be a violent explosion. When things calm down, the gases will have combined to form water.

7. Compounds are very unlike the elements that make them up.

8. Common table salt is a good example; you sprinkle it on your food daily, unaware that it is made of two deadly poisons: sodium and chlorine.

9. Sodium is a poisonous metal and chlorine is a poisonous gas sometimes used in war gases.
10. An element consists of atoms of one kind. The name of the element is the name of that particular kind of atom.
11. A compound is made up of two or more different kinds of atoms (or elements).
12. An atom consists of a core or nucleus, orbited by one or more rings of electrons.
13. The nucleus is made up of positively charged bodies called *protons*, and neutral bodies called *neutrons*.
14. In a normal atom the total number of electrons orbiting the nucleus is equal to the number of protons in the nucleus. The number of neutrons is not directly related to the number of protons and electrons. The "weight" of neutrons added to the "weight" of the protons determines the atomic weight (more correctly, the atomic mass).
15. The nucleus is held together by tremendous energies.
16. In the explosion of an atomic bomb, the nucleus is split apart and the binding energies are released in the form of heat, light, and other radiant energies.
17. The electrons in all of the orbits except the one at the greatest distance from the nucleus are also anchored firmly to the nucleus.
18. The electrons in the outermost orbit may be fairly tightly anchored to the atom or quite free, depending on the particular element.
19. *All electricity consists of electrons removed from the outer orbit of atoms (conductor atoms, generally).*
20. These outer-orbit electrons (called *valence* electrons) are forced through an electric circuit to produce light and heat, run electronic devices, and perform other work.

ATOMS AND ELECTRONS

Any discussion of electricity must begin with the electron and the atom to which it is attached. There are over 100 different kinds of atoms, a few of which are the by-product of humankind's tinkering with atomic energy. Of the 90 or so natural elements, only a few are found by themselves in nature. Most of them tend to combine with other elements to form compounds. When two or more elements combine into compounds, the observable properties of the compounds are almost always quite unlike the properties of either of the original elements. For example, hydrogen and oxygen are colorless, odorless gases that have an affinity for each other and combine to form the life-giving liquid, water.

Water is a benign liquid, but its formation is a violent thing. When oxygen and hydrogen are brought together in the presence of enough heat to trigger the reaction, they explode, releasing great quantities of heat and light. When it is all over, all that remains is a pool of water. Probably the most spectacular hydrogen-oxygen reaction to hit the news was that of the dirigible Hindenburg on May 6, 1937. The violence on that occasion not only ended the era of commercial lighter-than-air transportation, but until recently it also helped prevent hydrogen from being seriously considered as an energy source.

The formation of a chemical compound with its accompanying energy release is the result of a transfer of electrons between atoms. Ordinary batteries are based on chemical reactions where the electron exchange is controlled by making the electrons transfer through a conducting wire. Instead of uncontrolled quantities of heat and light, the electrons produce controlled and useful work in their travels between atoms.

The hydrogen fuel cell, designed for on-board power in space vehicles, combines hydrogen and oxygen to form usable water—the same reaction that occurred on-board the ill-fated Hindenburg. The difference is that the electron transfer in the fuel cell is controlled by forcing the electrons to travel by wire and perform energy-demanding tasks in the process.

The amount of energy involved in a chemical reaction varies, depending on the compound being formed or broken up, but all such reactions involve electron transfer and energy. The amount of energy involved is sometimes deceiving, particularly if the reaction progresses slowly.

A wag once described a particularly uneventful political campaign as being "as exciting as watching old cars rust in the dump." The reaction of iron combining with oxygen to produce rust certainly fails to conjure up images of a spectacular energy release. Yet the energy involved is significant. If the iron is powdered or spun into fine "steel wool," igniting it will provide a satisfactory spectacle. Spun magnesium wool is ignited in the presence of oxygen to produce an intense light for photographic flash bulbs.

The point of all this is that the transfer of electrons from atom to atom can be used to do work. Electricity is the controlled transfer of electrons. There are, of course, some uncontrolled forms of electricity such as lightning.

THE STRUCTURE OF THE ATOM

All atoms consist of a nucleus orbited by electrons arranged in orbits at varying distances from that nucleus. The only electrons of interest in electrical and chemical behavior are those in the outermost orbit. These outer-orbit electrons are called *valence* electrons. Electrons closer to the nucleus are so tightly bound to it that they become free only when the nucleus itself is in danger of being split. This is the province of nuclear bombs and nuclear power plants, not electronics. Figure 3-1 shows a 3-dimensional representation of an atom, and the standard schematic representation.

Electrical Stability

The nucleus consists of protons that have a positive charge and a mass some 1800 times that of an electron, and neutrons that are a neutral combination of a proton and an electron. Because unlike charges attract, it might be expected that the electrons would be drawn into the protons in the nucleus, forming a neutron lump. This is not what happens. If we turn a bucket of water upside down, we would expect the water to spill to the ground. If we spin the bucket in an orbit in a plane at right angles to the ground, with enough velocity, the water stays in the bucket, even when it is upside down!

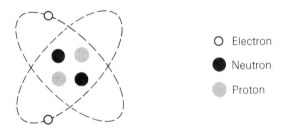

a. Three-dimensional representation of helium.

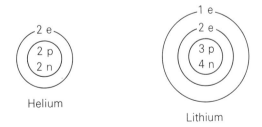

b. Schematic representation.

FIGURE 3-1. Representations of the atom.

In an atom, the motion of the electrons in their orbits balances out the attraction of the nucleus in the atom's planetary system. In all atoms there are exactly the same total number of orbiting electrons as there are protons in the nucleus.

For electrical stability, each orbit must contain a specific minimum number of electrons as shown in Figure 3-2. In any atom the inner shells will fill up first, so that each is complete. In the case of Shells M, O, and P, either 8 or 18 electrons constitute a complete shell. For Shell N, the shell is completely occupied in a stable condition with 8, 18, or 32 electrons. The goal of all atoms except hydrogen and helium is to have a complete outermost shell, with 8 electrons in most cases.

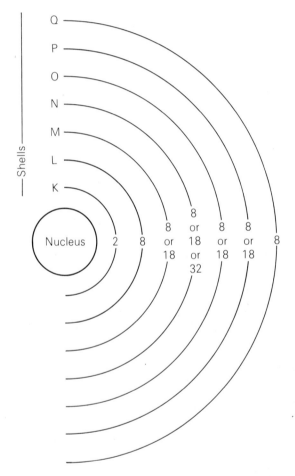

FIGURE 3-2. The number of electrons in each shell.

Only the noble gases are entirely stable and complete. They are too exclusive to interact with the peasantry of all the other elements. If they were humans, they would be a short-lived class because those gases do not react with each other either. As far as we know, they are completely inert (the ultimate leisure class) and do not form compounds in nature. Humankind's propensity for meddling with nature has yielded some *forced* compounds involving inert gases. These inert gases are electrically neutral, without human intervention or the intervention of such grand-scale natural phenomena as lightning or high-energy bombardment. Even under these "catastrophic" conditions, the reactions that happen are efforts to reestablish normal "noble" conditions. The six inert (noble) gases are helium, neon, argon, krypton, xenon, and radon.

All other elements are electrically unstable to varying degrees. The elements can be organized according to their chemical, physical, and electrical properties. Dmitri Mendeleev (1834–1907), a Russian scientist, published the first table showing such an organization. Since then there have been many changes, and elements not known to Mendeleev have been included. Table 3-1a is a modern version of the periodic table. The vertical columns in the periodic table list elements with similar, though not identical, properties. The lanthanide and actinide series are peculiar because all members of each series have properties so indistinguishable from each other that each series appears to be a single element.

Electron-transfer behavior, in both chemical reactions and electrical conductivity, is a function of the number of outer-orbit (valence) electrons. Eight electrons complete the outer shell and result in a nonconductive, chemically and electrically inert element. Only the inert gases have these characteristics. Half of the elements tend to give up electrons, leaving the next-to-outside shell complete. The other half tend to draw additional electrons to them. The maximum number of electrons in the outer shell is 8, and the elements with all 8 outer-shell electrons are the noble gases, which are inert. The very special elements with exactly 4 valence electrons form a group called *semiconductors*, the backbone of all modern electronic devices. These elements neither give up electrons nor accept them. Instead they participate in a complex electron-sharing behavior. The semiconductors are *not* inert. Those elements with outer shells containing less than 4 electrons give them up readily, yielding free electrons. The highly electrically conductive metals fall into this category. Elements with more than 4 electrons in the outer shell hang on tightly to them. These elements are poor electrical conductors (or good electrical insulators).

TABLE 3-1a PERIODIC CHART OF THE ELEMENTS

Shells		IA	IIA	IIIB	IVB	VB	VIB	VIIB		VIII		IB	IIB	IIIA	IVA	VA	VIA	VIIA	0
1	K	1 H 1.0079																	2 He 4.00260
2	L	3 Li 6.941	4 Be 9.01218											5 B 10.81	6 C 12.011	7 N 14.0067	8 O 15.9994	9 F 18.99840	10 Ne 20.179
3	M	11 Na 22.98977	12 Mg 24.305											13 Al 26.98154	14 Si 28.086	15 P 30.97376	16 S 32.06	17 Cl 35.453	18 Ar 39.948
4	N	19 K 39.098	20 Ca 40.08	21 Sc 44.9559	22 Ti 47.90	23 V 50.9414	24 Cr 51.996	25 Mn 54.9380	26 Fe 55.847	27 Co 58.9332	28 Ni 58.70	29 Cu 63.546	30 Zn 65.38	31 Ga 69.72	32 Ge 72.59	33 As 74.9216	34 Se 78.96	35 Br 79.904	36 Kr 83.80
5	O	37 Rb 85.4678	38 Sr 87.62	39 Y 88.9059	40 Zr 91.22	41 Nb 92.9064	42 Mo 95.94	43 Tc 98.9062	44 Ru 101.07	45 Rh 102.9055	46 Pd 106.4	47 Ag 107.868	48 Cd 112.40	49 In 114.82	50 Sn 118.69	51 Sb 121.75	52 Te 127.60	53 I 126.9045	54 Xe 131.30
6	P	55 Cs 132.9054	56 Ba 137.34	57 La* 138.9055	72 Hf 178.49	73 Ta 180.9479	74 W 183.85	75 Re 186.207	76 Os 190.2	77 Ir 192.22	78 Pt 195.09	79 Au 196.9665	80 Hg 200.59	81 Tl 204.37	82 Pb 207.2	83 Bi 208.9804	84 Po (210)	85 At (210)	86 Rn (222)
7	Q	87 Fr (223)	88 Ra 226.0254	89 Ac** (227)	104 (260)	105 (260)													

*Lanthanum Series

6	P	58 Ce 140.12	59 Pr 140.9077	60 Nd 144.24	61 Pm (147)	62 Sm 150.4	63 Eu 151.96	64 Gd 157.25	65 Tb 158.9254	66 Dy 162.50	67 Ho 164.9304	68 Er 167.26	69 Tm 168.9342	70 Yb 173.04	71 Lu 174.97

**Actinium Series

7	Q	90 Th 232.0381	91 Pa 231.0359	92 U 238.029	93 Np 237.0482	94 Pu (244)	95 Am (243)	96 Cm (247)	97 Bk (247)	98 Cf (251)	99 Es (254)	100 Fm (257)	101 Md (258)	102 No (255)	103 Lr (256)

TABLE 3-1b ELEMENT ABBREVIATIONS

Element	Symbol	Atomic Number	Element	Symbol	Atomic Number
Actinium	Ac	89	Mercury	Hg	80
Aluminum	Al	13	Molybdenum	Mo	42
Americium	Am	95	Neodymium	Nd	60
Antimony	Sb	51	Neon	Ne	10
Argon	Ar	18	Neptunium	Np	93
Arsenic	As	33	Nickel	Ni	28
Astatine	At	85	Niobium	Nb	41
Barium	Ba	56	Nitrogen	N	7
Berkelium	Bk	97	Nobelium	No	102
Beryllium	Be	4	Osmium	Os	76
Bismuth	Bi	83	Oxygen	O	8
Boron	B	5	Palladium	Pd	46
Bromine	Br	35	Phosphorus	P	15
Cadmium	Cd	48	Platinum	Pt	78
Calcium	Ca	20	Plutonium	Pu	94
Californium	Cf	98	Polonium	Po	84
Carbon	C	6	Potassium	K	19
Cerium	Ce	58	Praseodymium	Pr	59
Cesium	Cs	55	Promethium	Pm	61
Chlorine	Cl	17	Protactinium	Pa	91
Chromium	Cr	24	Radium	Ra	88
Cobalt	Co	27	Radon	Rn	86
Copper	Cu	29	Rhenium	Re	75
Curium	Cm	96	Rhodium	Rh	45
Dysprosium	Dy	66	Rubidium	Rb	37
Einsteinium	Es	99	Ruthenium	Ru	44
Erbium	Er	68	Samarium	Sm	62
Europium	Eu	63	Scandium	Sc	21
Fermium	Fm	100	Selenium	Se	34
Fluorine	F	9	Silicon	Si	14
Francium	Fr	87	Silver	Ag	47
Gadolinium	Gd	64	Sodium	Na	11
Gallium	Ga	31	Strontium	Sr	38
Germanium	Ge	32	Sulfur	S	16
Gold	Au	79	Tantalum	Ta	73
Hafnium	Hf	72	Technetium	Tc	43
Helium	He	2	Tellurium	Te	52
Holmium	Ho	67	Terbium	Tb	65
Hydrogen	H	1	Thallium	Tl	81
Indium	In	49	Thorium	Th	90
Iodine	I	53	Thulium	Tm	69
Iridium	Ir	77	Tin	Sn	50
Iron	Fe	26	Titanium	Ti	22
Krypton	Kr	36	Tungsten	W	74
Lanthanum	La	57	Uranium	U	92
Lawrencium	Lw	103	Vanadium	V	23
Lead	Pb	82	Xenon	Xe	54
Lithium	Li	3	Ytterbium	Yb	70
Lutetium	Lu	71	Yttrium	Y	39
Magnesium	Mg	12	Zinc	Zn	30
Manganese	Mn	25	Zirconium	Zr	40
Mendelevium	Md	101			

Properties of Valence Electrons

The Bare Facts

1. When the outer-orbit electrons are only lightly fastened to the atoms, they are called *free* electrons.
2. Any material with free electrons is an electrical conductor.
3. A conductor is capable of carrying an electric current.
4. When the outer-orbit (valence) electrons are tightly bound to the atom, they are not free and the material is a nonconductor or insulator.
5. A nonconductor (insulator) cannot carry an electric current.
6. Before valence electrons can leave the atom, a precise amount of energy is required to break them loose.
7. In the case of metals, the heat energy in the normal range of surface temperature on earth keeps the valence electrons disassociated from their respective atoms.
8. In insulators, the electrons cling to the atoms even at quite high temperatures.
9. In insulators, the temperature must often be raised to a point where the entire substance vaporizes, melts, or reacts chemically before electrons can be broken loose from the parent atom.
10. In semiconductors, the number of broken-loose electrons is dependent on the temperature. The higher the temperature, the more electrons will be disassociated from their atoms by thermal agitation.

ELECTRICAL CONDUCTION

In chemical reactions, elements that give up electrons easily combine with elements that freely accept them. In the process of electron transfer, electrons can be made to do useful work or to simply produce wasted heat and (or) light. In order to do useful electrical work, the electrons must move between atoms through some device that can extract the energy from the moving electrons. The electrons must be made to move, and this requires a force. The force that moves electrons is called *electromotive* force, and is abbreviated EMF.

Ions

When an electron leaves the valence shell and becomes completely free, the atom is left with one proton that has no electron to balance it. The atom is no longer neutral, but carries a 1 unit positive charge.

AN INTRODUCTION TO ELECTRICITY 67

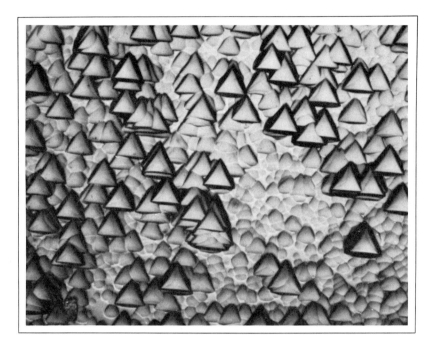

FIGURE 3-3. A silicon crystal magnified 150 times.

This atom is now a positive *ion*, exerting an electromotive force on any electron that strays close enough to be captured. A negative ion is formed when an excess electron is captured by a neutral atom. A neutral atom can exert a force on an electron if the atom has an incomplete valence shell, particularly when there are 5, 6, or 7 electrons in the valence shell. The negative ion is an atom with a negative charge. To summarize: *An ion results when a neutral atom gains or loses one or more electrons.*

In gases and liquids, both electrons and ions are mobile. Ions are roughly 2000 times heavier than electrons and move more ponderously than the lightweight electron, but they do move. In solids, the atoms have a permanent place and do not wander far, even when they become ions. Some scientists prefer not to call charged atoms in a solid ions, because of their limited mobility. However, it is more often convenient to give them the title anyway.

The conductive solids are mostly metals and are generally found with individual atoms arranged in a repeating geometric form called a *crystal*. Figure 3-3 shows a silicon crystal magnified 150 times. The geometric structure is called a *lattice*. The valence electrons between the atoms in a metal are so free at all ordinary earth temperatures that if we could see them, they would appear much like steam around the bodies in a turkish bath. The electrons in the inner shells would maintain their regular planet-like orbits around their special nucleus.

In an insulating solid, even the valence electrons would be following a regular orbital pattern, with only an infinitesimal number of apparently homeless, wandering valence electrons. There are no perfect insulators, nor can any conductor be classed as perfect. From a practical standpoint, however, most insulators and most conductors are very nearly perfect. Silver may be a slightly better conductor than gold, and gold slightly better than copper, and copper somewhat better than aluminum, and so on. We might make an analogy: A person who has 10 million dollars is wealthier than one who has only 8 million dollars, and a person who has 8 million dollars is wealthier than one who has an insignificant 1 million dollars. We would certainly class all of them as wealthy, as opposed to middle-class or poor! In the same fashion, one conductor may be better than another, but even the poorest conductor could not be properly classed as a partial insulator. Nor could the world's worst insulator legitimately be classified as any kind of conductor!

In the case of those very special, 4-valence-electron elements, carbon, silicon, and germanium, we would see valence electrons sharing nuclei in a pattern similar to the one shown in Figure 3-4c. In addition, there would be many more homeless wandering electrons than in an insulator, and far fewer than in a metal. The actual number of unbound electrons roaming more or less aimlessly would depend almost entirely on the internal temperature of the semiconductor in question.

Temperature has some effect on every natural phenomenon; the conductivity of metals and insulators is no exception. However, the effect of temperature on semiconductors is enormous in contrast to its relatively small effect on the conductivity of metals and insulators.

The Conductivity of a Metal

In metals, valence electrons are essentially homeless and free to wander. They are pushed around by thermal energy and are attracted to any area where the electron fog is less dense. In less-dense areas there will be more protons than electrons, and electrons will feel the force of those protons pulling them in the direction of the less-dense part of the swarm. The electrons are in constant random motion, being buffeted by a combination of thermal energy and electromotive forces produced by the positively charged protons. Randomness, by definition, has no direction, and although electrons are in constant motion there is no *net* current in an isolated chunk of metal.

AN INTRODUCTION TO ELECTRICITY   69

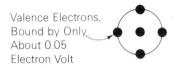

a. In lone silicon atoms the valence electrons are as lightly bound as they are in a conductor. Lone atoms are too rare to bear serious consideration.

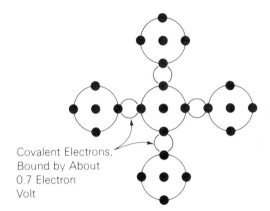

b. The extra bonding energy in covalent bonds accounts for semiconduction behavior.

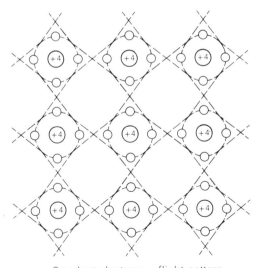

c. Covalent electrons—flight pattern.

FIGURE 3-4. Silicon semiconductor atoms.

Electrical Current Is Electron Movement

In the system we have been discussing, current flow does exist, but we cannot put it to work until we provide the proper motivation. Imagine a group of bricklayers milling around, waiting for the foreman to arrive and put them to work. Being bricklayers, the men are likely to handle the bricks, picking them up and setting them down at random. Work is being done, bricks are in motion, but no wall is getting built. The foreman arrives and provides the necessary motive force: "Build that wall if you want to get paid." The activity increases and the movement of the bricks becomes purposeful. In a chunk of metal there are plenty of electrons ready to do our bidding if we provide them with the proper motivation in the form of electromotive force.

At this point let us add to the definition of electricity, or electrical current: *Dynamic electricity is electrons in motion with a definite net direction.*

3-3 ELECTRICAL CURRENT

The Bare Facts

1. When we say that electricity flows, we really mean that valence electrons are leaving their respective atoms and moving through a conductor.
2. When electricity flows, valence electrons jump from atom to atom, moving in a direction toward the positive end of the wire.
3. Electricity (electrons) flows from negative to positive.
4. Electricity moves from an area of high electron density (negative) to an area of lower electron density (positive).
5. If we push an electron in one end of a wire, an electron will pop out of the other end in nearly the same time it would take for light to travel the same distance.
6. The electron that pops out will not be the same one we push in.
7. Imagine bumper-to-bumper traffic moving slowly down the freeway. The car at the head of the line stops suddenly. The last car bumps the car ahead . . . it in turn bumps the car in front of it. Almost instantly the car at the head of the line gets bumped ahead a little. Though the cars were moving slowly, the impact moves down the line very rapidly.

AN INTRODUCTION TO ELECTRICITY 71

8. Electron motion is analogous to the above example. Each electron moves fairly slowly, but the energy moves as though the electrons were moving with nearly the speed of light.

9. Electrons moving through a wire are called *current*, and current is measured in amperes.

10. An ampere is defined as a certain number of electrons, 6.25×10^{18} of them, moving past a point (in a wire) in 1 second.

11. Not all materials can carry an electric current.

A special note: We have defined electricity and electrical current as a movement of electrons from negative to positive. Although this is a valid definition there is another point of view: the concept of conventional current.

CONVENTIONAL CURRENT

As electrons move from negative to positive, the distribution of atoms missing valence electrons moves from positive to negative. This is called *conventional current*. By this definition, current is a movement of positive charges from positive to negative. People seem to have very strong feelings about which convention should be the standard—electron or conventional current flow. In this text we will generally talk about electron flow. There are, however, cases where the conventional-current view is more appropriate. In those cases we will use conventional current with proper notification. This author generally takes a neutral stand on the controversy, and will change sides whenever one convention is more convenient than the other. Advocates of both sides of the controversy can be very persuasive.

BASIC ELECTRICAL CONCEPTS

Electrical current is electricity. We will be discussing voltage and resistance as a part of electricity, but these exist only as a part of the moving-electron system. *Voltage* is another name for the electromotive force that moves electrons. It has only a potential existence without electrons and the movement of electrons. *Resistance* is a measure of the electrical friction that moving electrons encounter in nature's less-than-perfect conductors. Like any other friction, resistance has no meaning without some force acting to move some body. Again, resistance has no existence of its own. It is simply a manifestation of electrons moving against some opposition.*

---
* This is a somewhat oversimplified view, but is adequate for our purposes.

Electrical current is measured in amperes. An ampere (commonly referred to as an amp) is defined as a given number of electrons passing a point in 1 second. The number is rather large:

$$6{,}250{,}000{,}000{,}000{,}000{,}000$$

or, in the more compact scientific shorthand: 625×10^{16}, six hundred twenty-five with 16 zeros attached.

$$1 \text{ amp} = 625 \times 10^{16} \text{ electrons per second}$$

For the mathematically inclined:

---

1. 6.25×10^{18} is standard form.

$$^{625}/_{100} \times 10^{16} \times 100$$

2. Since: $100 = 10^2$

$$6.25 \times 10^{16} \times 10^2$$

3. Because exponents are added when multiplying:

$$10^{16} \times 10^2 = 10^{18}$$

In standard form, then,

$$6.25 \times 10^{18}$$

is the same numerical value as

$$625 \times 10^{16}$$

One of the reasons for using scientific shorthand is simply to make numbers of this size pronounceable. It is possible to pronounce the number 625 followed by 16 zeros as "six hundred twenty-five times ten to the sixteenth power." English numerical words for numbers larger than a million are limited and don't communicate well unless everyone knows exactly what they mean. Most people forget.

A "magical" box with 6.25×10^{18} electrons in it would contain a coulomb of electrons (like 1 ounce of cereal). Such a "magic" box actually exists and we will study about it later. It is called a *capacitor*.

Voltage

The Bare Facts

1. It takes a force and energy to move free electrons through a wire.

2. The force or pressure used to move the electrons through the wire (or an electrical device) is known as *electromotive force, potential difference*, or more commonly, *voltage*.

3. In an ordinary piece of wire, electrons (free electrons) are distributed evenly throughout the wire, and there is no *net* movement of electrons through the wire.

4. If some electrons can be pumped from one end of the wire to the other, this balance is upset and electrons move from the end of the wire that has more electrons to the end that has fewer electrons.

5. This movement of electrons is known as an *electric current.*

6. The difference between the number of electrons in one end of the wire and the other results in a potential difference, or voltage.

7. Voltage is potential energy. When you lift a weight off the floor it gains the potential energy to do work if you let it fall (preferably not on your foot).

8. Electrons do not move themselves. There must be an energy input.

9. Voltage is energy with the potential to do work.

10. Energy cannot be created or destroyed.

11. Other forms of energy can be converted into electrical energy and electrical energy can be converted into other forms.

12. Electrical current can flow only when this potential difference or voltage exists.

13. If we pump 10 or a million electrons from one end of the wire to the other, a potential difference will be created. (That is, we will have a voltage.) As soon as both ends have the same number of electrons again, there will be no more potential difference or voltage.

14. *All of the voltage is used up in the circuit regardless of characteristics of the electrical circuit or device.*

15. When a potential difference (or voltage) exists, the end with more electrons is called the *negative* end of the circuit, and the end with fewer electrons is called the *positive end.*

16. Electrons (current) always flow from the side of high electron density to the side of low electron density—*that is, from negative to positive.*

74 CHAPTER 3

17. Electrical pumps used to create potential differences are called *voltage sources*.

18. Some examples are batteries, generators, alternators, solar batteries, fuel cells, thermocouples, and piezoelectric crystals.

Electromotive force (EMF) is measured in volts (named after Alessandro Volta, 1745–1827). Voltage is always the result of two areas having different numbers of electrons. EMF is called *potential difference* because there are always some electrons in all regions in a conductor. The net force to move electrons is the difference between the number of electrons at two different locations.

In order to examine the nature of potential difference, suppose we make some sketches involving hypothetical "magic" boxes and see what happens when certain conditions exist. Let us imagine that we have reached into Box A in Figure 3-5a, grabbed 3 electrons, and dropped them into Box B. Box A has now lost 3 electrons, leaving 3

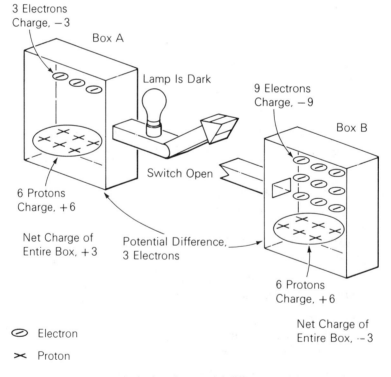

a. Isolated potential difference.

FIGURE 3-5. Illustration of potential difference.

protons that are not neutralized. Box A takes on a net (total) charge of +3 because there are 3 more positive charges (protons) in Box A than there are negative charges (electrons). In Box B there are now 9 electrons (negative charges) and only 6 protons (positive charges). The net charge in Box B is −3. There are 3 excess electrons. These intruder electrons in Box B are not particularly welcome and they are

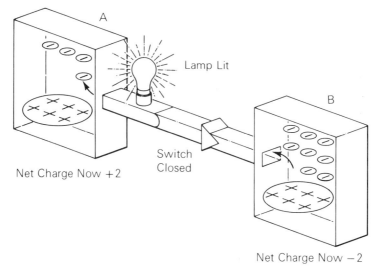

b. Moving electrons, doing work.

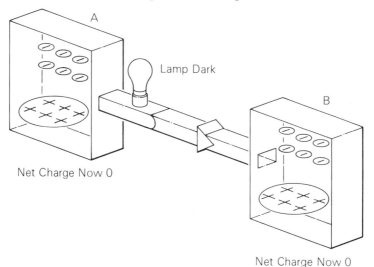

c. After three electrons have passed through the lamp.

FIGURE 3-5. (continued)

more than free to leave at any time, and good riddance. Those same electrons will be welcomed as prodigal sons if they can only get home to Box A. In Figure 3-5a the excess electrons are stuck, the switch is open, and there is no path home. In Figure 3-5b we create the path by closing the switch. The attraction between unlike charges draws an electron through the connecting wire. The proton is too heavy and well anchored to meet the electron partway, even though the electron pulls on it as strongly as it pulls on the electron. The homeward-bound electron does not get free passage in the example. It must work to earn its way back, by heating the little filament wire in the lamp.

No conductor is perfect; that is to say, all conductors offer some resistance to electron movement. The filament in the lamp is designed to have a high resistance. A lot of resistance means that there is considerable friction between moving electrons and atoms in the filament. As in other kinds of friction, energy of motion is converted into heat. The heat generated by the electrical friction in the high-resistance lamp filament is enough to bring the filament up to a white heat.

When a body is allowed to fall from a height under the influence of a gravitational force, it gains a great deal of speed in free fall. If it meets any resistance during the fall, the friction will slow the body down and convert energy of the moving mass into heat. A shooting star (meteorite) is a notable example. It accelerates toward earth, meeting little resistance in space. When it enters the earth's atmosphere, the friction heats it to visible brightness and slows it down. If the friction slows it down to something less than escape velocity, it falls to the earth. Upon impact it expends the remainder of its energy of momentum in digging a crater for the amazement of tourists.

The electron is accelerated by the attraction of the proton. In Box A, in Figure 3-5b, it gives up energy in the form of heat (and light), finally coming to rest with its energy expended back home in Box A. The electron has returned with no loss of charge, or mass; in fact it has not been changed at all by the journey. The system has given up some energy of motion to heat and light the lamp, but the electron itself hasn't undergone any changes. If the system has given up energy, then it must contain less energy than before and something about the *system* must have changed to reflect that energy loss.

An examination of Figure 3-5b provides the clue to what was given up in the process of producing heat and light. Suppose we stop the action just as it is shown (Figure 3-5b), with only one electron having left Box B and arrived at Box A. The difference between the conditions in Figures 3-5a and 3-5b are:

In Figure 3-5a there were 3 electrons in Box B that belonged in Box A, a potential difference of 3 units. In Figure 3-5b there are only 2 electrons in Box B that belonged in Box A, a potential difference of 2.

In lighting the lamp, the potential difference has dropped from 3 to 2. If we permit the action in Figure 3-5b to start again, electrons will continue to move from Box B to Box A until we have the conditions shown in Figure 3-5c. The lamp is dark. We have used up all of the potential difference; it has dropped to zero. There is no longer an electromotive force to drive electrons through the lamp.

EMF = Electromotive Force = Voltage = Potential difference = 0

EMF, electromotive force, voltage, and potential difference are interchangeable words.

In case we don't want to sit in the dark just because we have run out of potential difference, we can create some more. Actually, we cannot create it, but we can convert something else into it. Energy can neither be created nor destroyed, but its form can readily be changed. Our example was a bit unrealistic in that we just grabbed a handful of electrons from Box A and dropped them into Box B. If such a procedure were possible, it would involve converting human muscle energy into potential difference. In the real world it is possible to convert chemical, heat, mechanical, magnetic, and electromagnetic energy into potential difference.

Examples of electrical energy converters:

- Chemical: Batteries and fuel cells.
- Mechanical: Piezoelectric devices (some record-player tonearm cartridges convert the motion of the needle directly into electrical current).
- Heat: Thermocouples converting heat directly into electricity to operate electrical valves in natural-gas heaters.
- Magnetic: Motor-, steam-, or water-driven generators converting mechanical energy into magnetic fields to generate electricity.
- Photoelectric: Solar cells converting light directly into electricity.
- Friction: Lightning, and walking across a rug accumulating electrons on your body.

The practical unit of potential difference is the volt. The work done in moving 1 coulomb (6.25 × 10^{18} electrons) is defined as 1 joule. In the previous "magic-box" example, if we had begun by taking 6.25 × 10^{18} electrons out of Box A and placing them in Box B, we would have produced a potential difference between the boxes of 1 volt. The transfer of those electrons would have required 0.7376 foot-pounds of muscle work, the equivalent of lifting slightly less than 1 pound a distance of 1 foot against earth gravity. The metric equivalent of the foot-pound is the joule: 1 joule = 0.7376 foot-pound.

In the magic-box example, if we had transferred 1 coulomb (6.25 × 10^{18} electrons) from Box A into Box B, closing the switch would have resulted in 1 joule of heat and light before the potential difference became zero volts.

Resistance

The Bare Facts

1. Whenever electrons flow through a circuit, they meet some opposition.
2. In overcoming the resistance to electron flow, some energy must be lost from the circuit.
3. This loss is converted into heat and (usually) radiated into the surrounding air.
4. Current is never lost in an electrical or electronic circuit. In other words, *electrons* are not lost from the system.
5. The better the conductor, the less resistance to current flow.
6. The poorer the conductor, the greater the *resistance*.
7. Metals such as gold, silver, copper, and aluminum, etc., have very low resistance and are known as *conductors*.
8. Carbon, silicon, germanium, some metal alloys, and some liquids carry current but offer a higher resistance than metals to current flow.
9. Relatively high-resistance materials such as carbon and nichrome (an alloy of nickel and chromium) are used in commercial resistors and heating elements.
10. Resistance materials are useful for heating (and lighting). A high resistance converts a large amount of voltage into heat, and if the temperature is high enough, into light.

11. In the form of commercial resistors, these materials provide a means of reducing currents to lower values. This is the equivalent of having a lower applied voltage.
12. For example, a 6 volt car radio may be used in a 12 volt car if a resistor is used.

The practical unit of resistance is the ohm. *One ohm of resistance will restrict the current in a circuit to 1 amp when an electromotive force of 1 volt is applied.* In order to get a better picture of voltage, current, resistance, and their relationships, suppose we turn to an analogy.

The Skydiver

The skydiver in Figure 3-6 falls toward earth with parachute folded for as long as it is safe. The chute finally opens and his speed is reduced for a safe landing.

The Analogy

Electrical System	Skydiver System
Electromotive force (voltage)	Gravitational force
Current (electrons per second)	Velocity (feet per second)
Electrical resistance	Wind resistance

If our skydiver starts his free fall above the atmosphere, there will be no resistance. The gravitational pull (force) of the earth accelerates the diver's speed. The diver leaves the plane with no downward speed, but if left unresisted he will hit earth at a terrible speed. An electron in a vacuum will also accelerate under the influence of the attractive force of a proton. When the skydiver enters the atmosphere, he encounters wind resistance. He can then slow his fall by offering as much body surface as possible. For a given air density and body

80 CHAPTER 3

FIGURE 3-6. The skydiver.

area, the speed of his fall will depend entirely on the gravitational force and the wind resistance against his body. In physics this constant speed is called the *terminal velocity*. In like manner, the electron will slow as it encounters the resistance of atoms in its path.

With a steady stream of skydivers or electrons, we can define electron or skydiver current in terms of the number of skydivers or electrons passing a given point each second. In either case, current would increase with an increase in force (voltage or gravity), and current would decrease as the resistance increased. Figure 3-7 compares electrical and skydiver current.

Electrons are constantly speeding up and slowing down as they move between atoms. Individually, they never attain very high speeds, and yet electricity travels very fast indeed, far faster than the speed of individual electrons.

The Case of the Surprised Brakeman

In many books you will find a statement to the effect that electricity travels at nearly the speed of light. Such a statement seems to contradict our previous discussion about electron movement. The *effect* of electricity travels at speeds approaching that of light, even though the electrons themselves may be creeping along by comparison. In order to explain how this can be, let us resort to another analogy.

Imagine, if you will, a 31 car train stopped temporarily on a siding (see Figure 3-8, page 84). A new brakeman, who is just learning his job, has climbed up on a rear platform of the caboose while the train is at rest. Each car is 15 meters (50 feet) long, and the engineer is therefore 465 meters (1500 feet) from the caboose. When the engineer puts the engine into reverse and starts to move, the engine bangs the boxcar immediately behind, and ½ second later the novice brakeman is jarred off the rear platform of the caboose. The engine is barely moving, but the impact covered the 465 meters to the caboose in ½ second. A little calculation will show that the shock of that impact between the engine and the first car traveled at the rate of 930 meters (300 feet) per second, or 3348 kilometers (2000 miles) per hour! Assume that we noticed the chief brakeman walking beside the engine as it began to move. The average adult walks casually at the rate of about 6 kilometers (3.8 miles) per hour, so we could be certain that none of the cars were moving at more than about 6 kilometers per hour when the surprised brakeman was knocked off the caboose platform.

In this example no car moves at a speed greater than 6 kilometers (3.8 miles) per hour and yet the impact travels at a rate of over 3300 kilometers (2000 miles) per hour. How can a 3300-kilometer-per-hour impact speed be reconciled with our observations?

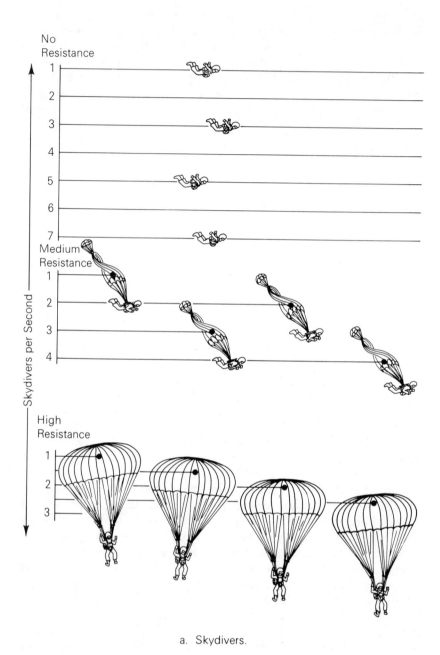

a. Skydivers.

FIGURE 3-7. Electrical and skydiver current compared.

AN INTRODUCTION TO ELECTRICITY   83

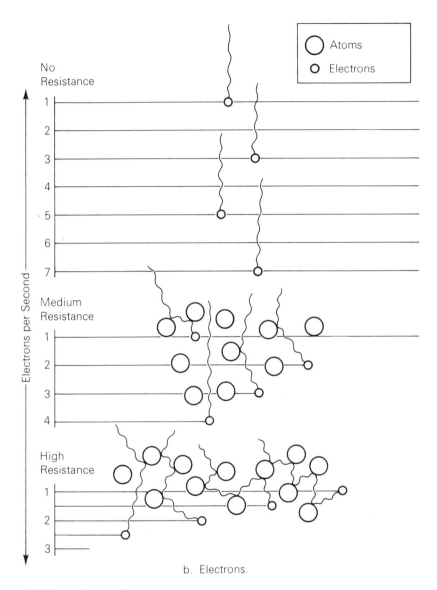

b. Electrons.

FIGURE 3-7. (continued)

As a matter of fact, we have been deceived. Assume that each hitch between cars has 2 centimeters of play; that is, the engine must move 2 centimeters before its half of the hitch makes contact with the front hitch on the car behind it, and so on. If there is 2 centimeters of gap in each of the 30 hitches (the rear hitch on the caboose doesn't count), the real movement required to jar the brakeman is only 60 centimeters total. The impact travels that 60 centimeters in one-half second,

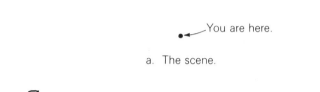

a. The scene.

b. The astonished brakeman.

FIGURE 3-8. The astonished brakeman. (Courtesy Kelly Graphics)

which is a speed of 120 centimeters per second or 4.32 kilometers (2.6 miles) per hour. This is a figure that does not conflict with our direct observations. Just because the mystery has been solved, let's not lose sight of the fact that the impact actually shook the surprised brakeman off the caboose 465 meters away in one-half second.

Electrons behave similarly, and when an electron is injected into one end of a conductor, an electron pops out of the other end almost instantly, but not the same electron that was injected.

If you are interested in the mathematics of the example, read the box at the top of the facing page.

3-4 ELECTRICAL ANALOGS

Electricity is dynamic. In any electrical system electrons are in constant motion, and a great many things are happening that we cannot see. As we progress through this book we will discover increasingly complex electrical behavior. We can learn the facts of what goes on in an electrical or electronic system, but it is just as important to have some intuitive *feel* for the dynamics of electronics. Animated pictures could be of great help, but in a book we are restricted to words and still-pictures.

> *Computing the apparent impact speed*
>
> The impact travels 465 meters in 0.5 second, or 930 meters per second.
>
> 1. To get meters per minute: 930 × 60 = 55,800
> 2. To get meters per hour: 55,800 × 60 = 3,348,000
> 3. To get kilometers per hour: 3,348,000 ÷ 1000 = 3348 kilometers per hour.
>
> *Computing the real speed*
>
> 30 hitches × 2 cm = 60 centimeters
> 60 centimeters in 0.5 second is a speed of 120 centimeters per second
>
> 1. To get meters per second: 120 centimeters ÷ 100 = 1.2 meters per second
> 2. To get meters per minute: 1.2 × 60 = 72
> 3. To get meters per hour: 72 × 60 = 4320
> 4. To get kilometers per hour: 4320 ÷ 1000 = 4.32 kilometers per hour.

We have been using a tool in this book to greatly increase the effectiveness of those words and still-pictures. This tool is the analogy. By relating electronic behavior to gears, wheels, the flow of water, and other familiar dynamic happenings, we can more easily visualize the invisible dynamic behavior in electronic devices and circuits. Without analogies, any real understanding of electronics is almost impossible. We are so dependent on our senses that it is very hard for us to really understand things when our senses cannot provide adequate input. Analogies provide us with a sort of second-hand set of sensory experiences. Much of the science of acoustics has been developed through the use of analogies with other, better-understood systems, such as mechanics and electronics.

The use of analogies in a book about electronics would probably result in more confusion than enlightenment if it were not for the fact that there are only 6 fundamental concepts in electricity . . . energy (voltage), flow (current), friction (resistance), mass (inductance), compliance (capacitance), and acceleration (changes in current). At this point *compliance* is probably an unfamiliar term, and *mass* may not seem to excite any ideas. But you are far more familiar with compliance and mass than you probably realize, as you will see later in the text. In this chapter we will examine the first 3 concepts and save the last 3 for later chapters.

For example, force is familiar to us in a number of contexts. Without it, the world's activity would cease. All movement requires force. The professor gives out grades to motivate students to study. Motivation is simply mild force and carries the connotation of motion.

Force is the ability to cause or change motion. The term *electromotive force* identifies the particular kind of force that moves electrons. Mechanomotive force identifies the kind of force required to move things in the mechanical world. Acoustomotive force identifies the force that moves air molecules to produce sound in our brain when the movements of air molecules force movements in the diaphragm of our ears. Table 3-2 describes force in electrical, mechanical, and acoustical systems.

TABLE 3-2 UNITS OF FORCE

Type of Force	Unit
Electromotive force	Volt*
Mechanomotive force	Pound or dyne
Acoustomotive force	Dyne, or gram per square centimeter

*The volt is more accurately a unit of work.

Conversions:

    980 dynes = 1 gram of force
    980,000 dynes = 2.2 pounds of force
    A larger unit of force in the metric system
      is the *newton*.
    9.8 newtons = 2.2 pounds of force

Some Definitions

Energy	The ability to do work
Work	Combination of force and motion—force moving something over a distance

3-5 DIRECT AND ALTERNATING CURRENT

The principal sources of electromotive force (voltage) fall into 2 basic categories: devices producing alternating current and those producing direct current.

DIRECT CURRENT

Batteries, solar batteries, thermoelectric generators, and fuel cells are direct-current devices. The principal source of alternating current is the electromagnetic generator of the sort used in automobiles, commercial power plants, and so on. There are also so-called direct-current electromagnetic generators that are actually alternating-current generators with integral reversing switches. Older automotive generators accomplished the switching by fastening switching segments on the rotating shaft. This array of segments is called a *commutator* and is contacted by fixed carbon brushes. Modern automotive generators use electronic switching devices called *diode rectifiers*.

Chemical batteries are a principal source of direct current. There are two classes: rechargeable and expendable batteries. Battery technology is still in its infancy in spite of the wide variety of available types. The demand for compact, lightweight, high-storage-capacity batteries for electric vehicles, cameras, and so on, is forcing the industry into new areas of research.

All sources of EMF are simply energy-conversion devices. Batteries convert the energy of chemical reaction into the EMF required to push electrons to do work. Generators convert mechanical energy into electromotive energy (force). We will examine how a generator works in Chapter 5.

Direct current is a continuous current that always flows in the same direction. Alternating current begins at zero amperes, rises to some maximum (peak) value, and returns to zero. The polarity of the voltage then reverses and the electrons are driven through the wire in the opposite direction. In the opposite direction, the current again passes through zero, rises to the same numerical peak value as before, and then returns to zero to begin the next cycle. A drawing is the best explanation, so let us examine Figure 3-9. Direct current is represented by a straight, unvarying line, in this case 10 volts above the zero reference line. Time is unimportant in direct current. It begins when the circuit is activated and continues until the current is shut off or the battery fails.

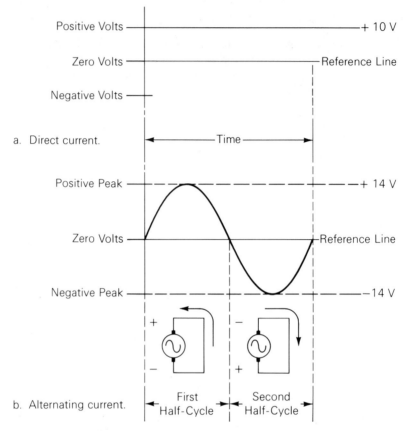

Note: A graph of current versus time would have the same shape as the voltage graph above.

FIGURE 3-9. Direct and alternating currents.

ALTERNATING CURRENT

In alternating current (AC), time is extremely important because of the cyclic nature of AC. One complete cycle is shown in Figure 3-9. The time required for one complete cycle is called the *period* of the wave.

The peak voltage of the AC is shown as 14 volts. This value was selected deliberately because the direct current (DC) and the alternating current illustrated will both do the same amount of work. The heat-producing capability of 10 volts DC is equal to the heat-producing capability of 14 volts (peak) of alternating current. The AC peak voltage is equal to about 140 percent of the same value of steady-state DC voltage.

Effective AC Voltage (AC Sine Wave)

The effective value of an AC voltage (its direct-current equivalent) is equal to about 70 percent (70.7 percent) of the AC voltage's peak value. This 70 percent value is called the *RMS* value and is what is always meant by a figure such as 120 volts AC, unless a peak or peak-to-peak value is specifically mentioned.

1. To find the equivalent direct-current voltage for an alternating current:

$$AC_{RMS} = 70.7 \text{ percent of the peak voltage}$$

Mathematically:

$$\boxed{AC_{RMS} = 0.707 \; AC \text{ peak}}$$

2. To find the value of the peak AC voltage required to equal a given direct-current value:

$$AC_{peak} = \frac{AC_{RMS}}{0.707}$$

Or:

$$AC_{peak} = 1.4 \times AC_{RMS}$$

Mathematically:

$$\boxed{\begin{array}{c} AC_{peak} = \dfrac{1}{0.707} \times AC_{RMS} \\ \text{The reciprocal of } 0.707 = 1.4 \\ \text{(Divide 0.707 into 1)} \end{array}}$$

The curve shown in Figure 3-9 is known as a *sine wave*. It is a graphical description of the ocean's incoming and outgoing tides, for example, or the trace a swinging pendulum would make if a strip of paper were slid along beneath it. Many of nature's cycles follow this pattern. In nature the most visual example of the sine-wave shape is that of ripples radiating out from a pebble dropped into a calm lake. Even here our usual view does not clearly reveal the sine-wave shape. The sine-wave shape becomes apparent in the cross section shown in Figure 3-10.

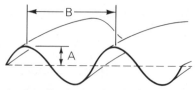

A: Amplitude (peak value)
B: One Complete Cycle

a. As we normally see them.   b. Shown in cross section.

FIGURE 3-10. Water waves.

Frequency and Period

The frequency is defined as the number of complete cycles the sine wave passes through in 1 second. The unit, cycle per second (cps), has been named *hertz* after Heinrich Rudolf Hertz (1857–1894), the discoverer of radio waves, which are also sine waves. The abbreviation for hertz is Hz.

$$\text{Cycles per second} = \text{hertz} = \text{Hz}$$

Frequency and period are described in Figure 3-11. For the mathematically inclined:

Period and frequency are reciprocals.

$$T = \frac{1}{f} \quad \text{and} \quad f = \frac{1}{T}$$

where
T = the time for one period
f = the frequency, in hertz (cps)

Designating Points on the Sine Curve

Because electrical sine waves were first produced by a rotating generator, positions in time along the curve are designated in degrees. The output voltage of the generator varies in a sine-wave fashion as the shaft rotates. The output voltage at any instant depends on the position of the generator shaft at that instant. The sine wave is divided into 360 degrees (360°), the number of degrees in a circle (one complete shaft rotation). This is illustrated in Figure 3-12.

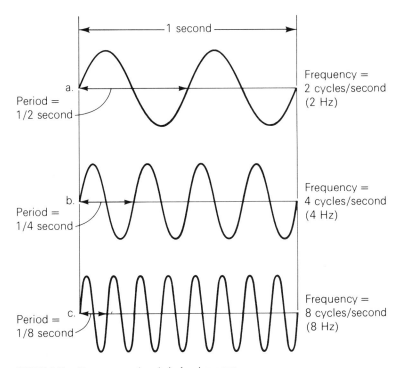

FIGURE 3-11. Frequency and period of a sine wave.

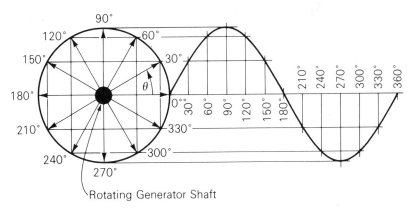

FIGURE 3-12. The sine wave measured in degrees.

The voltage at specific points along the curve can be found by using Table 3-3. Because the voltage is varying with time, the value at a certain number of degrees is called the *instantaneous* value. To use Table 3-3, simply multiply the value taken from the chart times the peak voltage.

TABLE 3-3 FINDING THE INSTANTANEOUS VOLTAGE

Number of Degrees				Multiply the Peak Voltage by:
0	180	360	—	0
30	150	210	330	0.5
45	135	225	315	0.87
60	120	240	300	0.707
90	270	—	—	1

Note: Affix a minus sign to voltages that are found beyond 180° (but not over 360°).

Example:

An alternating-current sine wave has a peak voltage of 168 volts:

1. Find the instantaneous voltage at 45°:
   From the chart in Table 3-3, we get 0.707.
   Instantaneous voltage = 168 × 0.707 = 119 volts

2. Find the instantaneous voltage at 210°:
   From the chart in Table 3-3, we get 0.5.
   $e_{inst.}$ = 168 × 0.5 = 84 volts

For the mathematically inclined:

$$e_{inst.} = e_{max}(\sin \theta)$$
where θ is the angle of rotation.
See Figure 3-13.

The values in Table 3-3 are sin (sine) values for the angles shown. Values for other angles can be found in trigonometry tables or with scientific calculators.

Figure 3-13 summarizes voltages, currents, and resistances. Figure 3-14 summarizes alternating-current peak, peak-to-peak, and RMS values.

AN INTRODUCTION TO ELECTRICITY 93

a. The standard measurement of electron quantity (a standard "scoop" of electrons).

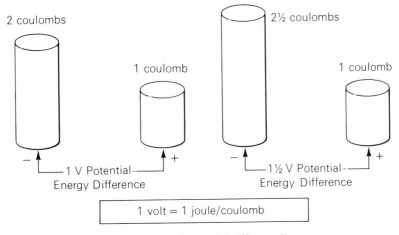

b. Voltage (potential difference).

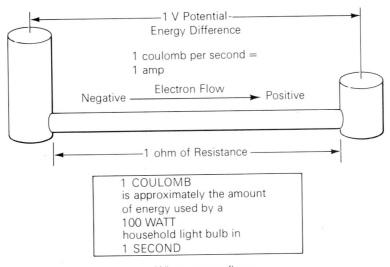

c. When current flows.

FIGURE 3-13. Voltage, current, and resistance summary.

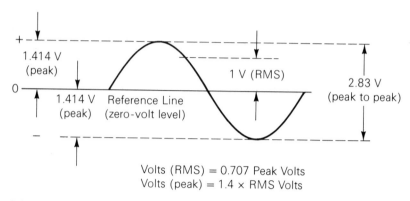

FIGURE 3-14. Alternating-current voltage summary.

Study Problems

Atoms and Electrons
1. There are approximately how many different atoms?
2. What is the difference between an atom and a compound?
3. Do two or more of the *same* kind of atom often combine to form compounds?
4. Does energy absorption or release always accompany the formation of a compound?
5. Is there always an energy exchange when a compound is broken up into its constituent atoms?
6. Suppose we are about to form a new compound, using two elements whose properties are very well known and understood. Is it fairly simple to predict in advance what properties the new compound will have?
7. Describe the basic construction of an atom. Make a labeled sketch of it.
8. Electrons carry what kind of charge?
9. Protons carry what kind of charge?
10. What is the approximate difference between the weight of a proton and that of an electron?
11. What is the relationship between the total number of electrons in the orbits and the number of protons in the nucleus of an atom?
12. In electricity, the only electrons involved in our work are those that reside in a specific orbit; which orbit is that?
13. What title do we give the electrons in that special orbit?
14. What is electricity (dynamic, not static)?
15. In terms of valence electrons, what is the difference between insulators and conductors?

16. In terms of valence electrons, how do the number of valence electrons available for electrical conduction differ between conductors and semiconductors? What specific kind of energy is involved in this difference? (If you don't understand this question, skip it. Continue answering questions until it does make sense).
17. True or false: In metals, nearly all valence electrons are free to participate in electrical conduction at any normal temperature on earth.
18. True or false: In the case of insulators, the statement in Question 17 is true because temperature has only a modest influence on the electrical behavior of metals and insulators.
19. True or false: In semiconductors, the number of valence electrons free to participate in conduction is almost entirely governed by temperature. (Go back to Question 16 if you were unable to answer it. If you did answer it, you may wish to modify your answer.)

Electrical Current
20. Define electrical current.
21. Which of the following statements is correct?
    a. Electrons (current) flow from negative to positive.
    b. Electrons (current) flow from positive to negative.
22. True or false: Electrons move from areas of high electron concentration to areas of lower electron concentration in a conductor.
23. Explain why electrons appear to move through a conductor at nearly the speed of light, even though the electrons actually move at a vastly slower speed. (Please do not use a train as part of your explanation).
24. When there are 6.25×10^{18} electrons flowing past a point in a circuit each second, how much current is flowing?

Voltage
25. What kind of force is required to move electrons through a circuit?
26. Voltage has two other common names. What are they?
27. Define potential difference.
28. How is potential difference produced?
29. Define voltage.
30. Define current.
31. If there are 6.25×10^{18} more electrons at Point A in a circuit than there are at Point B, what is the potential difference, in volts, between Points A and B?

Resistance
32. True or false: Better conductors have lower resistances.

33. True or false: In some applications high-resistance, poor conductors are more satisfactory than low-resistance, good conductors.
34. If Question 33 is true, suggest such an application.
35. In an electrical circuit, which of the following is "used-up" by resistance in the circuit?
    a. Current (electrons).
    b. Voltage (potential difference).
36. One of the quantities in Question 35 is converted by resistance into a form of energy. What form is it?

Chapter 4

CIRCUITS WITH RESISTORS AND VOLTAGE SOURCES

4-1 INTRODUCTION

Electronic components can be divided into 2 basic categories: devices that amplify and those that do not. Amplifying devices such as transistors and vacuum tubes are called *active* devices, while devices that do not amplify are called *passive* devices.

Passive devices include resistors, capacitors, and inductors (coils). There are only 3 members of the passive class, but each of the 3 is available in a wide variety of sizes, shapes, and electrical values. Any reasonably complex electronics system is likely to use a variety of each of the passive components. The parts list for one color TV model lists 160 resistors, 90 capacitors, and 51 coils (inductors), not including an estimated 30 small coils in the tuner.

In this chapter we will examine resistive circuits and commercial resistors. We will take a look at inductors and capacitors in Chapter 6.

Important note: Resistors behave the same way with both alternating and direct current. When resistors using alternating current are discussed in the text, the voltages and currents will usually be RMS values. Capacitors and inductors behave differently with the 2 kinds of current.

4-2 RESISTORS

The Bare Facts

1. Resistors are made of carbon, resistance wire wound on a ceramic form, or certain more exotic materials.

2. The value of carbon resistors is indicated by colored bands instead of printed values.

3. Other resistors are marked with printed values.

4. The carbon-composition resistor is by far the most common, and it is therefore important to learn the color code.

5. Resistors work in exactly the same way whether the voltage is alternating or direct current.

6. Resistors convert potential energy difference into heat, and limit current flow in a circuit.

A resistor is a unit with a specified number of ohms of resistance. Its function in a circuit is to reduce a voltage (potential difference) by some specific amount and to control the flow of current. Because a resistance converts voltage into heat, on the unit you will find the resistance specified as well as a value for the amount of heat that the resistor can dissipate safely. The resistance is specified in ohms and the heat dissipation is given as the electrical heat equivalent in watts. We will study electrical power later in this chapter.

Early resistors were made by winding a specific length of high-resistance wire around a ceramic tube. The same construction technique is still used for resistors that must handle high power levels where high temperatures are expected. For lower wattages (1 watt or less), carbon composition, thin resistance films, and ceramic-metal compositions are used. Figure 4-1 illustrates several of the more popular resistor types.

WATTAGE RATING

The power dissipation capabilities in a resistor are proportional to its physical size. Power resistors are manufactured only in a few standard wattages, and it is possible to determine the power rating simply by looking at the resistor. Many larger units have the rating

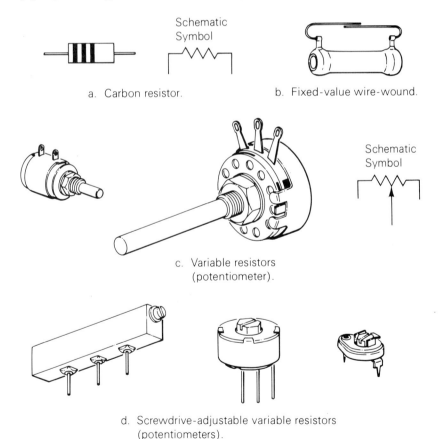

FIGURE 4-1. Commercial resistors.

printed on the body. In the case of smaller resistors, there is little space for printing, and size will be your only guide. Figure 4-2 shows the sizes of common carbon resistors. (W is the abbreviation for watts, and mm is the abbreviation for millimeters.)

THE RESISTOR COLOR CODE

Carbon-composition resistors of the type shown in Figure 4-2 use a series of color bands to mark the resistance. Each color stands for one of the digits between 0 and 9. The colors, their significance, and a memory aid follow.

Color Code and Memory Aid

0	Black	Big	BAD
1	Brown	Boys	BOYS
2	Red	Race	RAPE
3	Orange	Our	OUR
4	Yellow	Young	young
5	Green	Girls	GIRLS
6	Blue	But	BUT
7	Violet	Violet	violeT
8	Gray	Gray	GIVES
9	White	Wins	wiLLiNGLY

Figure 4-3 is a complete summary of the color code. The fifth band is used only in military-rated resistors. Resistors are very reliable devices, even by military standards, so the fourth band is nearly always

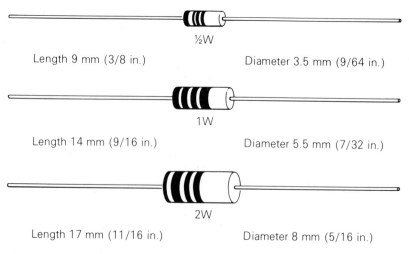

FIGURE 4-2. Carbon-resistor sizes.

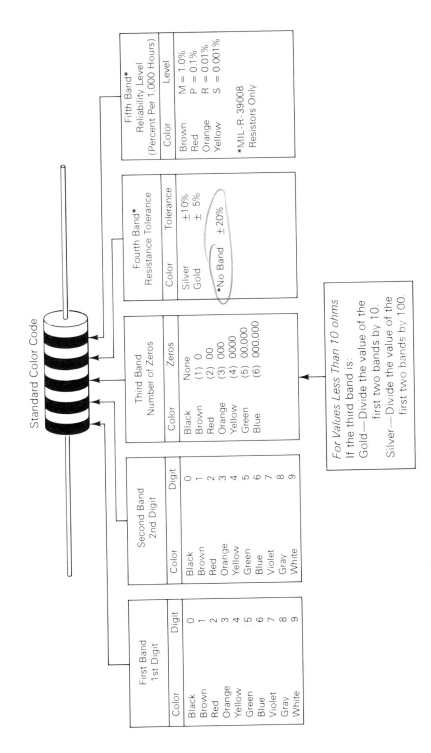

FIGURE 4-3. The resistor color code.

102 CHAPTER 4

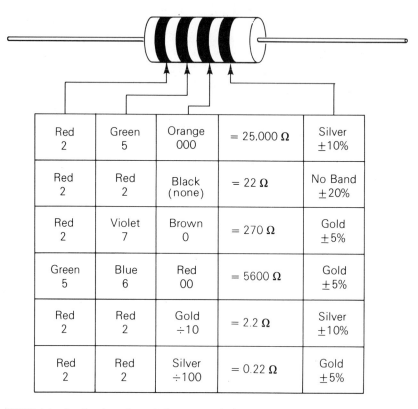

FIGURE 4-4. Reading the color code (some examples).

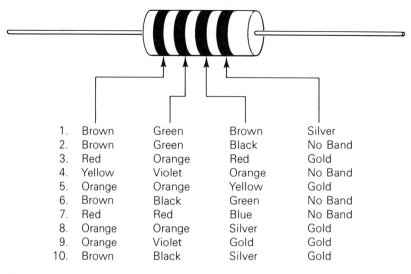

1.	Brown	Green	Brown	Silver
2.	Brown	Green	Black	No Band
3.	Red	Orange	Red	Gold
4.	Yellow	Violet	Orange	No Band
5.	Orange	Orange	Yellow	Gold
6.	Brown	Black	Green	No Band
7.	Red	Red	Blue	No Band
8.	Orange	Orange	Silver	Gold
9.	Orange	Violet	Gold	Gold
10.	Brown	Black	Silver	Gold

FIGURE 4-5. Some study practice.

yellow. Figure 4-4 provides some examples and Figure 4-5 is for student practice. The Greek letter omega (Ω) used in Figure 4-4 is the standard abbreviation for ohms.

Learning to read color-coded resistors is made easier by the fact that you will never see most of the possible color combinations. Resistors come in standard values, and nonstandard values are virtually nonexistent. This means a limited number of frequently occurring color patterns. An old hand can read color patterns as a whole in the same way you read words without struggling through, one letter at a time. Table 4-1 lists resistor manufacturers' standard values at each tolerance rating. More accurate resistors are available, but they are more expensive. Values for precision resistors are stamped rather than color-coded.

When greater precision than ±5 percent is required, the price takes a considerable jump. For example, a ±1 percent resistor might cost from 5 to 8 times as much as a ±5 percent resistor. Precision resistors are also available in nonstandard values at still higher prices. Carbon resistors are readily available up to 22,000,000 ohms (22 megohms). For higher values the appropriate number of zeros is added to the two digits given in Table 4-1.

RESISTANCE IN CIRCUITS

An ordinary water-supply valve makes an excellent analogy for discussing resistance. Like the commercial resistor, its purpose is to provide a "high" resistance in a "package." The valve is essentially a variable resistor to control water flow, just as an electrical resistor is a high resistance in a package used to control current flow. In each case the device causes a drop in energy (or pressure)—electromotive energy or voltage for the resistor, and water pressure for the valve. Figure 4-6 illustrates the analogy.

Wires and water pipes also have some resistance, but it would take a lot of either to provide very much resistance. If you did not have a valve to reduce the water flow, a few thousand feet of pipe would reduce it, but it would hardly be a practical solution to the problem. In the case of an electrical circuit a resistor might be replaced by several thousand feet of wire, but again this is not a practical approach.

Table 4-2a lists wire resistances for several typical copper-wire sizes used in electronics. Table 4-2b compares the resistance of other metals to that of copper. Nichrome and tungsten are used in heating elements, lamp filaments, and other places where a high-resistance metal is required.

TABLE 4-1 RETMA STANDARD RESISTOR VALUES

Name of Series	±20%	±10%	±5%
Percent Step Size	40%	20%	10%
Values in Each Series	10	10	10
	—	—	11
	—	12	12
	—	—	13
	15	15	15
	—	—	16
	—	18	18
	—	—	20
	22	22	22
	—	—	24
	—	—	—
	—	27	27
	—	—	—
	33	33	33
	—	—	36
	—	39	39
	—	—	43
	47	47	47
	—	—	51
	—	56	56
	—	—	62
	68	68	68
	—	—	75
	—	82	82
	—	—	91
	100	100	100

Note: Each value is available in multiples of 10, for example, 15, 150, 1500. . . .

CIRCUITS WITH RESISTORS AND VOLTAGE SOURCES 105

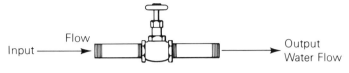

a. Water valve (fluid resistor).

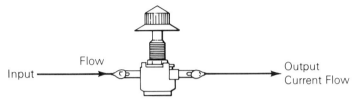

b. Variable resistor.

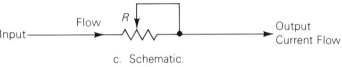

c. Schematic.

Notes:
1. In each case, assume a fixed input voltage or water pressure.
2. The water or current flow will be less for a higher resistance and greater for a lower resistance.

FIGURE 4-6. Water valve and its electrical equivalent.

TABLE 4-2 RESISTANCES OF MATERIALS

a. Resistances of Copper Wire.

Wire Gauge No.	Length (ft)	Resistance (ohms)
25	100	4
22	50	1
22	100	2

b. Resistances of Other Metals Compared to Copper. (Copper = 1.)

Metal	Resistance
Aluminum	1.59
Brass	4.40
Gold	1.38
Iron	6.67
Lead	12.76
Nichrome	60.
Nickel	7.73
Platinum	5.80
Silver	0.92
Steel	8.62
Tin	8.2
Tungsten	3.2
Zinc	3.62

4-3 CIRCUITS AND ANALOGIES

One of the most common words in electronics is *circuit*. As the word implies, electrons are driven from some "pool" of electrons through various devices and back to the source (pool). As with the water supply, the pool holds only so much, and unless it is replenished it dries up. When a continuous flow of electrons is required they must be returned to the pool after they have done their work. Water analogies using pipes and garden hoses can sometimes be deceptive because the circuit, though closed, is a complex one. Rainfall runs down rivers, pipes, irrigation ditches, and so on until eventually it arrives at lakes and oceans. The sun, serving as a pump, lifts water by evaporation and it forms clouds that result in more rainfall. The circuit is closed. For our water analogies we will stick to systems with fewer complications such as the system in Figure 4-7.

Analogs

Hydraulic	Electrical
Water flow	Current flow
Water (energy)	Pressure = Voltage
Resistance to water flow	Resistance to current flow
Water valve	Electrical (variable) resistor

The pump provides the energy to move the water around the circuit. When the valve is more nearly closed, it offers *more* resistance, reducing the flow. A more open valve offers a lower resistance, allowing a greater flow. The resistance of the valve is analogous to electrical resistance, in that both are frictional and both convert energy into heat.

Because of the excellent heat-radiating capability of the metal valve and the water's capacity to absorb the heat and carry it along, we rarely experience a warm valve, nor do we notice that the water

CIRCUITS WITH RESISTORS AND VOLTAGE SOURCES 107

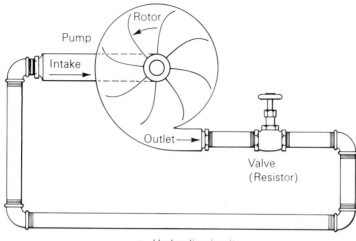

a. Hydraulic circuit.

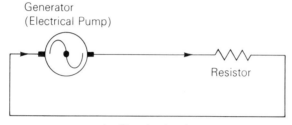

b. Electric circuit.

FIGURE 4-7. Complete circuits.

leaving the valve is very slightly warmer than when it entered the valve. In many electronics circuits, we can also touch resistors without being able to detect any heat because the resistor is radiating heat efficiently and because the electrons also carry heat away. In Figure 4-8 let us make a gap in the circuit to allow us to peek into the closed system and examine the relationships among pump force, valve resistance, and water flow.

Figure 4-8a shows a high-resistance circuit with the valve more closed than open. A high resistance allows only a small flow. In the hydraulic circuit we had to break the circuit and show water flowing into a bucket to illustrate the amount of flow. We have broken the electrical circuit for the same purpose, but because we cannot see electrons we must insert an instrument into the break to measure the current flow. Current is measured in amperes, so the instrument is called an *ammeter*.

Water flow is measured in liters per minute and electrical current is measured in electrons per second (1 amp = 6.25×10^{18} electrons

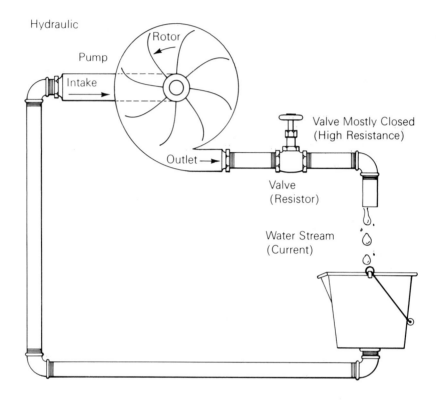

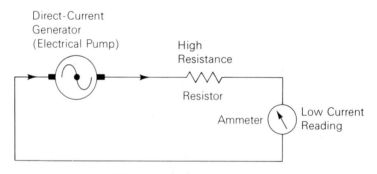

a. High resistance.

FIGURE 4-8. The effect of resistance on current flow.

per second). The circuit in Figure 4-8b is a low-resistance circuit with the valve opened wider. Notice the greater water flow and the corresponding increase in the ammeter reading in the low-resistance electrical circuit in Part b of the figure as compared to the reading in

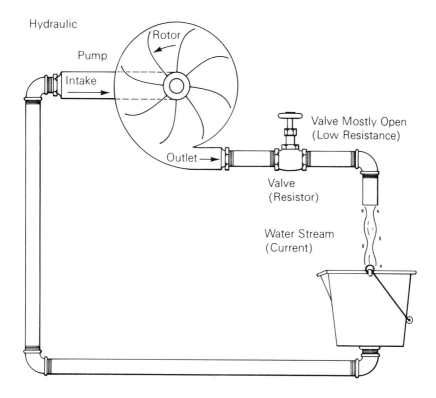

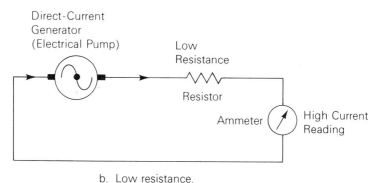

b. Low resistance.

FIGURE 4-8. (continued)

Part a. For a given pump pressure or generator voltage, an increase in resistance reduces the flow (of current or water). A decrease in resistance results in an increase in the flow. That is the first important relationship.

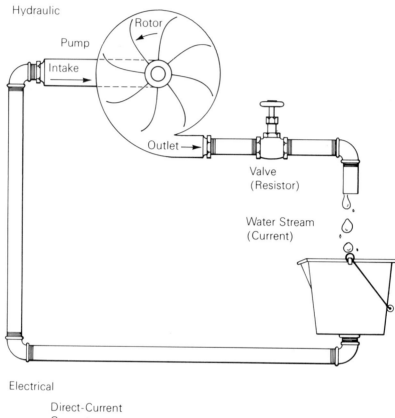

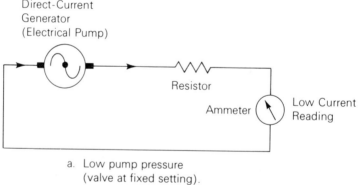

a. Low pump pressure
(valve at fixed setting).

FIGURE 4-9. The effect of pump pressure on flow.

Suppose we set the valve to a particular opening and fix the value of the resistor in the electrical circuit, and then increase pump pressure and generator voltage. What happens to the flow? We intuitively know that the flow will increase as shown in Figure 4-9b.

Hydraulic

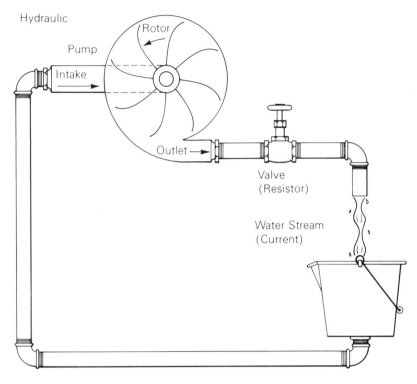

Electrical

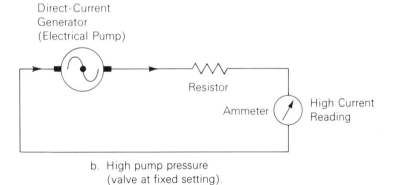

b. High pump pressure (valve at fixed setting).

FIGURE 4-9. (continued)

1. An increase in pressure forces a larger *volume* of water through a given valve resistance.
2. An increase in voltage forces a larger *current* through a given electrical resistance.

This is the second important relationship among voltage (pressure), resistance, and current (water flow).

4-4 OHM'S LAW

Georg Simon Ohm developed a practical way of dealing with these three interrelated factors in 1826, and was ridiculed by his colleagues at the University of Cologne for his efforts. Ohm was so right in his formulation that we now call it *Ohm's Law*, not *Ohm's Theory* or *Principle*. We now recognize Ohm's Law as one of the fundamental electrical relationships.

Interestingly, Ohm developed his law by analogy (not our water analogy) partly because the electron had not yet been discovered. Ohm had become fascinated by Fourier's studies in heat flow in which he saw heat flow as the analog of current flow, thermal resistance as the analog of electrical resistance, and temperature difference as the analog of voltage. Ohm was also interested in acoustics, the science of sound, and again he leaned heavily on analogies between electricity and mechanics. The analogy became one of the principal mathematical tools for acoustical engineers for the next 140 years. Acoustical engineers and scientists have been primarily responsible for developing the ideas of analogies into a legitimate science.

The Bare Facts

1. Ohm's Law expresses the interdependent relationships among voltage, current, and resistance.

2. When the values of 2 of the variables are known, the third may be found by simple multiplication or division.

3. Ohm's Law stated in words:
   One volt of potential difference will force a current of one ampere through a resistance of one ohm.

4. Ohm's Law stated in mathematical symbols:
   a. When the voltage is unknown:
   $$E = I \times R \quad \text{or} \quad V = I \times R$$
   b. When the current is unknown:
   $$I = E/R \quad \text{or} \quad I = V/R$$
   c. When the resistance is unknown:
   $$R = E/I \quad \text{or} \quad R = V/I$$

CIRCUITS WITH RESISTORS AND VOLTAGE SOURCES

where:
> E or V is the potential difference in volts
> I is the current in amps
> R is the resistance in ohms

Both symbols, E (for electromotive force) and V (for voltage) are used. The symbol E was more common for many years, but in the last few years the symbol V has become the preferred symbol.

Using Ohm's Law we can always find the current by simple division if we know the voltage and resistance.

$$\text{Current} = \frac{\text{Voltage}}{\text{Resistance}}$$

Or, in mathematical shorthand:

$$\boxed{I = \frac{V}{R}}$$

When we know the current and resistance, we can find the voltage by simple multiplication.

$$\text{Voltage} = \text{Current} \times \text{Resistance}$$

Mathematically:

$$\boxed{V = I \times R}$$

And finally, if we know the voltage and current, we can find the resistance by:

$$\text{Resistance} = \frac{\text{Voltage}}{\text{Current}}$$

Mathematically:

$$\boxed{R = \frac{V}{I}}$$

Because it is easy to get the symbols out of order, Figure 4-10 provides some help for the unreliable memory.

Study Problems

1. If 10 volts is applied to a circuit with a resistance of 100 ohms, what current will flow?

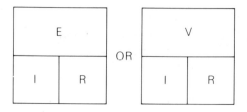

a. Memory aid.

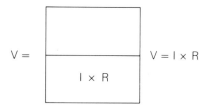

b. To find V.

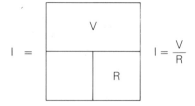

c. To find R.

$I =$ [V over R box] $I = \dfrac{V}{R}$

d. To find I.

FIGURE 4-10. Ohm's Law memory aid and examples.

2. If the voltage is increased to 20 volts, what current will flow?
3. If the voltage is increased to 100 volts, what current will flow?
4. If the voltage is reduced to 1 volt, what current will flow?
5. If 1 amp of current flows when the voltage is 10 volts, what is the resistance?

6. What resistance must be used if 2.5 amps of current are to flow when 100 volts is applied?
7. What resistance must be used if 0.05 amp of current is to flow with 100 volts applied?
8. What is the value of the resistor when 250 volts causes 0.1 amp of current to flow?

4-5 SERIES AND PARALLEL CIRCUITS

There are two basic circuits with which Ohm's Law can be used: series and parallel. Before we look at some practical applications of Ohm's Law we should examine the nature and properties of the series circuit. We will examine the parallel circuit and combinations of the two in later sections.

SERIES CIRCUITS

The Bare Facts

1. A series circuit is like a single train track with each element behind the other.
2. The same amount of current always flows through every element in the circuit.
3. If one element fails (opens), all current stops. This is known as an *open circuit*.
4. An open circuit does not cause sparking and heating, but quietly stops the current flow.
5. An on-off switch is connected in *series* with the rest of the circuit. If it is open, all current stops.
6. A fuse is a device designed to burn open before anything else in the circuit. Since it is wired in series with the rest of the circuit, it stops the current if it burns open.
7. To find the current in a series circuit, the resistance values of all elements are added. The total value is used for the resistance figure, and the current is computed using Ohm's Law.
8. Adding more resistive elements in series increases the total resistance and decreases the current.
9. If a fuse is inserted in a series circuit (fuses are always in series), current flowing through the fuse converts a small amount of voltage to heat. When the current exceeds the rated value in amps, the heat generated is enough to melt the metal strip in the fuse, causing an *open* circuit. (The fuse has some resistance.)

10. A switch in the *off* position will cause an open circuit.
11. A circuit breaker is a thermal or electromagnetic device that opens the circuit when its current rating is exceeded.
12. A short circuit is a "leak" with zero ohms of resistance (or nearly zero ohms).
13. Short circuits tend to be violent, with sparks, overheating, fire, or blown fuses and tripped circuit breakers.
14. When any 2 of the 3 values—voltage, current, resistance—are known, Ohm's Law can be used to find the remaining value.
15. The sum of the voltage *drops* in a series circuit is equal to the applied voltage.

The series circuit is illustrated in Figure 4-11 in both electrical and hydraulic forms.

Having set the two valves for some specified flow, we could remove one valve and adjust the remaining one to duplicate the original (two-valve) flow. The remaining valve must be further closed, adding more resistance, to duplicate the original flow. We must add resistance in that remaining valve by exactly the same amount that was taken out of the circuit when we removed one valve. Resistances in series are additive. If the two resistances in Figure 4-11b are 100 ohms and 200 ohms, they could be replaced by a single 300 ohm resistor and we would still get the same current. This addition process is the rule no matter how many resistors we put in the series.

Mathematically, for series resistors:

$$R_t = R_1 + R_2 + R_3 \ldots$$
$$R_t \text{ means } R_{total}$$

A very important point: on examination of the hydraulic circuit in Figure 4-11a it seems obvious that the same amount of water must flow through both valves. What is slightly less obvious but just as true is that the same amount of current must flow through both resistors in the electrical circuit in Figure 4-11b. Obvious or not, it is a very important thing to remember.

One might wonder at this point why we would want to replace two resistors with their single equivalent, or why we might have two resistors in the first place when one would do. We will find out shortly.

CIRCUITS WITH RESISTORS AND VOLTAGE SOURCES 117

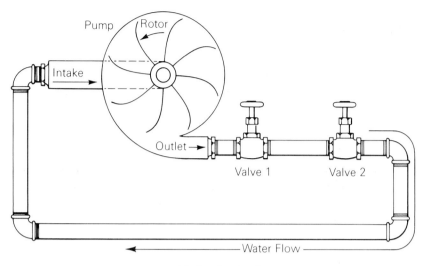

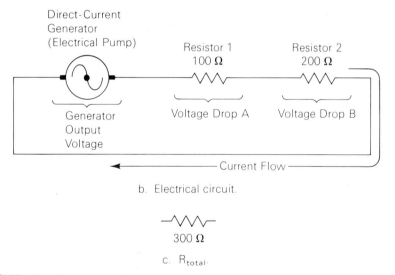

FIGURE 4-11. The series circuit.

A second very important point: a less obvious concept is hidden in the Ohm's Law relationship. In any series circuit the voltage (or pressure) drops around the loop (circuit) will add up to the generator voltage (pump pressure). By the time the electrons get back to the source, all of the original force provided by the pump or generator has been spent in forcing those electrons through the resistance opposing the flow. As the river water stops running when it reaches the ocean, so electrons have their energy of movement used up when they have returned to the generator. In Figure 4-11b, you will see the voltage drop (reduction) indicated for each resistor. The sum of the voltage drops A + B must equal the generator output voltage. This is true no matter how many resistors are involved. The reason for this behavior is the nature of current flow as it was described in the magic-box example in the previous chapter (see Figure 3-5).

EXAMPLES

Suppose we look at 3 typical examples of how Ohm's Law can be used to solve problems one might encounter in everyday life.

Example 1

Being on a very limited budget, but in need of a radio for your car, you have purchased from the local junk yard the only one you could afford that was in working order. At that price, as you would expect, there is a small hitch. The radio, according to the nameplate, was designed to be used on a 6 volt system at a current of 3 amps. Your car, of course, has a 12 volt system. Can we use a resistor to solve the problem? Indeed we can, but what value?

We can look at the problem in either of two ways:

1. How can we reduce the 12 volts available from the car's electrical system from 12 volts to 6 volts when there is a current of 3 amps flowing?

2. How can we make a circuit with the car radio as part of it that causes a current of 3 amps to flow with an applied voltage of 12 volts?

Both views of the problem are simply two *apparently* different ways of saying the same thing. The view that one takes depends mostly on which one seems simpler at the time. Either approach will yield the same results.

A car radio is obviously much more complex than a resistor, but we can treat it as simply a black box. The term *black box* in electronics jargon refers to some device or circuit where we are not really concerned about the circuitry inside it. We are interested only in what we can learn from the nameplate or things that can be measured in the connections that lead from the box to the outside world. As in any problem, we begin the solution by starting with what we know.

CIRCUITS WITH RESISTORS AND VOLTAGE SOURCES 119

Known:
1. Radio proper operating voltage = 6 volts.
2. Radio proper operating current = 3 amps.
3. Available voltage from car's electrical system = 12 volts.

The Problem:
How do we get rid of a surplus of 6 volts at a current of 3 amps?

Whenever a voltage reduction is required, a resistor is the most obvious choice. So the problem is one of finding the proper value. The problem is shown schematically in Figure 4-12.

First Solution:
The radio requires 6 volts and 3 amps to operate properly. Because the same current must flow through both radio and resistor, the resistor current must be 3 amps. The car generator (which actually produces direct current) provides us with 12 volts. The car radio requires 6 volts, which means we must drop a surplus of 6 volts with the resistor.

We have two known values pertaining to the resistor:

$$3 \text{ amps}$$
$$6 \text{ volts}$$

Using Ohm's Law, we can find the resistance value required:

$$R = \frac{V}{I} = \frac{\text{volts}}{\text{amps}} = \frac{6}{3} = 2 \text{ ohms}$$

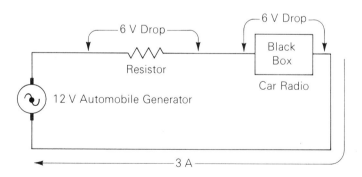

FIGURE 4-12. Circuit for the car-radio problem.

Second Solution:

This second approach to the problem might be more appropriate if the value of the car radio's internal resistance had been given on the nameplate. But no matter, we can find it with Ohm's Law:

Radio voltage = 6 volts
Radio current = 3 amps

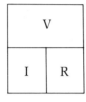

$$R = \frac{V}{I} = \frac{\text{volts}}{\text{amps}} = \frac{6}{3} = 2 \text{ ohms}$$

This value proves to be the same as the value required for the extra resistor. This is pure coincidence, in the sense that the voltage consumed by each was 6 volts. Had the voltages differed, the resistances would have been unlike.

We hang on to that value and begin the actual solution. The problem: What resistance is in a circuit with 12 volts applied and a current of 3 amps?

$$R = \frac{12 \text{ volts}}{3 \text{ amps}} = 4 \text{ ohms}$$

This is the resistance of the entire circuit. Two ohms of that is already a part of the circuit in the radio. The remaining 2 ohms is the value of the extra resistor. $R_{dropping} = 2$ ohms.

$$R_{dropping} = R_{total} - R_{radio}$$

In the foregoing example, the radio and the dropping resistor both use the same amount of energy. The radio does some useful work while the dropping resistor converts the energy into waste heat. The battery must supply twice the energy actually doing useful work. In most cases the amount of energy converted into waste heat by resistors is small enough to be of little or no concern in energy conservation.

Example 2

You have purchased a portable heater for your cubbyhole in the garage, for use on those cold nights when you feel like tinkering. The heater is connected to your single light-bulb socket with an adaptor. The circuit breaker turns off all the lights in the house every time you plug it in. It has been back to the dealer, who argues that there is nothing wrong with the heater. He suggests that you are probably overloading the circuit. You are doubtful. It doesn't seem like that little dinky heater should overload any circuit. Is the dealer handing you a line?

Let's look at some facts. The heater nameplate says that the heater is rated at 10 ohms at 120 volts. The circuit breaker is set to break at 20 amps. You have estimated that the typical load from lights on in the house is 15 amps.

The Problem:

There are 15 amps of load on a 20 amp circuit breaker. That leaves a good 5 amps for your little heater. The heater nameplate does not specify the current, so you can't tell whether it should draw more than 5 amps or not. With Ohm's Law, of course, you can find out.

The Solution:

The nameplate says:

$$\text{Current} = \frac{\text{volts}}{\text{ohms}} = \frac{V}{R} = \frac{120}{10} = 12 \text{ amps}$$

Apparently you should not have lost your temper with the dealer. You are indeed overloading the circuit by 7 amps. Even turning out some lights in the house probably won't solve your problem. The solution is not a happy one. Either wear heavy clothing or run a line from another circuit that can carry the current.

Example 3

You have always wanted a surplus AC/DC Universal Widget. At long last you have found one. The nameplate reads: 5 amps AC or DC, 24 ohms. The wire to supply power to the Widget has no plug on it. You know from experience in Widget hunting that they come in 24 volt, 120 volt, and 220 volt models.

CHAPTER 4

The Problem:

Can you put a household plug on this one, plug it into the 120 volt outlet and expect it to work?

$$\text{Known:} \quad \text{5 amps} \\ \text{24 ohms}$$

$$\text{Find:} \quad \text{The voltage (V)}$$

The Solution:

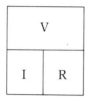

$$V = IR = 5 \times 24 = 120 \text{ volts}$$

Here you were luckier than you were with the heater purchase; the Widget is a 120 volt model.

Study Problems

1. Suppose you are going to use the radio from the previous example on your yacht, which has a 24 volt electrical system. What value of dropping resistor would be required? (Any but the most well-to-do yatchtsman can tell you that the purchase of a yacht soon reduces one to haunting junk yards for accessories.)
2. In Example 3 above, assuming that the 220 volt model requires 3 amps and the 24 volt model requires 18 amps, what resistance value would be found on the nameplate of (a) the 220 volt model; (b) the 24 volt model?
3. A Christmas-tree-light *series* string draws 0.1 amp. There are 12 lamps in the string. It is designed to operate on a 120 volt outlet. One bulb is burned out. It is December 20. Bulbs must be ordered from a catalog, which takes 6 weeks. What is the value of the resistor required to replace the lamp to make a functional string of 11 lights?
4. What is the resistance of a toaster rated at 120 volts, 8 amps?
5. A bulb for an automobile tail lamp is rated at 1 amp and has a resistance of 12 ohms. Is it intended for use in a 6 volt or 12 volt automobile system?
6. A bulb intended for use on a Hungarian tractor is rated at 2 amps and has 12 ohms of resistance. What voltage system is used in the tractor?
7. A 220 volt European bowling-ball cleaner has a resistance of 120 ohms. How much current will it draw when plugged into:

a. A 220 volt outlet?
b. A 110 volt outlet?

PARALLEL CIRCUITS

The Bare Facts

1. A parallel circuit consists of several resistances connected directly across the voltage source.

2. A parallel circuit may be considered as several independent circuits supplied by a common voltage source.

3. Each branch of the circuit is independent and does not affect other branches of the circuit in any way.

4. If one circuit (branch) fails (opens), the other branches continue to function normally.

5. Adding more resistances in parallel increases the current taken from the source.

6. A parallel branch with a resistance near zero ohms is said to be a short circuit.

7. A short circuit results in sparks, excessive heating, and either damage to the system or a blown fuse, whichever happens first.

The parallel electrical circuit and its hydraulic analogy are shown in Figure 4-13. It is evident in 4-13a that changing the resistance of Valve 1 has no effect on the stream from Valve 2, and likewise the stream from Valve 1 is independent of the setting of Valve 2. The total volume of water leaving the pump and returning to it is the sum of the 2 individual stream flows.

In the electrical circuit the value of Resistor 1 has no effect on the current through Resistor 2.* The 2 currents (I_1 and I_2) are independent of each other. The total current from, and back to, the generator is the sum of the 2 currents ($I_1 + I_2$).

Household appliances and many other devices are connected to the mains in parallel (see Figure 4-14). Each device adds to the total current taken from the mains but each device is independent of the others in its operation. The current through each device can be found by Ohm's Law ($I = E/R$); the total current is simply the sum of the individual branch currents.

---
*Except in the event of a short circuit.

124 CHAPTER 4

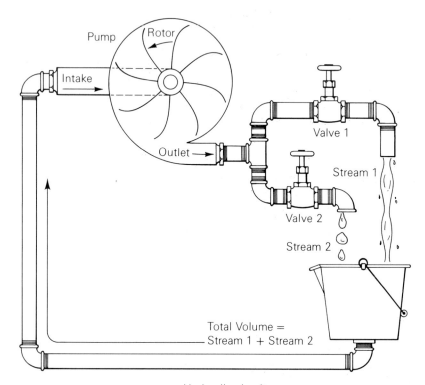

a. Hydraulic circuit.

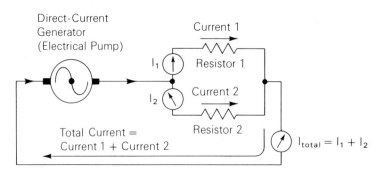

b. Electrical circuit.

FIGURE 4-13. The parallel circuit.

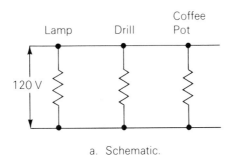

a. Schematic.

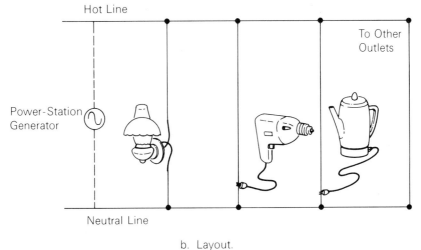

b. Layout.

FIGURE 4-14. Household appliances connected in parallel.

4-6 SHORT CIRCUITS AND OPEN CIRCUITS

The Bare Facts

1. A short circuit has a resistance of nearly zero ohms.
2. In a short circuit the current is excessive and can cause excessive heating and possible fire danger.
3. A short circuit can prevent the operation of all devices on the line.
4. Fuses and circuit breakers *open* the circuit, stopping the current flow when a short circuit occurs.
5. A fuse is a low-melting-point metal-alloy link that melts when a short causes excessive current to flow.

6. A circuit breaker is a heat-operated switch that opens the circuit when excessive current causes heating of a special bimetallic element.
7. Fuses must be replaced after a short occurs. Circuit breakers can be *reset* after the short has been cleared.
8. An open circuit stops current flow.
9. An open circuit causes no heating or fire danger.
10. A fuse or breaker *opens* the circuit to remove the dangers of a short.
11. An open circuit in a particular appliance or other device will prevent it from working, but will not affect other devices on the same line (in a parallel circuit).
12. A switch is an "on-purpose" open circuit in the *off* position.

To the layman, every problem in an electrical system is a short. A short circuit is the greatest hazard because it means that some failure has placed nearly zero ohms of resistance across the generator. This means a nearly infinite current, excessive heating, and a serious fire hazard.

Fuses and circuit breakers are placed in series in the circuit to open the circuit and stop the flow of current before anything serious can happen. The fuse is a link of metal with a low melting point and some (small) resistance. Excessive current heats the fuse link until it melts, interrupting the current flow. Current does not flow in an open circuit.

Switches are always put in series to turn devices on and off. The circuit breaker is a switch that is automatically switched to the open (*off*) position when the current through it exceeds a preset value. A strip of laminated metal called *bimetallic* strip bends as it heats, due to excessive current, and trips a switch. An accidental open circuit can also cause a device to be inoperative. But an open circuit does not draw any current, does not blow fuses or trip breakers, does not interfere with the operation of other devices on the line, and causes no fire hazard. Figure 4-15 illustrates open circuits, short circuits, fuses, and switches.

4-7 RESISTORS IN SERIES AND PARALLEL

Resistors are often connected in series in a line like cars on a one-lane highway, in parallel like parallel parked cars, or in a combination of the two. To find the total resistance in a circuit like those in Figure 4-16, simply add the individual values.

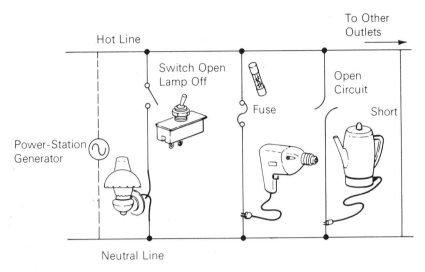

FIGURE 4-15. Switches, fuses, and open and short circuits.

When resistors are connected in parallel, the arithmetic is a little more difficult. There is a simple procedure when several resistors have identical values. Assume that the 3 resistors in Figure 4-17 are 300 ohms each.

Rule

Divide the common value by the number of resistors.

Examples:

1. For three 300 ohm resistors in parallel, as shown in Figure 4-17:

$$R_{equivalent} = {}^{300}/_3 = 100 \text{ ohms}$$

2. When there are 2 resistors of unequal value, as shown in Figure 4-18, use the following formula:

$$\boxed{R_{eq} = \frac{R_1 \times R_2}{R_1 + R_2}}$$

Using the values $R_1 = 25$ ohms and $R_2 = 100$ ohms, we get

$$R_{eq} = \frac{R_1 \times R_2}{R_1 + R_2} = \frac{25 \times 100}{25 + 100} = \frac{2500}{125} = 20 \text{ ohms}$$

The equivalent value of a group of parallel resistors is always less than the value of the lowest resistance in the group.

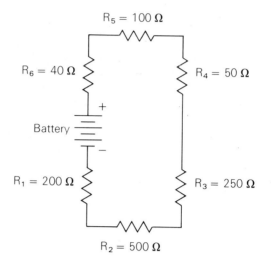

Total Resistance = 200 + 500 + 250 + 50 + 100 + 40 = 1140 Ω

a. Resistors.

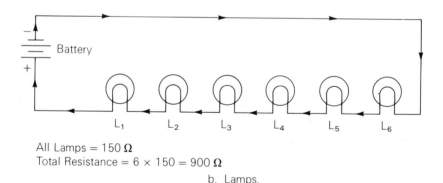

All Lamps = 150 Ω
Total Resistance = 6 × 150 = 900 Ω

b. Lamps.

FIGURE 4-16. Series-circuit examples.

The formula in Example 2 is good only for 2 resistors at a time. When there are more than 2, the formula can be used several times. Two values can be used the first time. The second repetition can be the *result* (R_{eq}) of the first pair used as R_1, and the third resistor can be R_2, and so on. This method is very time-consuming. However, there is a universal formula that can be used in all cases. Without a calculator it too is tedious, but with a little calculator help it is not difficult.

CIRCUITS WITH RESISTORS AND VOLTAGE SOURCES 129

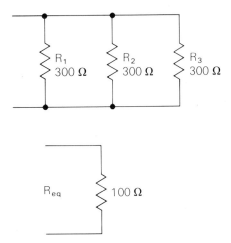

FIGURE 4-17. Resistors in parallel.

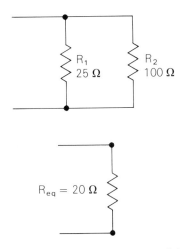

FIGURE 4-18. Unlike values in parallel.

The universal formula for resistors in parallel:

$$R_{eq} = \frac{1}{\dfrac{1}{R_1} + \dfrac{1}{R_2} + \dfrac{1}{R_3} \cdots}$$

The formula is monstrous-looking, but it can be presented in a simpler form. The 3 dots (called an *ellipsis*) means you can include as many more resistors as you like.

With a calculator you can use the following procedure. Your calculator manual can probably show you how best to reduce the effort, depending on the features of your calculator.

1. Divide R_1 into 1. Record the result.
2. Divide R_2 into 1. Record the result.
3. Repeat for each R.
4. Add the results of all divisions (Steps 1, 2, and 3).
5. Divide the sum into 1 to get the final answer.

A simplified presentation of the universal parallel-resistance formula:

Add:

$$R_{eq} = \frac{1}{R_1} + \frac{1}{R_2} + \frac{1}{R_3}$$

Procedure (hand):

1. Substitute resistor values. Using the common denominator, add the fractions.
2. Turn the answer from Step 1 upside down and divide.

Example:

Find the equivalent resistance of a 25, a 50, and a 100 ohm resistor in parallel:

$$R_{eq} = 1/25 + 1/50 + 1/100$$

Finding the common denominator:

$$R_{eq} = 4/100 + 2/100 + 1/100 = 7/100$$

Inverting* and dividing:

$$100/7 = 14.3 \text{ ohms}$$

Unfortunately, parallel-resistor solutions are time-consuming. Tables 4-3 and 4-4 can be used for a quick lookup for many situations. To use the tables, find the value of one resistor on the left-most column.

---

*The inverting procedure is a shortcut method for finding the reciprocal of a fraction.

TABLE 4-3 ARITHMETIC-SAVER PARALLEL-RESISTANCE CHART 1

First Value → 100 ← ... Equivalent Resistance

	10	12	18	22	27	33	39	47	56	68	100
10	5.00	5.45	6.42	6.875	7.29	7.67	7.96	8.25	8.48	8.72	9.09
12	5.45	6.00	7.20	7.76	8.30	8.80	9.18	9.56	9.88	10.20	10.71
18	6.42	7.20	9.00	9.90	10.80	11.65	12.32	13.01	13.62	14.23	15.25
22	6.875	7.76	9.90	11.00	12.12	13.20	14.06	14.98	15.79	16.62	18.03
27	7.29	8.30	10.80	12.12	13.5	14.85	15.95	17.15	18.22	19.33	21.33
33	7.67	8.80	11.65	13.20	14.85	16.50	17.875	19.39	20.77	22.22	24.81
39	7.96	9.18	12.32	14.06	15.95	17.875	19.5	21.31	22.99	24.78	28.06
47	8.25	9.56	13.01	14.98	17.15	19.39	21.31	23.50	25.55	27.79	31.97
56	8.48	9.88	13.62	15.79	18.22	20.77	22.99	25.55	27.79	30.70	35.99
68	8.72	10.20	14.23	16.62	19.33	22.22	24.78	27.79	30.70	34.00	40.48
100	9.09	10.71	15.25	18.03	21.23	24.81	28.06	31.97	35.99	40.48	50.00

↑ Second Value

TABLE 4-4 ARITHMETIC-SAVER PARALLEL-RESISTANCE CHART 2

	39	47	56	68	100
120	29.43	33.77	38.18	43.40	54.54
180	32.05	37.27	42.71	49.35	64.28
220	33.13	38.73	44.64	51.94	68.75
270	34.08	40.03	46.38	54.32	72.97
330	34.88	41.14	47.87	56.38	76.74
390	35.45	41.94	48.97	57.90	79.59
470	36.01	42.73	50.04	59.40	82.46
560	36.46	43.36	50.91	60.64	84.85
680	36.88	43.96	51.74	61.82	87.18
1000	37.54	44.89	53.03	63.67	90.91

(Top header label: First Value; Left column label: Second Value; Body label: Equivalent Resistance)

Find the value of the second resistor in the top-most row. Draw an imaginary vertical line down the column from the value in the top-most row and an imaginary horizontal line from the value in the left-most column. Where these lines intersect, you will find the equivalent value of the two parallel resistors.

Example:

Find the equivalent value of a 27 ohm and a 39 ohm resistor connected in parallel. Table 4-3 yields a value of 15.95 ohms. This value appears in 2 places in the table, depending on which of the 2 values is found in the left-most column and which is found in the top-most row. The table is built around the standard values in the ±10% family.

Study Problems

Find the equivalent resistance of each pair of parallel resistors:
1. 100 ohms and 56 ohms
2. 33 ohms and 33 ohms
3. 47 ohms and 68 ohms
4. 10 ohms and 100 ohms
5. 33 ohms and 100 ohms

VALUES ABOVE 100 OHMS

The usefulness of Table 4-3 can be extended by simply multiplying all values by 10, 100, and so on.

Example:
Find the equivalent value of two 1000 ohm resistors in parallel.

Solution:
Mentally add a zero to each number you use. Find 100 + (0) = 1000 in the top-most row of the table, and 100 + (0) = 1000 in the left-most column. The value found at the intersection is 50; adding a zero gives us 500 ohms, the correct value.

You may have noticed that the table does not include a very large range of values. For most purposes the table is quite adequate because the lower-value resistor becomes dominant when paralleled with much higher values. For example, a 10 ohm resistor paralleled with a 1000 ohm resistor yields an equivalent resistance of 9.9 ohms. This is a change of 1 percent in the original 10 ohm value, smaller than the probable tolerance error of the 10 ohm resistor itself. (It could be off by as much as 10 percent.)

COMBINED SERIES AND PARALLEL CIRCUITS

In electronics it is common to find series and parallel circuits combined in the same system. Problems involving series-parallel circuits can be solved by ordinary Ohm's Law and a little simplification provided by combining resistors. Suppose we look at an example.

Example:
You have just purchased a new stereo system. A fast-talking salesman has convinced you that several speakers for each channel were far better than a single speaker for each channel. As a remarkable coincidence, he happened to have a number of last year's models that he could let you have at near cost.

The instruction manual makes the following statement: "Connect the right and left channel speaker terminals to a 16 ohm loudspeaker. (*Note:* Less than 16 ohms can damage the unit; more than 16 ohms will cause a loss of power output.)" Upon unpacking the speakers, you find that you have purchased:

> Four 8 ohm speakers
> Four 16 ohm speakers
> Two 4 ohm speakers

The manual also states: "Multiple speakers may be used, providing they are connected in series-parallel to equal 16 ohms." For each

134 CHAPTER 4

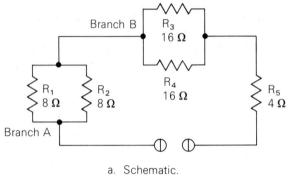

a. Schematic.

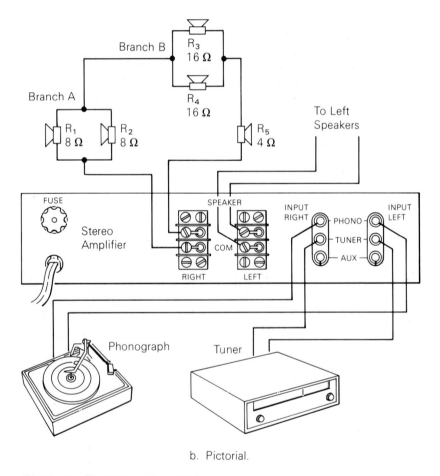

b. Pictorial.

FIGURE 4-19. The stereo-speaker example.

channel you have two 8 ohm speakers, two 16 ohm speakers, and one 4 ohm speaker.

The Problem: How can these speakers be connected to total 16 ohms? After several obviously faulty attempts, you have come up with the circuit in Figure 4-19. Is this a proper solution to the problem? To find out, we combine series and parallel values, as we discussed in the previous section, until we have a single equivalent resistance. For any but the most horrendous series-parallel circuits, the following procedure will work:

1. Find the equivalent resistance of all parallel branches. Redraw the circuit.
2. Add the series values in the redrawn circuit.

If we follow that procedure for the circuit in Figure 4-19 we can reduce Branch A to a single 4 ohm resistance as shown in Figure 4-20a.

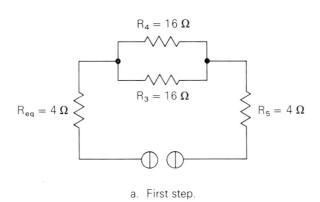

a. First step.

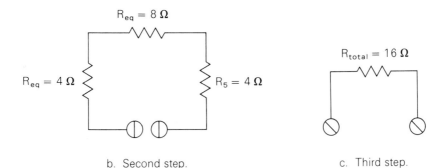

b. Second step.   c. Third step.

FIGURE 4-20. Solution to the speaker problem.

The Math:

$$R_{eq} = \frac{R_1 \times R_2}{R_1 + R_2} = \frac{8 \times 8}{8 + 8} = \frac{64}{16} = 4$$

Because the 2 values R_1 and R_2 are equal, the math is really unnecessary.

The second step in the procedure is to reduce Branch B to its single resistance equivalent. Figure 4-20b shows that result.

In the third and final step, we add the 3 series resistances in Figure 4-20b to get the final result of 16 ohms as shown in Figure 4-20c

The attempt in this case was successful.

Study Problem

Can you find any other arrangements of the speakers that will provide the required 16 ohms?

4-8 POWER

The Bare Facts

1. To light, heat, or do mechanical work requires power.

2. Electrical power is measured in watts.

3. Power is computed by multiplying the operating voltage of a device by the current flowing through the device—that is, volts times amps.

4. 746 watts is equal to one horsepower.

5. The power rating of a device is the power consumed, not the work done.

6. For example, a 1 horsepower electric motor would require a little more than 746 watts to do a full horsepower's work due to resistance and mechanical friction.

7. Some devices are very efficient and the power used is almost equal to the work done. Other electrical devices are very inefficient.

8. A DC electric motor is typically a high-efficiency device, often ranging from 75 percent to 95 percent efficient.

9. A motor that is 95 percent efficient would do 95 watts of work for each 100 watts consumed.

10. The incandescent lamp is a very inefficient device, with efficiencies ranging from about 6 percent to 30 percent.

11. In a lamp using 100 watts, at best only 30 watts of light energy would be produced and some 70 watts would be radiated in the form of heat.

12. The average incandescent lamp is therefore a better heater than it is a light source (it is frequently used as a heater).

13. You can compare relative efficiencies of a lamp and an electric motor by simply feeling the amount of heat given off by each.

14. Whenever one kind of energy is converted to another—for example, electrical to mechanical or electrical to light—some of the energy is lost in heat. This loss is due to friction in a mechanical system and resistance in an electrical system.

15. A resistor in an electronic circuit is an intentional waste of electrical energy. Voltage is deliberately given up in the form of heat to obtain a proper operating voltage from one that is too high.

16. Resistors are rated in watts as well as ohms. The wattage rating tells how many watts of electrical power the resistor can convert to heat without overheating itself. Therefore, a higher-wattage resistor may always be used as a substitute for a lower-wattage resistor, but not vice versa.

17. In the above example, the ohms (resistance value) should remain the same.

Most of what you need to know about power up to this point is contained in The Bare Facts. What remains to be said involves some examples and a little arithmetic. Power in watts is defined as the product of the voltage and current used by a device, and is the rate of work per unit of time.

$$\text{Power} = \text{volts} \times \text{amps}$$

$$\boxed{P = V \times I}$$

Electrical power in watts is related to mechanical power in hosepower:

1 horsepower = 746 watts = 550 foot-pounds per second

1 kilowatt (1000 watts) is equal to 1.34 or about one and one-third horsepower. One horsepower is approximately ¾ kilowatt.

Table 4-5 shows the power requirements of some common electrical and electronic devices. The power shown in the table is power consumed, not the power produced in the form of work output. For example, the 50-watt-per-channel stereo amplifier produces 100 watts of audio power, but consumes 150 watts. The horsepower figures are provided for a little perspective. The table illustrates the fact that devices with resistive heating elements are very power-hungry. This accounts for the fact that the little toaster can overload an already loaded circuit, tripping the circuit breaker, while a much more impressive automatic washer often will not.

The no-iron fabrics contribute to energy conservation by taking that 1000 watt hand iron off the line in millions of homes, a savings of as much as 8,000,000 kilowatt-hours of energy per week, nationwide.

POWER CALCULATIONS

Power calculations involve only simple arithmetic. When the voltage and current are both known, it is only necessary to multiply those two figures together:

$$\text{Volts} \times \text{amps} = \text{watts}$$

If resistance is known instead of either voltage or current, Ohm's Law can be used to find the missing value.

Example:

Find the power consumption of a toaster with a resistance of 10 ohms designed to operate from the common 120 volt line.

Known: Voltage
Resistance

Needed: Amps
Watts

$$I = \frac{V}{R} = \frac{120}{10} = 12 \text{ amps}$$

Power = V × I = 120 volts × 12 amps = 1440 watts

There are three formulas that provide a one-step solution without having to use Ohm's Law.

CIRCUITS WITH RESISTORS AND VOLTAGE SOURCES   139

TABLE 4-5 TYPICAL POWER RATINGS OF ELECTRICAL AND ELECTRONIC DEVICES

Device	Power Consumed (watts)	Equivalent Horsepower
Television receiver	50–150	0.067–0.2
Radio receiver	1–50	0.001–.067
Electric clock	2	0.002
Ultraviolet lamp	385	0.38
Electric fan	100	0.13
Portable electric heater	1000	1.3
Record changer	75	0.1
Electric blanket	200	0.27
Heating pad	60	0.08
Electric shaver	12	0.016
Sewing machine	75	0.1
Refrigerator	150	0.2
Coffee maker	1000	1.3
Hand iron	1000	1.3
Ironer	1650	2.2
Floor lamp	300	0.4
Vacuum cleaner	125–300	0.17
Electric skillet	1100	1.5
Mixer	100	0.13
Toaster	1500	2.0
Broiler	1500	2.0
Waffle iron	1000	1.3
Washer	700	0.9
Baker	900	1.2
50 W stereo amplifier	150	0.2
Automotive headlights	250	0.3
¼ inch electric drill motor	250	0.3

1. When voltage and current are known:

$$P = V \times I$$

2. When voltage and resistance are known:

$$P = \frac{V^2}{R}$$

3. When current and resistance are known:

$$P = I^2 R$$

If you don't remember the appropriate formula, you can always use Ohm's Law to find the missing voltage or current.

Study Problems

$$\text{Power (in watts)} = \text{volts} \times \text{amps}$$

1. A resistor is connected across a 10 volt battery; 0.02 amp of current flows. How much power is consumed?
2. A lamp draws 0.11 amp on a 110 volt line. What is the wattage rating of the lamp?
3. If 746 watts = 1 horsepower, what is the approximate horsepower rating of a 110 volt motor when it draws 3.6 amps running?
4. A 10 ohm resistor is connected across a 10 volt battery; how much power does it dissipate? (*Hint:* find I first.)
5. A 10 kilohm (10,000 ohm) resistor is connected across a 150 volt source; how much power is radiated as heat (electrical watts)?
6. A current of 2 amps flows through a 100 ohm resistor; how much power is dissipated by the resistor? (*Hint:* find the applied voltage.)
7. A current of 0.1 amp flows through a 1000 ohm resistor; how much power does it consume?
8. A 100 watt lamp is designed to operate at 120 volts; what is its resistance?

Chapter 5

MAGNETISM AND MAGNETIC MACHINES

5-1 INTRODUCTION

The electron is a bundle of mysteries. Is it a particle or a wave? Why is it always accompanied by a magnetic field? What is it made of, if it is made of anything at all? Within the atom the electron stores energy and releases it in the form of light or other radiant energy; what is the relationship between electrons and radiant energy? Don't expect answers to these questions. They are being investigated but are still shrouded in mystery. An important part of the mystery is our ignorance of the nature of light and magnetism. It is remarkable that we have developed such a sophisticated technology when we have only begun to get to the solution of these very basic mysteries. Even without a full understanding of their hidden nature, we can study the behavior and applications of electrons and magnetism. In this text we will try to explain what can be explained without losing the excitement of the many remaining mysteries.

We know that a moving electron is always accompanied by a magnetic field. We know that a direct current through a wire produces a magnetic field that will attract objects made of iron, and to a lesser extent some other materials. We know that if we push electrons back and forth through a wire with an alternating current, the magnetic field will fly mysteriously through space. Under the right circumstances this will cause a current of the same frequency to flow in a wire far from the original radiating source. This is the technique used in radio communication.

We know that if we make the electrons move back and forth rapidly enough, the electromagnetic wave radiated into space becomes visible. Our eyes see different colors corresponding to different vibrating frequencies of the moving electrons.

We don't know just what an electron is. It does have some mass (weight), so it is not pure energy. However, at times it behaves as a wave with properties much like those of light. The electron microscope takes advantage of this property. Electromagnetic energy can exert a slight force and often behaves as though it consisted of particles. Light-driven sailing ships are currently being designed by NASA to rendezvous with Haley's comet on its next passing. Electronic controls will trim the sails so the vessel can tack in the solar wind.

There is growing evidence that the mysteries of the electron are somehow intimately tied to the fundamental nature of space. Space may not only be the next frontier for pioneer explorers; it may also be the next hunting ground for the theoretical physicists who are currently in pursuit of *quarks, charm,* and other elusive subatomic entities.

Magnetic fields always seem to exist with complementary poles

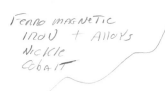

Ferro magnetic
Iron + Alloys
Nickle
Cobalt

MAGNETISM AND MAGNETIC MACHINES 143

designated *north* and *south*. There is a strong feeling among many scientists that magnetic poles should be able to exist in isolation.

There is a group of equations, developed by James Clerk Maxwell in 1862, that summarize the interactive behavior of electrical charges. These equations still describe nearly all we know about electricity with great accuracy, but in Maxwell's mathematical analog, magnetism is simply a by-product of electricity. In 1931, P.A.M. Dirac altered Maxwell's equations. In the Dirac revision, these equations require 2 magnetic charges that are exact analogs of the electron and the proton, a North and a South magnetic charge. Dirac's version may not be correct, but Maxwell's version provides no direction for exploration, nor any clue to the solution of the mystery of magnetism.

In 1975, scientists from the University of Houston and the University of California at Berkeley found what they believed to be the tracks of a single isolated magnetic charge. No isolated magnetic charge had ever been found before, although the search has been going on for more than 40 years. The isolated magnetic charge is called a *magnetic monopole*. The magnetic monopole seems to be among the rarest of things in the subatomic zoo. Some question must naturally arise about the importance of magnetic monopoles, particularly when their existence is still in doubt. It would be nice to be able to give some detailed answers, but the best that is possible is some educated speculation.

To begin with, if the magnetic monopole does exist, it must have an attraction for a monopole of the opposite polarity estimated at about 137 times as great as the attraction of an electron for a proton. One of the most exciting potential applications so far is a magnetic "bottle" to control the super-hot plasma in a fusion nuclear generating plant. Plasma control has been a major technological problem in fusion reactors. However, the enormous attractive force between monopoles could conceivably provide a totally new source of power—perhaps another science fiction idea becoming fact.

Now it is time to settle down to learning about some things that we know in the magnetic realm. We will examine light and lasers and radio in subsequent chapters.

5-2 MAGNETS

The Bare Facts

1. One of the unexplainable properties of the electron is the fact that whenever an electron is in motion, a small magnetic field is set up around the electron.

2. Since electrons move in orbits around the atom, magnetic fields are set up around each orbiting atom, and the atom itself sometimes becomes a miniature magnet.

3. A magnetic field is directional, and two magnetic fields may either cancel each other or add together to form a field of double strength.

4. In an unmagnetized piece of material, the random fields tend to cancel each other so the material seems to have no overall magnetic field.

5. Some materials may have nearly all the fields (atomic magnets) lined up as a result of applying an outside field from another magnet.

6. When the magnetizing field is removed, the atomic magnets may remain lined up to form a permanent magnet. The magnet may last anywhere from a fraction of a second to centuries.

7. Magnetic *current* (called flux) flows outside of the magnet from north to south in a complete 360° loop.

8. The word *flow* as applied to magnetic lines of force is only a *convenience* word. There is no scientific evidence that a steady state magnetic field requires any time at all to propagate, a necessary condition for something to actually flow.

9. How long a magnet lasts depends on how strongly it is magnetized and the material out of which it is made.

10. The earth is a natural magnet.

11. The central core of the earth is somewhat fluid. Friction between the core and crust as the earth rotates on its axis strips off electrons and causes a current to flow, producing a magnetic field. This phenomenon is known as the *magnetohydrodynamic effect*. (Any movement of electrons constitutes a current flow.)

12. Natural magnets are sometimes formed under the influence of the earth's magnetic field. These natural magnets are called *lodestones*, composed of a substance known as *magnetite* to the mineralogist.

ELECTRONS AND MAGNETISM

A magnetic field exists in the vicinity of any electron that is in motion. No electron has ever been found at rest, and scientists speculate that any electron that did come to rest would cease to exist as an electron. If electrons should slow down in their orbits, they would spiral into the nucleus like a spent space vehicle spiraling into the earth's atmosphere. Should such an "impossible" thing happen, that atom would cease to exist as the kind of matter we know.

An electron in motion is always accompanied by a magnetic field. Magnetic fields are never detected unless associated with moving electrons. Electrons are never found at rest. As nearly as we can determine, everything is surrounded by magnetic fields. However, not all substances exhibit obvious magnetic properties.

We have somewhat oversimplified the orbital behavior of valence electrons, and have failed to mention electron spin. Electrons not only orbit the nucleus, but also, like the earth, rotate about their axis. The axial spin, as well as the orbital rotation, produces a magnetic effect. All materials, elements, and compounds are affected to some small extent, at least, by external magnetic fields. Electron motion generates a magnetic field, and a magnetic field influences electron motion.

The electron produces simple magnetic fields when it moves, and when it moves in more complex patterns, it produces electromagnetic waves that travel through space. The entire electromagnetic spectrum, including light and the longer wavelengths we use for radio communications, is the result of electrons in motion.

We don't really understand why an electron in motion generates a magnetic field. Even more mind-boggling is how both an electric and a magnetic field leave a moving electron to travel through space at the speed of light. This electromagnetic energy travels through space and is capable of doing work at distances we can talk about, but not really comprehend. To further complicate the situation, the electron not only produces all known wavelengths but seems to have a special wavelength of its own. The electron microscope takes advantage of this peculiarity. Later we will examine electromagnetic waves in more detail. For now we will look at the electron's simpler magnetic manifestations.

MAGNETISM

One of the most mysterious aspects of nature is magnetism. Magnetism in all of its shapes and forms is a manifestation of the electron. Because of our ability to harness the electron to perform so many tasks, we often assume that it is well understood. Mathematical equations have been formulated that allow us to predict how electrons will behave under various circumstances and what sort of magnetic fields will be associated with them in each case.

The first permanent magnet seems to have been discovered in

China in the form of the mineral magnetite, an iron oxide. The first practical application of magnetism was as a navigation aid, and the magnetic compass is still the most widely used navigation aid.

The earth is a natural magnet with its north magnetic pole located near the geographical North Pole. Its south magnetic pole is near the geographical South Pole. Present theory views the earth as an electromagnet, so we will come back to the magnetic earth after a look at smaller-scale electromagnets. The early application of magnets to the compass has left us with the designation of north-seeking and south-seeking poles for the opposite ends of all magnets.*

Magnetic Materials

Magnetic materials fall into three general categories: those that become permanently magnetized, those that rapidly lose their magnetic property when an external magnetic field is removed, and those that do not exhibit significant magnetic behavior. Iron is the principal magnetic material, along with some iron compounds. Pure iron loses its magnetism rapidly. The addition of 0.1 to 2 percent carbon, resulting in the alloy called *steel*, causes the iron to keep its magnetism almost indefinitely.

Magnetic oxides are used to coat the tape in tape recorders, and refractory iron oxides called *ceramics* are used to form magnets in a variety of shapes. Nickel and cobalt are the only other highly magnetic materials, and nickel alloy magnets are common (alnico, for example).

In addition to orbiting the nucleus of the atom, electrons also spin about their axis. The two spin directions that electrons can take establish minute magnetic fields of opposite direction. In unmagnetized material, atoms are oriented randomly in all directions and their tiny fields tend to cancel each other. In magnetic materials, groups of atoms with the same magnetic polarity join together to form miniature magnets called *domains*. Under the influence of an external magnet these domains tend to align with the external field. Domain behavior is illustrated in Figure 5-1.

In soft iron, when the external field is removed, the domains are soon shaken back into a random condition by thermal vibrations. The iron ceases to be a magnet. In steel the domains remain aligned after the external field is removed and the bar becomes a permanent magnet.

Magnetic fields form a closed loop, with the magnetic current (flux) flowing from the north to the south pole as shown in Figure

---

*Because opposite magnetic poles attract, the earth's north magnetic pole may be considered to be located at its geographical South Pole.

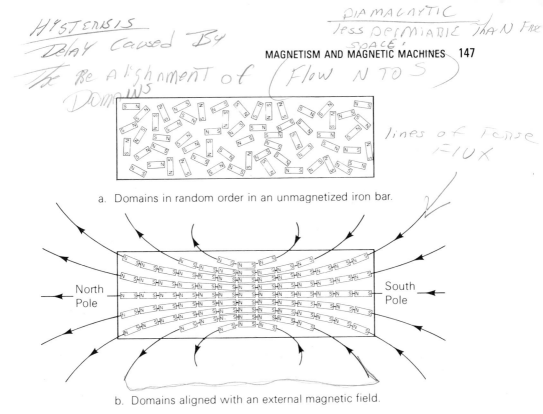

a. Domains in random order in an unmagnetized iron bar.

b. Domains aligned with an external magnetic field.

FIGURE 5-1. Magnetic domains.

5-2a. Figure 5-2b shows iron filings that have been shaken onto a piece of paper placed over a bar magnet. When a permanent magnet is broken into pieces, each piece is also a magnet. The domain theory accounts very well for this phenomenon, as shown in Figure 5-3. Remember that the domains are microscopic in size. They are greatly exaggerated in the figure.

Hysteresis

Because magnetic domains have inertia, some time is required to align or realign them. This delay is called *hysteresis*. The phenomenon is particularly important in electromagnets that are operated by alternating or some other varying current. When the electromagnet is operated by a rapidly changing current, the domains lag behind the changing field because they can't move fast enough to keep up.

Magnetic Attraction and Repulsion

In magnetism as in electricity, opposite signs attract and like signs repel. This attractive and repulsive action is utilized in electric motors and other electromechanical devices. Figure 5-4 illustrates the rule and shows an iron-filing representation of the fields for both attraction and repulsion.

148 CHAPTER 5

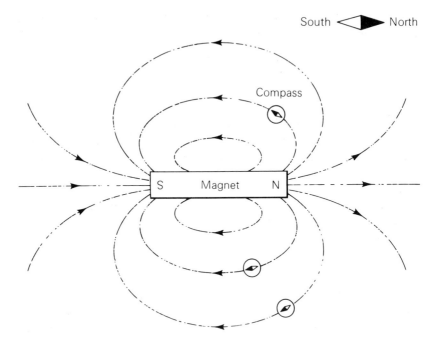

a. The magnetic field around a bar magnet.

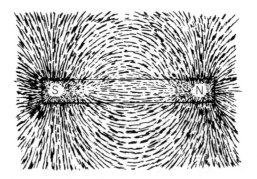

b. Iron filings used to demonstrate field pattern.

FIGURE 5-2. Magnetic current (flux) flow.

5-3 ELECTROMAGNETS

Conductors, whether or not they have magnetic characteristics, become temporary magnets when current flows through them. The moving electrons establish a magnetic field around the current-carrying wire as described in Figure 5-5. The hand grasping the wire

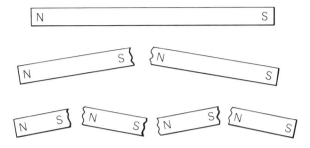

a. Bar magnet broken in two and then into four pieces.

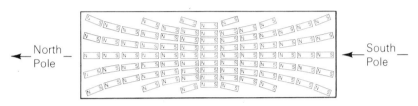

b. Bar magnet showing domain alignment.

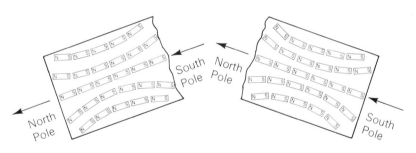

c. Domains when a magnet is broken.

FIGURE 5-3. What happens when a magnet is broken.

illustrates a memory aid known as the *left hand rule*. Using the left hand as shown in the illustration, the thumb points in the direction of electron flow and the fingers point in the direction of the magnetic field.* The attractive force is broadside to the direction of current flow, as illustrated by the deflection of the compass.

The magnetic field around a current-carrying wire is a closed loop and has no external poles unless the loop is opened. Figure 5-6

---
*The right hand rule is used when one is talking about conventional current.

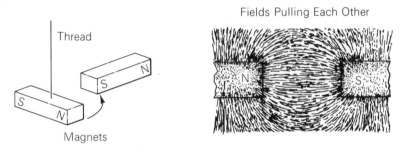

a. Opposite poles attract.

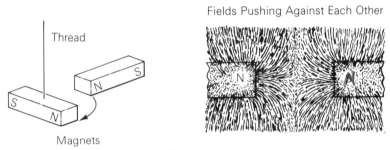

b. Like poles repel.

FIGURE 5-4. Magnetic attraction and repulsion.

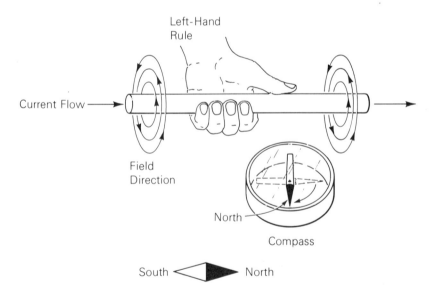

FIGURE 5-5. Magnetic field around a current-carrying conductor.

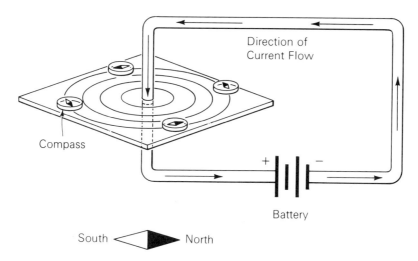

FIGURE 5-6. Magnetic field around current-carrying wire.

illustrates this situation using 4 magnetic compasses to partially interrupt the field to form north and south poles. Electrical current flows from negative to positive, while magnetic "current" flows from north to south.

SOLENOIDS

A strong electromagnet called a solenoid can be formed by winding a long wire into a coil. This concentrates the many little fields along the wire into a compact space so that the little fields add up to a total large field in a concentrated area. The field pattern formed is the same as that produced by a bar-shaped permanent magnet. The solenoid and its associated field are shown in Figure 5-7.

Adding an Iron Core

The field strength of a solenoid can be greatly increased by inserting a soft iron core. When the current is started through the coil, the current-carrying coil sets up a magnetic field. The coil's field also aligns the domains in the iron in a direction that aids the electromagnet's field. We then have two magnets working together: the electrical one and an iron one. When the current is interrupted, the electromagnetic field collapses and the domains relax into a random state. The entire unit exhibits no magnetism until current again starts to flow. Figure 5-8 illustrates the situation.

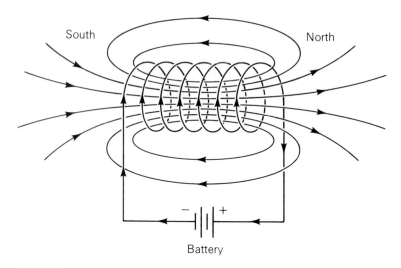

FIGURE 5-7. The solenoid.

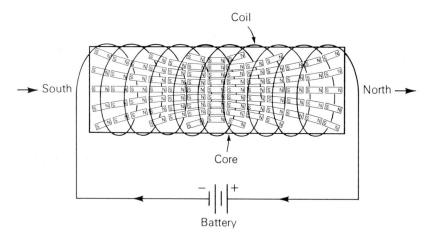

FIGURE 5-8. Solenoid with an iron core.

Study Problems

1. Explain why a broken magnet becomes two small magnets.
2. Complete the following statements:
   a. Like poles _____.
   b. Opposite poles _____.

c. Magnetic "current" flows from _____ to _____.

d. An electron in _____ produces a magnetic field.

e. _____ makes the best temporary magnet.

f. Steel makes a (more ____) (less ____) permanent magnet than iron.

g. Besides iron and steel, the metal _____ is also magnetic.

3. Explain what a domain is.
4. How do domains act when placed in a magnetic field?
5. In what direction are the domains pointing in an unmagnetized piece of iron?
6. Explain why inserting an iron core into a solenoid makes it a much stronger magnet.
7. Where are the poles in the magnetic field surrounding a current-carrying wire?

MAGNETIC CIRCUITS

The Bare Facts

1. All magnetic systems are complete circuits.

2. The space (including air) around a magnet offers a high resistance to magnetic flux (current).

3. Air, wood, water, and nearly all other substances except iron, nickel, and cobalt are magnetic insulators.

4. Iron is the best magnetic conductor.

5. Resistance in magnetic circuits is called *reluctance*.

6. Current in a magnetic circuit is called *flux*.

7. The equivalent of voltage (EMF) in a magnetic circuit is magnetomotive force (MMF).

8. "Magnetic Ohm's Law" is the same as electrical Ohm's Law when reluctance is substituted for resistance, magnetomotive force for electromotive force, and flux for electrical current.

There are 3 basic elements in magnetic circuits that are directly analogous to voltage, current, and resistance in electrical circuits.

Analogies

	Electrical	Magnetic
Voltage	Electromotive force (EMF), V	Magnetomotive force (MMF)
Current	Current (amperes), I	Flux, ϕ
Resistance	Ohms, R	Reluctance, $\mathcal{R}$

In magnetic circuits, magnetic current always exists because nothing is a perfect magnetic insulator. Space has a very high reluctance (magnetic resistance), as does everything else except iron, its oxides, nickel cobalt, and certain alloys. Nickel and cobalt are considered medium-reluctance materials. The reluctance of all materials (except iron, nickel, cobalt, and a few special alloys) is roughly 10,000 times the reluctance of iron.

Examples:

1. High-reluctance materials
    a. Vacuum
    b. Air
    c. Wood
    d. Water
    e. Plastic

2. Low-reluctance materials
    a. Iron
    b. Iron oxides
    c. Special alloys

3. Medium-reluctance materials (the "semiconductors" of the magnetic world)
    a. Nickel
    b. Cobalt

The Air Gap

There are a number of magnet shapes; the bar-shaped magnet is the one with the greatest reluctance between poles. The reluctance be-

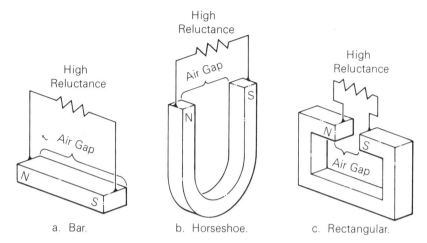

FIGURE 5-9. Magnetic shapes and air gaps.

tween poles is dependent on the distance between them. The space between poles is known as an *air gap* and is a high-reluctance path between poles. In the same way that a longer wire has more resistance than a shorter one, a longer air gap has higher reluctance than a shorter one (see Figure 5-9).

The horseshoe and rectangular forms allow for any desirable air gap, and are the more common shapes found in electronics. When an iron object is to be lifted with a magnet it is placed in the air gap. The shorter the air gap between the poles and the object, the greater the flux through the object and the greater the lifting power.

The air gap is a high-reluctance path, while an iron (or other magnetic) object placed in the gap provides a low-reluctance path. This can be viewed as analogous to parallel resistances, illustrated in Figure 5-10. The reluctance of the iron is so much lower than the air reluctances that it draws most of the magnetic flux (current) through it. As in any parallel circuit where one resistance (reluctance) is very low compared to all others in the circuit, the low-reluctance path becomes the dominant path. The stray air flux in a magnetic circuit is frequently referred to as *leakage* flux. In the hydraulic analogy, high resistances with a fixed value are formed by crimping the pipe to restrict the flow. The valve in Figure 5-10c is assumed to be nearly wide open.

Magnetic Ohm's Law

The magnetic-circuit relationships are the same as those of Ohm's Law and are often called Ohm's Law of Magnetism even though they

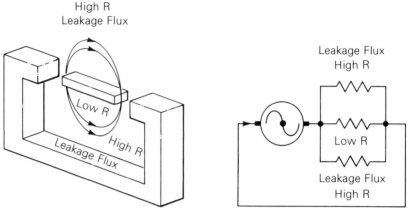

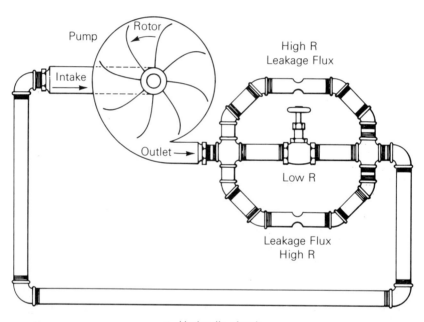

c. Hydraulic circuit.

FIGURE 5-10. Magnetic-circuit analogies.

were not formulated by Ohm. Figure 5-11 is a memory aid for magnetic Ohm's Law. We will not get involved any deeper in magnetic laws or the mathematics of magnetic Ohm's Law. It is a highly specialized subject that most technicians and even many engineers do not get much involved in.

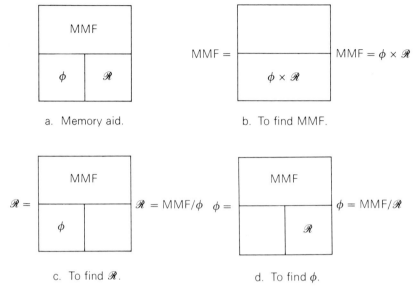

FIGURE 5-11. Magnetic Ohm's Law memory aid and examples.

Study Problems

1. List several high-reluctance materials.
2. List the low-reluctance materials.
3. Complete the following table of analogs:

Electrical	Magnetic Analog
Voltage	
Current	
Resistance	

4. Write the 3 forms (variations) of magnetic Ohm's Law. (*Hint:* see Figure 5-11.)
5. Define air gap.

5-4 MAGNETIC MACHINES

The Bare Facts

1. A solenoid with part of its iron core fixed to the coil and part free to move forms a simple reciprocal motor.

2. The relay is an electromagnetically actuated switch.

3. The moving part of a solenoid or relay is called the *armature*.

4. An electric (rotary) motor consists of:
    a. A magnetic-field assembly (a permanent magnet or electromagnet, called the *field*).
    b. A rotating electromagnet connected to a shaft, called the *armature* or *rotor*.
    c. A reversing switch called a *commutator* mounted on the shaft for DC motors, or a pair of slip rings for AC motors.
    d. A pair of brushes (generally made of carbon because of carbon's self-lubricating and conductive properties). The brushes rub against the commutator or slip rings, forming a moving contact.

5. Electromagnets can also be used to provide the magnetic field necessary to make a motor work.

6. When electromagnet fields are used, there are 2 ways in which the field coil can be connected to the armature coil: in series, or parallel with it.

7. When field and armature windings are connected in series, the motor is called a *series-wound* motor.

8. When field and armature coils are connected in parallel, the motor is called a *shunt-wound* motor.

9. Both series-wound and shunt-wound motors can have their shaft directions reversed by exchanging the field wires.

10. The series-wound motor can be used with alternating current. The shunt-wound motor cannot normally be used with alternating current.

11. Most home-shop power tools and small home appliances use series-wound motors.

12. A DC generator is constructed exactly as a motor.

13. Any DC motor, properly connected, can be used as a generator, and vice versa.

14. Many AC motors use slip rings and many DC motors have a commutator instead.

THE SOLENOID

Moving electrons in a coil produce a magnetic field whose strength is proportional to the current and the number of turns of wire in the coil. A greater current produces stronger fields around the wire carrying it, and the number of turns is a measure of how many individual fields are concentrated between the poles of the solenoid.

$$MMF = amps \times turns$$

Earlier in this chapter we examined the effect of winding a solenoid around a soft iron core. If that same arrangement is used, but the coil is wound around a nonmagnetic cylinder in which the iron core is free to move, we have the simplest form of electric motor.

Assume that a portion of the core is partway out of the coil, and another part of the core is fixed to the coil as shown in Figure 5-12. When current flows through the coil, producing a magnetic field around it, the coil's field aligns the domains in the iron core, making it a temporary magnet. The core is split, with one part fixed to the coil and the other part free to move. The 2 pieces of the core attract each other. Such a device is called a *solenoid*, and it can be used to pull a lever or perform other mechanical functions. For example, a solenoid is used to shift gears to change the tub speed in automatic washing machines. A permanent magnet can also be used as the movable core, but it is rarely used in commercial solenoids due to the added cost and slow release. Figure 5-13 shows some examples of commercially available solenoids.

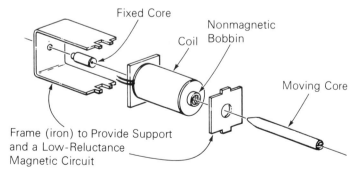

FIGURE 5-12. Construction of a solenoid.

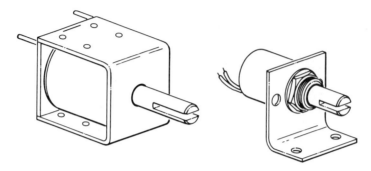

a. Direct-current types.

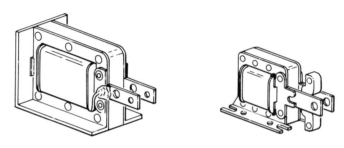

b. Alternating-current types.

FIGURE 5-13. Typical commercial solenoids.

The *relay* is basically a solenoid that actuates a switch when a current flows through the coil. Sometimes switch contacts are mounted on a modified solenoid as shown in Figure 5-14a, but more often the movable part of the core is in the form of a flat bar acting as a lever to move one contact against another. The movable bar is known as the *armature* (it moves like an "arm"). Figure 5-14b shows the construction of the typical relay.

THE ELECTRIC MOTOR

If a permanent magnet is used in a solenoid and the current is periodically reversed, the core can be made to move with a reciprocating motion. Such a reciprocating motor is rarely needed in practice, but the same principle can be used to produce rotary motion. Figure 5-15 is a drawing of a simple direct-current electric motor. The motor uses a permanent magnet with the pole ends shaped to form a small air gap to minimize the reluctance in the magnetic circuit. The shaft that delivers power to a rotating load has a rotating electromagnet and a reversing switch attached to it. The rotating electromagnet is called the *rotor* or *armature* and is powered by the battery.

MAGNETISM AND MAGNETIC MACHINES   161

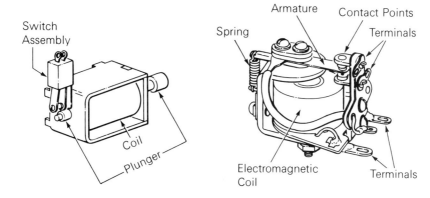

a. Combination relay/solenoid.   b. Typical relay construction.

FIGURE 5-14. The relay.

The current-reversing switch is called the *commutator;* it consists of copper segments that are insulated from each other and connected to the two armature wires. Current flows from the battery through the two carbon brushes that rub against the commutator segments, carrying current through the coil of the armature electromagnet.

With the armature in Position 1, shown in Figure 5-15a, the upper end of the armature is its south pole and the bottom end is its north pole. The 2 pairs of poles attract each other, turning the armature in a clockwise direction. Ordinarily the armature would stop moving as soon as it reached Position 2. However, at this point the brushes slide off the copper segments of the commutator, interrupting the current momentarily. This allows the armature to coast slightly beyond the point where the 2 poles are in exact alignment.

As the armature coasts for a few degrees beyond Position 3, the brushes again slide onto the copper segments on the commutator, causing current to flow again through the armature coil. Now, however, the battery connections to the coil are reversed and the current is flowing in the opposite direction. The armature pole that was previously south now becomes north, and the former north pole becomes south. The repulsion of like poles drives the armature further in a clockwise direction. At Position 4 attraction again takes over. At Position 5 the commutator again interrupts the current to the armature. At Position 6 the current is again reversed and the armature poles are interchanged to start a new round.

The permanent magnet in Figure 5-15b is called the *field* magnet. The field magnet in many motors (and generators) is also an electromagnet. Until recently, permanent-magnet technology had not

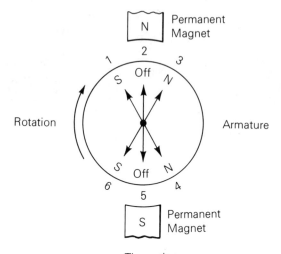

a. The action.

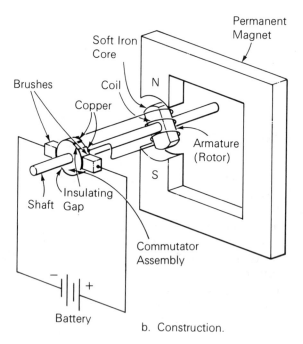

b. Construction.

FIGURE 5-15. The electric motor.

advanced to the point where any but the smallest motors could be built with permanent magnetic fields. Recent improvements in magnets now permit motors of up to a few horsepower to be built with such fields.

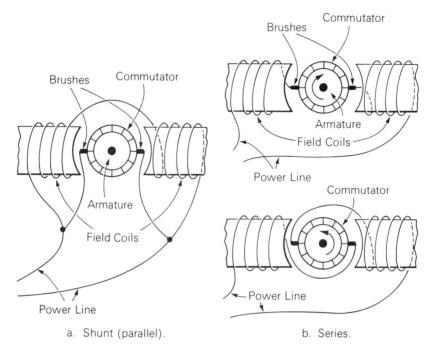

FIGURE 5-16. Shunt- (parallel) and series-wound motors.

In motors with an electromagnetic field, the field coil can be designed to operate either in series or in parallel (shunt) with the armature winding. Figure 5-16 shows the two winding arrangements. The shunt-wound motor can be reversed by exchanging field-coil wires, but cannot be used with alternating current. The series-wound motor can also be reversed by rearranging the leads shown in Figure 5-16b. The series-wound motor *can* be used with alternating current, and is found in electric drill motors, vacuum cleaners, electric mixers, and most other home-shop power tools. Figure 5-17 shows a cutaway view of an automobile starter motor.

THE GENERATOR

When a magnet is moved in and out of a coil, electrons in the wire of the coil are driven first in one direction as the magnet is pushed in, and then in the opposite direction as the magnet is pulled out. When the magnet is at rest within the coil, no electrons move and there is no current flow. While this reciprocating arrangement can be used to generate an alternating current, it is seldom used.

Rotary generators work on the same principle. It is necessary only to apply mechanical force to move a coil relative to a magnetic field.

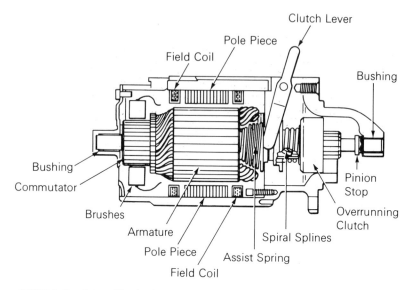

FIGURE 5-17. Automobile starter motor.

The coil can be stationary and the magnet movable or the other way around, as long as there is relative motion between the two.

Electric generators are constructed exactly like the electric motors we have examined. In fact, any electric DC motor can be connected to operate as a generator. All generators are basically alternating-current-producing devices, but a commutator and brush assembly can be used to switch the output voltage to produce a pulsating direct current. Figure 5-18 shows a schematic diagram of an alternating-current and a direct-current generator. The only real difference between the two is the use of continuous slip rings to carry the current from the armature of the AC generator to the load, and a segmented commutator used in the DC generator for the same purpose.

The electric generator produces an alternating current that can be taken either from slip rings for AC, or from a commutator to reverse the output polarity for pulsating DC. Figure 5-19 shows the relationship between armature position and output voltage for AC and DC generators.

Study Problems

1. What is a commercial solenoid? Give some examples where one might be used.
2. What is a relay?

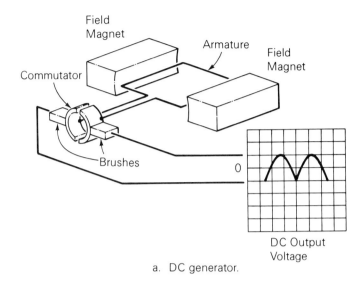
a. DC generator.

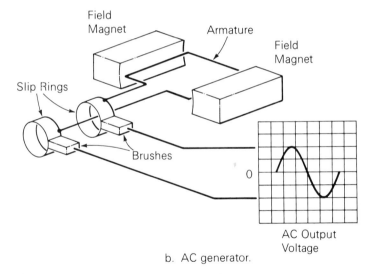
b. AC generator.

FIGURE 5-18. Alternating-current and direct-current generators.

3. Explain how an electric motor works.
4. Motors are classified as either series-wound or shunt-wound. What is meant by these classifications?
5. What is the stationary magnet in an electric motor called?
6. What is the rotating part of an electric motor called?
7. What is the purpose of slip rings in an AC motor or generator?

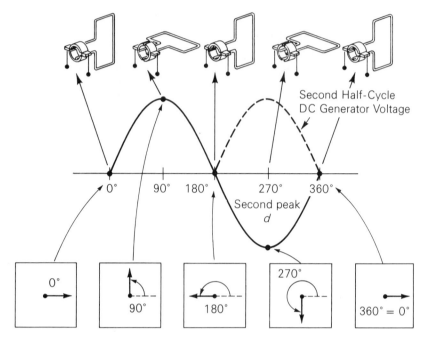

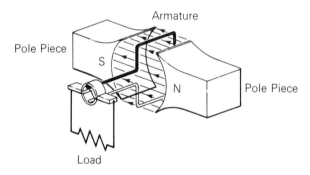

a. Armature rotation and output voltage.

b. Armature and pole-piece relationship.

FIGURE 5-19. Generator output voltages.

8. A commutator in a DC motor or generator serves two purposes. What are they?
9. Which of the two kinds of motor (series or shunt) is commonly used with ordinary household power?
10. What kind of output voltage (shape) does an AC generator produce?
11. What kind of direct-current voltage does a DC generator produce? (*Hint:* see Figures 5-18 and 5-19.)

FRACTIONAL-HORSEPOWER AC MOTORS

The Bare Facts

1. The most common fractional-horsepower motors designed for operation with alternating current only are called *induction* motors.
2. Most fractional-horsepower induction motors use a simple rotor called a *squirrel cage*.
3. The squirrel-cage rotor has no electrical connection to the power line.
4. No brushes, commutators, or slip rings are found in motors using a squirrel-cage rotor.
5. The squirrel-cage "winding" is a closed circuit.
6. Current in the cage is electromagnetically induced from the field (stator) coil.
7. Squirrel-cage motors are essentially single-speed motors.
8. The speed is approximately determined by the AC line frequency (generally 60 hertz) and the number of poles in the stator.
9. Squirrel-cage motors require an extra coil to start them rotating.
10. Resistors or capacitors are often included in the starting-coil circuit to increase the starting torque.
11. A centrifugal switch is generally used to disconnect the starting circuit when the motor gets up to speed.
12. The efficiency of squirrel-cage motors is low compared to that of DC motors.

Many appliances, such as phonographs, vending machines, and washing machines, use an induction motor, which runs on AC only. This type of motor is cheap to manufacture, reliable, and runs at a fairly constant speed. Induction motors are different from those we have discussed so far, because there is no electrical connection to the rotor. The rotor gets its current by electromagnetic induction. The device is basically a transformer with a movable secondary. (See Chapter 7 for a discussion of transformers.)

The most common induction motor in consumer applications is the squirrel-cage motor, shown in its simplest form in Figure 5-20. The squirrel cage is actually a winding with the equivalent of a single turn of very heavy wire. In the rotor detail drawing in Figure

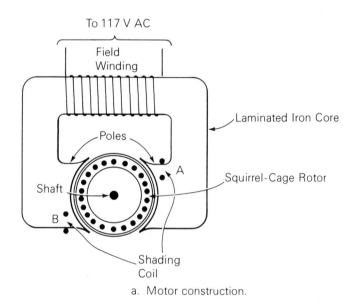

a. Motor construction.

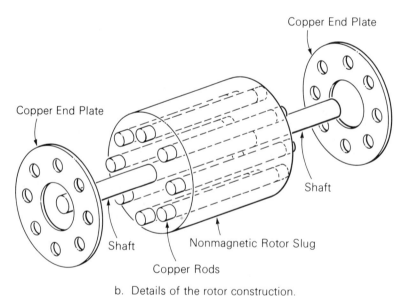

b. Details of the rotor construction.

FIGURE 5-20. The simple squirrel-cage motor.

5-20b, the copper bars are insulated from the nonmagnetic rotor slug where they pass through it. Two copper rings are pressed onto the copper bars to interconnect them electrically. As the alternating current flows through the field winding, the changing magnetic field cuts the bars in the squirrel cage, causing a current to flow in the squirrel cage. The current flowing in the squirrel-cage bars sets up a magnetic field around the cage. The squirrel cage follows the effectively rotating magnetic field produced by the field coil, with its changing AC current. The squirrel-cage magnetic field must always be slightly behind the magnetic field set up by the field coil. This lag is called *slip*, and it is the slip that allows the field coil's magnetism to effectively pull the squirrel cage behind its rotating field. When the motor is running normally, a fan blade or other load maintains the slip. However, the motor is not self-starting and requires some special starting method.

Induction Motor Starting

All of the various starting methods depend on the use of a special winding that produces an extra field that lags behind the main field. This can be accomplished by electrical phase shift. We will study phase shift in detail in a later chapter. Basically, currents or voltages in an alternating-current circuit can be delayed in time by using appropriate resistances and capacitances in conjunction with the motor starting coil.

Figure 5-21 illustrates the 3 most common starting circuits for squirrel-cage motors. The arrangement in Figure 5-21a is often found in motors used in washing machines and similar appliances. These motors are generally rated at ⅓ horsepower or less. The method is not practical for larger motors or motors that must start under heavy loads.

The resistor and starting winding are switched out of the circuit when the motor reaches about 75 percent of its normal running speed. The contacts on a centrifugal switch are actuated by a weight system fastened to the motor shaft. As the motor gains speed, the weights move outward from the shaft against a spring. When the shaft is turning at about 75 percent of running speed, the weights will have moved far enough to open the contacts by way of a simple mechanical linkage.

The capacitor starting arrangement in Figure 5-21b is able to start the motor under heavier loads than is practical with resistance starting. Capacitor starting is also used with motors larger than ⅓ horsepower. The capacitor is switched out of the circuit at about 75 percent of normal running speed.

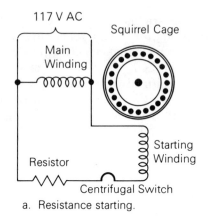

a. Resistance starting.

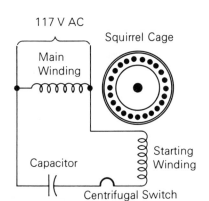

b. Capacitor starting (with switch).

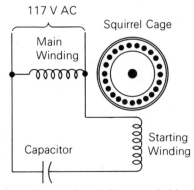

c. Capacitor starting (without switch).

FIGURE 5-21. Starting the squirrel-cage motor.

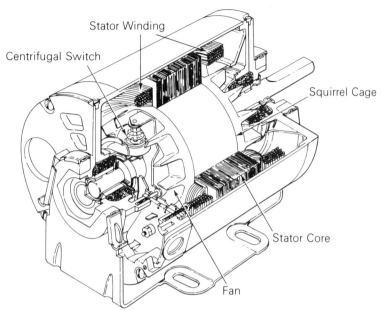

d. Cutaway view of a typical squirrel-cage motor.

FIGURE 5-21. (continued)

The circuit in Figure 5-21c eliminates the need for the centrifugal switch, but is practical only for small motors of the kind that might be found in high-quality record turntables. The size of the capacitor increases with the size of the motor. When the capacitor must remain in the circuit continuously, it must be much larger than a capacitor used for starting only. It is generally costly to use the arrangement in Figure 5-21c for larger motors.

The Shaded-Pole Motor

The simple motor shown in Figure 5-20 uses an induction starting winding that is left connected in the circuit. Slots or holes are punched in the laminated core near each pole and a couple of turns of heavy copper wire are wound through them. The windings are shorted and current is induced in these small *shading* windings in the same fashion as in the squirrel-cage bars. This kind of motor is common in inexpensive phonographs, vending machines, and small fans. The shaded-pole starting arrangement is not practical for motors larger than about $1/20$ horsepower.

Some larger induction motors use a rotor that consists of several windings of small wire instead of the squirrel cage. These so-called *wound-rotor* motors operate on the same principle as the squirrel-cage machines and use the same basic starting techniques. Wound-rotor machines are comparatively expensive and more suitable for industrial uses than for consumer devices.

Study Problems

1. How is current caused to flow in a squirrel-cage rotor?
2. Does the squirrel-cage induction motor require slip rings or a commutator?
3. What is the purpose of a shading pole?
4. Why are capacitors often used in induction-motor starting circuits?
5. What is the purpose of a centrifugal switch in an induction motor?
6. Can the squirrel-cage motor be used with direct current?

5-5 ELECTRICAL METERS

The Bare Facts

1. Moving-coil meters are close relatives of the DC electric motor.
2. They use a permanent-magnet field and a wound-coil rotor.
3. No brushes, slip rings, or commutators are required because the rotation of the rotor (armature) is limited to less than 180°.
4. The iron core in the rotor field does not rotate.
5. The rotor coil turns in an air gap between field poles and the rotor core.
6. The rotation of the rotor coil is restrained by a calibrated spring.
7. A pointer fastened to the rotor coil is deflected by an amount proportional to the current flowing through the armature (rotor) coil.
8. A numbered scale provides the readout.
9. This kind of meter is called an *analog* meter.
10. These meters can also be used for measuring volts and ohms with extra circuitry.
11. Multirange meters can be constructed by providing switches that switch in different resistance values.
12. An ohmmeter circuit requires an in-circuit battery or other voltage source.

MAGNETISM AND MAGNETIC MACHINES 173

13. The current range of a moving-coil meter can be increased by adding a resistor in parallel with the meter coil. This resistor is known as a *shunt*.

14. The moving-coil meter is a *DC-only* meter. To measure AC, the alternating current is first converted to direct current by a meter *rectifier*.

The basic principle of the DC permanent-magnet electric motor is used in measuring instruments commonly known as *analog meters*. The most common of these meters is the permanent-magnet moving-coil meter shown in Figure 5-22. The coil is wound on a

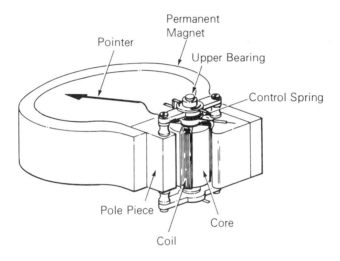

a. The complete movement.

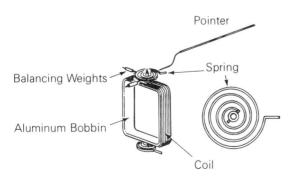

b. The moving coil in detail.

FIGURE 5-22. The moving-coil-meter movement.

lightweight aluminum bobbin that is suspended between 2 jewel bearings. Balancing weights allow the meter to be used in any position. The spiral-spring tension can be adjusted to set the pointer at the proper place on the scale (not shown). An adjustment screw is usually available to the operator for this purpose. The core is fixed and the coil moves in the air gap between the core and the poles. The meter travel is limited by 2 adjustable stops.

When current flows through the coil, its electromagnetic field interacts with the permanent magnet's field, turning the coil and moving the pointer. The coil moves against the pull of carefully calibrated spiral springs. The deflection of the pointer is proportional to the current flowing through the coil.

VOLTMETERS AND OHMMETERS

The moving-coil meter is basically a current device, but it can also be used to measure voltage or resistance. The trick is to take advantage of Ohm's Law. The current in a series circuit depends on the applied voltage and the resistance. If a known resistance value is placed in series with the meter coil, the amount of current that will flow for different voltages can be calculated. The scale on the meter can then be marked in volts. A similar procedure can be used to construct an ohmmeter when a battery of known voltage is built into the circuit. Ohmmeter scales are nonlinear in these instruments, but this causes no problem in most applications.

Multipurpose instruments (multimeters) using the moving-coil movement are very common. The functions (voltage, current, and resistance) are selected by turning a knob on a rotary switch. Another switch selects the proper wired-in resistances for various ranges. The face of the meter is marked with several different scales that are keyed to settings on the range and function switches. Figure 5-23 is a photograph of a typical multimeter.

THE WATT-HOUR METER

The watt-hour meter is a form of induction motor that is used to keep track of your monthly use of electricity. You can find one of these somewhere outside your house. The construction of a typical watt-hour meter is shown in Figure 5-24. The simple aluminum rotor in this meter is not able to produce much torque. It can, however, produce enough to turn the counter that registers your accumulated electricity use for billing purposes. You will not be likely to find this "flypower" motor in use anywhere else.

MAGNETISM AND MAGNETIC MACHINES 175

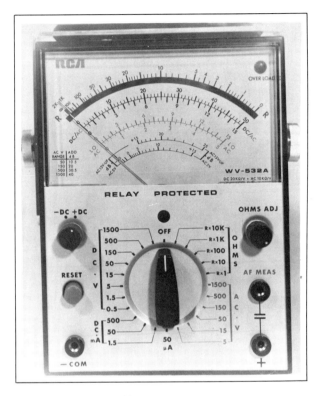

FIGURE 5-23. Typical multimeter.

Because power (watts) is the product of voltage and current, the motor is driven by two windings. One winding is in series with all of the electrical devices in operation in your home at any given time. A second coil monitors the line voltage as it arrives at your home. The disc turns at a rate determined by the actual voltage and current at any given instant. This ensures a true measure of the energy actually used.

The shaft of the motor turns a mechanical counter that keeps a running total of the power used. The permanent magnet acts as a drag brake to slow the disc down. The drag is made large enough so that variations in temperature effects and friction in the counter have minimal influence on the rotor speed. A friction-type brake would be impractical because wear and temperature variations would make it all but impossible to keep the meter accurate over long periods. The magnet also prevents the disc from coasting when the power consumption is reduced.

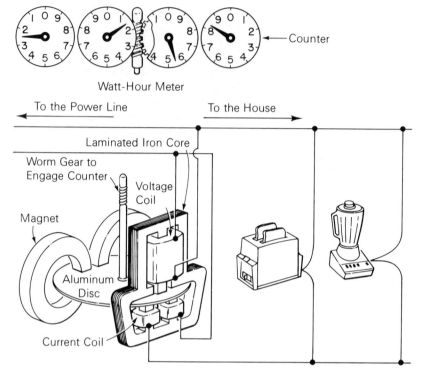

FIGURE 5-24. The induction watt-hour meter.

Study Problems

1. The moving-coil meter is a relative of what kind of motor?
2. The watt-hour meter is related to what kind of motor?
3. What is the purpose of a shunt? (*Hint:* see Bare Facts.)
4. Is the moving-coil meter classed as an analog or a digital meter?
5. Why is the simple aluminum-disc rotor used in watt-hour meters rarely used elsewhere?
6. Why does the watt-hour meter have two field coils?

5-6 ELECTRICAL POWER DISTRIBUTION

About 30 percent of the power generated in this country is generated in hydroelectric plants. About 65 percent of our power is produced by steam-driven generators using coal or oil for fuel. The balance is produced by diesel and other internal-combustion engines, and nuclear power. The household power line is usually referred to as the *120 volt main*, although the voltage varies with the locality. 110, 115, and 117 volt lines are common.

At the power plant, the generator produces a much higher voltage, in the neighborhood of 13,000 to 18,000 volts. Where power must be sent over long distances a transformer steps the voltage up to 130,000 to 345,000 volts or more.

Resistance losses in the transmission lines increase as the current increases. It is more efficient, for example, to send 1000 watts along a transmission line as 1000 volts and 1 amp than it is to send it as 100 volts and 10 amps.

> Power = volts × amps
> 1. Power (watts) = 1000 volts × 1 amp
> 2. Power (watts) = 100 volts × 10 amps

In each case the power is 1000 watts.

Suppose the transmission line has a resistance of 2 ohms. Let's calculate the power loss. Power = I^2R is the formula for power where current and resistance are known.

> For the 1000 volt, 1 amp system, the power loss in 2 ohms of resistance is:
> $P = I^2R$
> $P = 1^2 \times 2 = 2$ watts (power loss)
>
> For the 100 volt, 10 amp, system:
> $P = I^2R$
> $P = 10^2 \times 2 = 200$ watts (power loss)

Strategically located substations use transformer banks to step the 130,000 to 345,000 volts down to about 4000 volts for residential distribution, and to about 13,000 volts for industrial distribution. Additional transformers near the home, office, or store step the voltage down again to about 120 volts.

Manufacturing plants, large office complexes, and so on often have their own substations and receive power at the 13,000 volt level. These substations are often located in underground vaults beneath the building. Figure 5-25 summarizes the production and distribution of power.

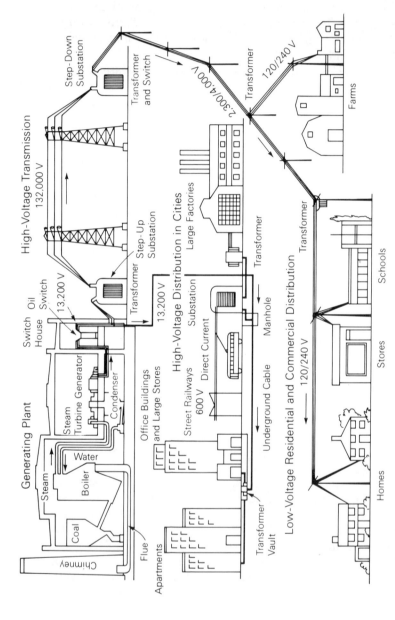

FIGURE 5-25. Power production and distribution system. Coal is the energy source. (Bureau of Labor Statistics, U.S. Department of Labor.)

DELIVERY OF POWER TO THE HOME

A transformer on the pole (or underground) near your home has two windings, each of which provides 120 volts with respect to a common (neutral) line. This makes 220 volts available for air conditioners and other power-hungry units.

Figure 5-26a shows the voltages out of the transformer and 5-26b shows the equivalent circuit. (Remember that V is the abbreviation

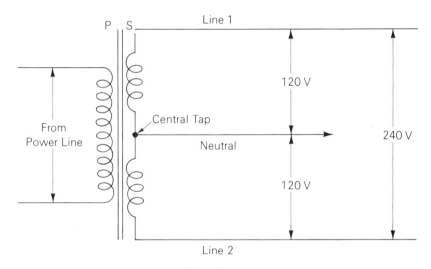

a. Transformer.

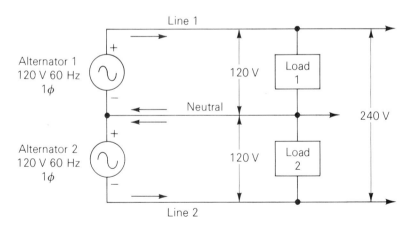

b. Equivalent circuit.

FIGURE 5-26. Voltages delivered to a house.

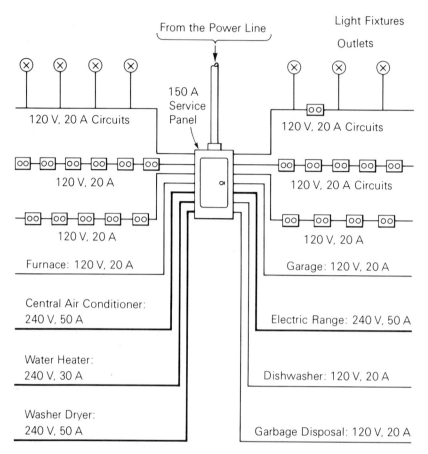

FIGURE 5-27. Power distribution in a typical home.

for volts, and Hz is the abbreviation for hertz.) The neutral line is grounded to the earth; that's where we get the term *ground*.

Distribution in the House

Figure 5-27 shows how the power is distributed in a typical house. (The letter A is the abbreviation for amperes.) The service panel contains circuit breakers for the individual circuits leaving the service panel (older houses may use fuses). Modern circuit breakers serve both as protective devices and as master switches to turn the power off for specific circuits. Figure 5-28 shows typical house wiring techniques, and Figure 5-29 shows some typical standard house-wiring hardware.

MAGNETISM AND MAGNETIC MACHINES 181

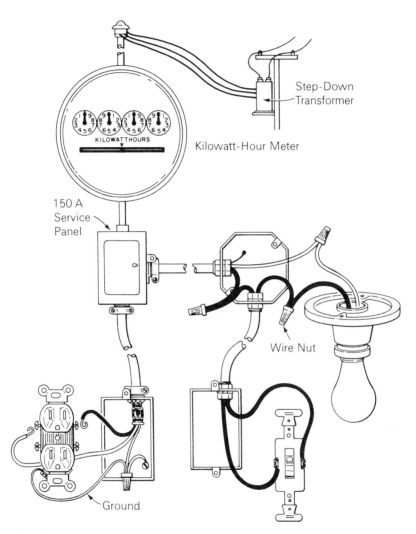

FIGURE 5-28. House wiring.

182 CHAPTER 5

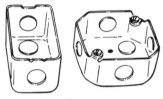

a. Boxes.

b. Fittings.

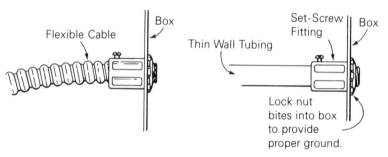

c. Conduit.

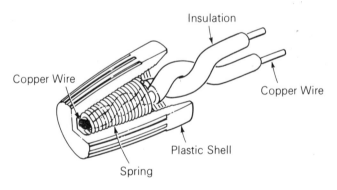

d. The wire nut.

FIGURE 5-29. House-wiring hardware.

Reading the Kilowatt-Hour Meter

Two different kinds of dials are shown in Figure 5-30a. The dial on the left is easier to read but less common than the one on the right. Figure 5-30b shows how to read the meter. A reading between two digits is read as the smaller. Note that the dials don't all turn in the same direction.

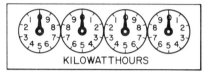

a. Digital type (left) and dial type (right).

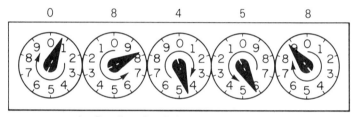

b. Reading the dial-type meter.
The reading is 08458 kilowatt-hours.

FIGURE 5-30. Reading the kilowatt-hour meter.

Chapter 6

Courtesy of Kelly Williams

INDUCTANCE AND CAPACITANCE

6-1 INTRODUCTION

The last two passive devices we will look at are capacitors and inductors. They are both energy-storage devices. The inductor is nothing more than a coil with or without a magnetic core. In the preceding chapter we were concerned with the magnetic properties of coils. Here we will be concerned with the electrical properties of coils. We will study their ability to store energy in their magnetic fields and to deliver that energy back to the circuit when the field collapses.

The capacitor is also an energy storage device but it works on a different principle. Capacitors and inductors *mirror* the electrical behavior of each other. They can be used separately or combined to do such special jobs as tuning in a specific radio or television station and rejecting all others.

6-2 CAPACITORS

The Bare Facts

1. The capacitor is an energy-storage device.

2. It consists of 2 metal plates separated by an insulator.

3. The amount of energy a given capacitor can store increases with the area of the plates, and decreases as the thickness of the insulating material increases.

4. The insulating material is called the *dielectric*.

5. The storage capacity of a capacitor can be greatly increased by using a high-quality dielectric material.

6. Air is a poor dielectric. Some plastics, glass, and ceramic materials are better.

7. Capacitors are classified according to the dielectric material: mylar, paper, ceramic, and so forth.

8. The unit of capacitance is the farad.

9. A 1 farad capacitor can store enough energy to deliver one amp for 1 second when charged to 1 volt.

10. 1 ampere-second is 1 coulomb, about enough energy to light a 100 watt household lamp for 1 second.

11. The farad is too large a unit for most electronics applications. The microfarad, abbreviated μF and meaning 1/1,000,000 of a farad, is more often used.

12. Capacitors in parallel increase the effective plate area. Therefore a 1 microfarad capacitor in parallel with a 2 microfarad capacitor yields an effective total of 3 microfarads.
13. Capacitors in series must be treated like resistors in parallel to find equivalent values.
14. Capacitor analogs are springs, compressed air in a tank, deformable diaphragms, and so forth.
15. A capacitor charged from a direct-current source builds an increasing counterforce as it charges.
16. The rate of current flow into a capacitor diminishes as it charges.
17. All charging and discharging capacitors follow a special curve known as the *universal-time-constant* curve.
18. Charging voltage and current are out of step (out of phase) in time in a capacitor circuit.
19. The maximum current occurs before the maximum voltage (earlier in the charging cycle).
20. The combination of a resistor, a capacitor, and a DC voltage is called a *timing* or *time-constant circuit*.
21. One time constant is the time required to charge a capacitor through a resistor to 63 percent of the applied DC voltage.
22. A capacitor is considered to be fully charged after 5 time constants have elapsed. Time constants are measured in seconds.

    The capacitor is an energy-storage tank. The basic capacitor consists of 2 plates separated by an insulating material called the *dielectric*. The dielectric can be any insulating material including vacuum, air, plastic, glass, ceramic materials, and so on. The dielectric is analogous to the core in a solenoid; the effective size of the capacitor can be greatly increased by careful selection of the dielectric material.

    Figure 6-1 shows the structure of the simplest capacitor. The storage capacity increases as the area of the plates increases, and decreases as the distance between them increases. The capacity is also increased by using a better dielectric material.

    The figure of merit for dielectric materials is called the *dielectric constant* (K).* Better dielectrics have larger numerical values for K.

---

*The figure of merit is a ranking number that describes how much better one material is than another with respect to a vacuum.

INDUCTANCE AND CAPACITANCE 187

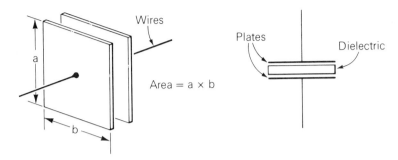

FIGURE 6-1. The capacitor.

Vacuum, the poorest of dielectric "materials," has a dielectric constant of 1. Air is very nearly as poor a dielectric as vacuum and has a constant of approximately 1. A material with a K of 10 would be 10 times as effective as a vacuum (or air). Table 6-1 lists some dielectric materials and their dielectric constants. Capacitor types are primarily classified according to the dielectric material used. Paper has been replaced by mylar (Saran wrap® is a mylar plastic). Mylar capacitors are rolled into cylinders in order to get a fair amount of plate area into a compact space. The mylar capacitor is illustrated in Figure 6-2.

TABLE 6-1 DIELECTRIC CONSTANTS

Material	Dielectric Constant (K)
Vacuum	1
Air	1
Paper	2–5
Oil	2–5
Mica	3–8
Glass	6–8
Mylar	5–8
Ceramics	80–1200

$$\text{Capacitance} = K \times \frac{\text{Area}}{\text{Distance}}$$

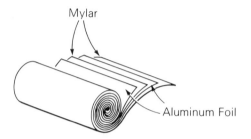

FIGURE 6-2. Mylar capacitor construction.

Figure 6-3 illustrates some of the more common commercial capacitor styles. Each type comes in several electrical and physical sizes. The unit of capacitance is the farad. One farad can store 1 coulomb (6.25×10^{18}) of electrons with a charging potential of 1 volt. The microfarad (0.000,001 or 10^{-6} farad) and the picofarad (0.000,000,000,001 or 10^{-12} farad) are more practical units. As a reminder, the abbreviation for microfarad is μF, and for picofarad, pF.

CAPACITORS IN SERIES AND PARALLEL

Capacitors come in standard values with the same digits as resistors. Placing capacitors in parallel adds the individual capacitances, effec-

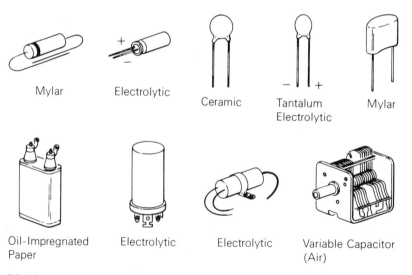

FIGURE 6-3. Commercial capacitors.

tively increasing the total plate area. Capacitor values in series combine like resistor values in parallel. The parallel-resistance charts in Chapter 4 can be used for finding equivalent values for series capacitors. Figure 6-4 summarizes the equivalent values of capacitors in series and parallel.

CAPACITOR ANALOGS

Figure 6-5 shows several capacitor analogs. Each device illustrated is capable of storing energy that can do work when released. In Figure 6-5a, as the springs are stretched they store energy that can exert a force in the direction of the upward arrows. Figure 6-5b shows a compression-type spring that is compressed by the cam. The energy stored in the spring closes the valve when the cam lobe moves away from the valve stem. Compressed air in a pressure tank (Figure 6-5d) and water in a tank (Figure 6-5e) are also excellent capacitor analogs.

A principal characteristic of all capacitor-like systems is an increasing counterforce as more energy is stored in the "capacitor." If you have ever pumped up a bicycle tire or football, you have experienced this phenomenon. The pumping is easy at first, but as the tire takes on more air, the pumping gets more difficult. A point is finally reached where you can no longer exert enough force to push any additional air into the tire. The compressed air is exerting a counterforce that increases with the amount of air pumped in. This increasing counterforce is also characteristic of all kinds of springs. You can feel it when winding a watch, for example. Suppose we examine that counterforce and what it means in electrical capacitors.

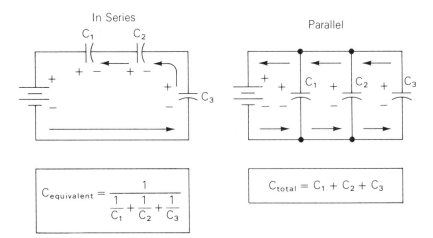

FIGURE 6-4. Capacitors in series and parallel.

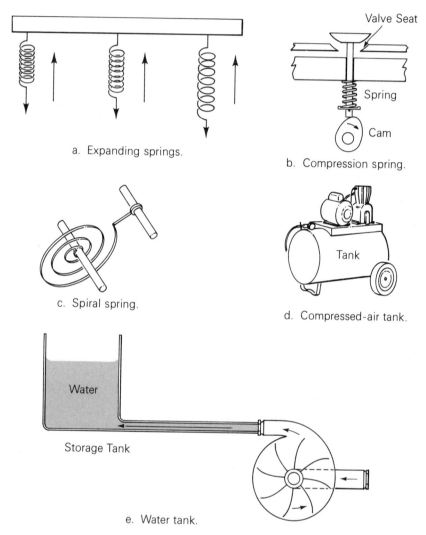

FIGURE 6-5. Capacitor analogs.

COUNTER-VOLTAGE IN CAPACITORS

The capacitor is an energy-storage tank, but it behaves in a special way. In Figure 6-6b the battery forces electrons from Plate A onto Plate B. This leaves Plate A deficient in electrons and Plate B with a surplus. The result is a potential difference between the 2 plates equal to the battery voltage. When Switch A is opened (Figure 6-6c), the electrons are trapped and the potential difference remains across the capacitor. When Switch B is closed (Figure 6-6d), the electrons

INDUCTANCE AND CAPACITANCE 191

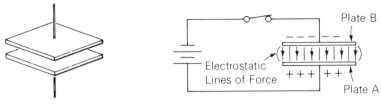

a. The basic capacitor.

b. A charged capacitor.

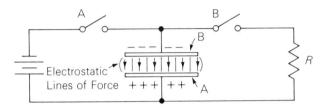

c. Capacitor holding (storing) charge.

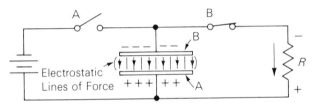

d. Capacitor discharging.

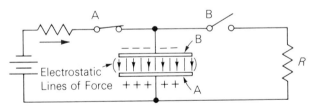

e. Circuit with charging resistance.

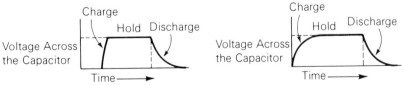

f. Charge-discharge graph without charging resistor.

g. Charge-discharge graph with charging resistor.

FIGURE 6-6. Charging and discharging a capacitor.

are free to flow through Resistor R until the initial neutral uncharged condition is reestablished.

The current through the resistor is not constant as it would be if connected to a battery with no capacitor in the circuit. As electrons move through the resistor from Plate B back to Plate A, the potential difference drops toward zero. At first, when the potential difference is at its highest, a relatively large current flows. As the potential difference is reduced, so is the current (Ohm's Law), so the capacitor discharges more slowly as time goes on. Figure 6-6f is a graph that illustrates this slowing down of the discharge rate with time. The straight vertical rise on the charge part of the graph (Figure 6-6f) shows that the capacitor was charged with very little resistance in the circuit. The charging was nearly instantaneous.

In reality there is no such thing as a zero-resistance system. Perpetual-motion machines depend on a zero-resistance (friction) system, and that is the reason the U.S. Patent Office will not issue patents for such machines. In Figure 6-6e we have added a charging resistance to conform to reality. Figure 6-6g shows a graph of the charge-hold-discharge cycle with the charging resistance added. In order to help you visualize why this rate of charge and discharge behaves as it does, we will examine the analogy in Figure 6-7.

The storage tank (capacitor) is divided into segments. Water weighs 1 kilogram (kg) per liter (l). The pump produces a force (voltage) of 100 kilograms. The valve is set for 100 liters per minute (current) with a force of 100 kilograms.

We will use Ohm's Law to compute the flow in each stage of the example. We have specified the pump force (voltage) as 100 kilograms and the flow (current) as 100 liters per minute. For this example we need to know the "hydraulic resistance" of the valve. By Ohm's Law:

$$\text{Resistance} = \frac{\text{Force}}{\text{Flow}} = \frac{\text{Volts}}{\text{Current}}$$

$$\text{Resistance} = \frac{100 \text{ kilograms}}{100 \text{ liters/minute}}$$

$$\text{Resistance} = 1 \text{ "hydraulic" ohm}$$

To compute the flow at each stage of the example we will use Ohm's Law again in the form:

$$\text{Flow} = \frac{\text{Effective force}}{1 \text{ ohm}}$$

Equivalent to:

$$I = \frac{E}{R} \text{ in electrical terms}$$

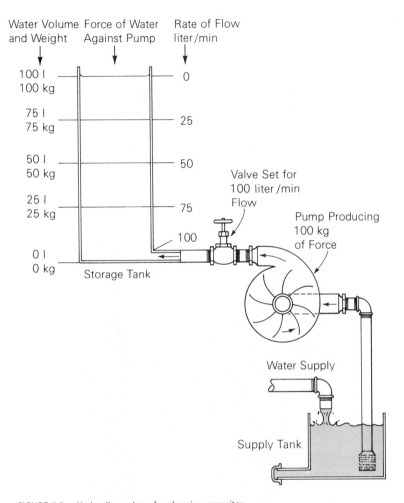

FIGURE 6-7. Hydraulic analog of a charging capacitor.

1. Starting with an empty storage tank, the pump is delivering 100 liters per minute.
2. When the tank is one-quarter full, the water pushes down against the pump with a counterforce of 25 kilograms. The effective pump pressure is now only 75 kilograms (100 kilograms − 25 kilograms). The flow has been reduced to 75 liters per minute.

$$\text{Flow} = \frac{\text{Effective force}}{1 \text{ ohm}}$$

$$\text{Flow} = \frac{75 \text{ kilograms}}{1 \text{ ohm}} = 75 \text{ liters/minute}$$

3. By the time the tank is half full, the pump is working against 50 kilograms of counterforce and its effective force is reduced to 50 kilograms. This produces a flow of 50 liters per minute, half the flow rate when the storage tank was empty.

$$\text{Flow} = \frac{\text{Effective force}}{1 \text{ ohm}}$$

$$\text{Flow} = \frac{50 \text{ kilograms}}{1 \text{ ohm}} = 50 \text{ liters/minute}$$

4. When the tank is three-quarters full, the effective force is 100 kilograms opposed by 75 kilograms (100 kilograms − 25 kilograms), or 25 kilograms.

$$\text{Flow} = \frac{\text{Effective force}}{1 \text{ ohm}}$$

$$\text{Flow} = \frac{25 \text{ kilograms}}{1 \text{ ohm}} = 25 \text{ liters/minute}$$

5. When the tank has reached the top of the uppermost section (the 100 liter, 100 kilogram mark) in the drawing, the water weighs 100 kilograms and pushes against the pump with 100 kilograms of counterforce. The pump must keep working to produce its 100 kilograms of force just to keep the water level where it is. The flow at this point is zero. If we were to pour water in the top of the tank, water would flow out of it, either flowing past the pump vanes or driving the pump backward until the level was back down to the 100 kilogram mark.

If you will study the drawing (Figure 6-7), it will become evident that as the tank fills, the *rate* at which it fills decreases. The fuller it becomes, the more downward force there is, pushing against the pump and causing a reduction in its flow. Each higher level takes longer to reach than the one before it. If we plot a graph of flow versus time, and force stored in the tank versus time, we get the curve shown in Figure 6-8.

Of course, there are not enough points shown in Figure 6-7 to plot so sophisticated a curve, but the curve involved has proven to be one of nature's favorites; its shape is called the *natural logarithm* curve. It is found in nature in population growth and most other natural growth. Here, the curve is the shape of current and voltage growth and decay in capacitors and inductors. In electronics it is called the *universal-time-constant* curve. Figure 6-9 shows a capacitor charge-discharge circuit called a *time-constant* or *timing* circuit, and the curve from Figure 6-8 appropriately labeled.

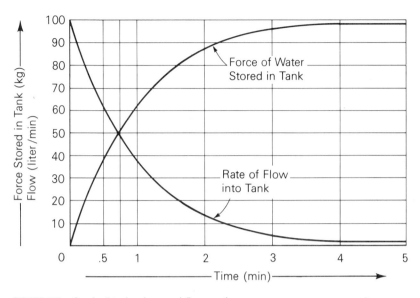

FIGURE 6-8. Graph of tank volume and flow vs. time.

TIME CONSTANTS

Most electronic timing devices use a charging capacitor to mark time. Photographic timers, time-delay circuits for alarm systems, garage lights, and so on are common examples. A resistor-capacitor time constant is determined by the simple product of the 2 values. The resistor-capacitor time-constant formula is:

Time constant = Resistance × Capacitance

$$tc = RC$$

In the example (Figure 6-9b) C = 1 microfarad (μF) and R = 1 megohm (MΩ); the time constant is 1 second(s). The universal-time-constant charts tells us the voltage stored in the capacitor at the end of 1, 2, 3, . . . time constants. At the end of 1 time constant (1 second in this case) the capacitor, according to the chart, will have a charge of 63 percent of the battery voltage, or 63 volts. At the end of 2 time constants the capacitor will have taken on a charge of about 86 volts (adding 63 percent of the 63 volts).

The universal-time-constant chart can be used for any values of resistance and capacitance, any time period, and any voltage. The capacitor is always considered to be fully charged for all practical

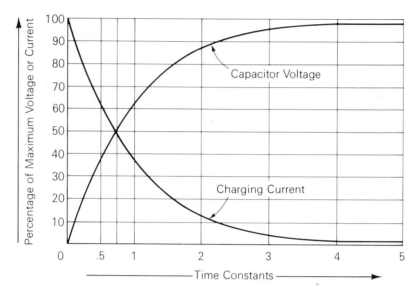

a. Graph.

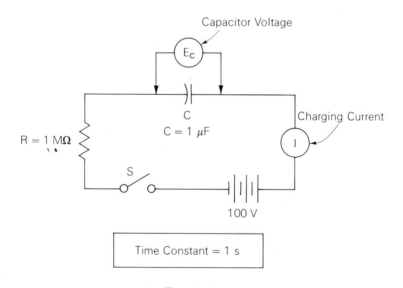

b. The circuit.

FIGURE 6-9. The capacitor universal-time-constant chart.

purposes at the end of 5 time constants. If the capacitor is being discharged, the universal chart can be used to find the voltage left in the discharging capacitor by simply changing the label Charging Current in Figure 6-9a to Capacitor Voltage, Discharging.

TIME Constans are = To R×C

T = R×C

INDUCTANCE AND CAPACITANCE   197

Study Problems

$$C = 1 \text{ microfarad}, R = 1 \text{ megohm}$$

1. In the circuit in Figure 6-9b, what is the voltage stored in the capacitor if the switch is opened after:
    a. 0.5 time constant has elapsed?
    b. 1.5 time constants have elapsed?
    c. 3 seconds have elapsed?
2. If the capacitor value is changed to 2 microfarads, what is the voltage stored in the capacitor after:
    a. 2 seconds have elapsed?
    b. 1.5 seconds have elapsed?

6-3   THE CAPACITOR AND ALTERNATING CURRENT

The Bare Facts

1. In an alternating-current capacitive circuit, current leads the voltage by 90° (¼ cycle). This is called *phase shift*.

2. A capacitor presents an opposition to the average current flow in an AC circuit.

3. This opposition is called *capacitive reactance*.

4. Reactance opposes current flow without power loss.

5. The reactance of a capacitor becomes smaller as the value of the capacitor gets larger.

6. The reactance of a capacitor becomes smaller as the alternating-current frequency gets higher.

7. Alternating-current Ohm's Law for capacitors takes the same form as Ohm's Law for resistors.

$$V = I \times X_c; \quad I = \frac{V}{X_C}; \quad X_C = \frac{V}{I}$$

where X_C = Capacitive reactance.

8. When there are both capacitors and resistors in a circuit, the combined opposition to current flow due to R and X_C is called *impedance*.

9. Impedance can be determined graphically. X_C and R cannot simply be added.
10. They cannot simply be added because of the phase difference caused by the capacitor.

The universal-time-constant chart in Figure 6-9 contains an important key to capacitor behavior in alternating-current circuits. Notice that the charging current is 100 percent of maximum at the start of the charging operation, while the capacitor voltage is still at zero percent of maximum. Later, the current has dropped to near zero while the capacitor voltage has risen to near 100 percent of maximum. The voltage and current are out of *time* with each other.

In a circuit using only resistors, the voltage and current are always in step. When the voltage is at its maximum, the current is also at its maximum. Likewise, when the voltage is zero, so is the current.

Important Point:

1. In a capacitor, the maximum charging current flows when the capacitor voltage is near zero. The current maximum occurs first, and current is said to *lead* the voltage in time.
2. In a capacitor circuit, the capacitor voltage is minimum when the charging current is maximum, and capacitor charge voltage is maximum when charging current is minimum.

Figure 6-10a shows a part of the universal-time-constant chart for direct-current capacitive circuits. The drawing in Figure 6-10b shows the slightly modified universal-time-constant curve that results when alternating current is applied to a capacitive circuit. The shape of the curve is slightly different because of the varying voltage in an alternating current as opposed to steady, unchanging direct current. The important point is that the current reaches its maximum sooner than the voltage in an AC circuit. This time displacement results in a situation where the current in an alternating-current capacitive circuit is one-quarter of a cycle (90°) ahead of the voltage.

> In an alternating-current *capacitive* circuit the current leads the voltage by ¼ cycle.

There is a convenient memory aid to help you remember what leads what in circuits with capacitance or inductance. As we will see

INDUCTANCE AND CAPACITANCE 199

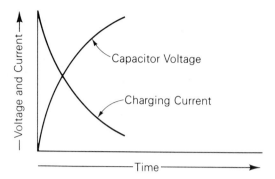

a. The universal-time-constant curve with direct current in a capacitive circuit.

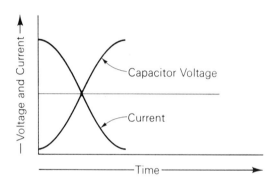

b. The universal-time-constant curve with alternating current in a capacitive circuit.

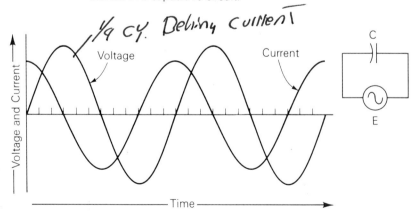

c. The alternating-current waveforms showing the voltage-current time displacement in a capacitive circuit.

FIGURE 6-10. Voltage-current time displacement in a capacitive circuit.

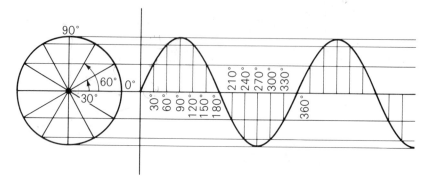

FIGURE 6-11. How the passage of time is measured for sine waves.

in a following section, voltage leads the current in an inductive circuit. Current leads the voltage in a capacitive circuit. To use the memory aid we must use E for voltage. The memory aid is:

ELI the ICE man

where:

E = voltage
L = inductive
C = capacitive
I = current

ELI reads: Voltage (E) leads the current (I) in an inductive (L) circuit.

ICE reads: Current (I) leads the voltage (E) in a capacitive circuit (C).

The time displacement in a capacitive circuit is shown by the waveform drawing in Figure 6-10c.

Positions in time on a sine wave are measured in degrees as shown in Figure 6-11. As a result, the current in an AC capacitive circuit is said to lead the voltage by 90°. The time displacement in AC circuits is called *phase shift*.

CAPACITIVE REACTANCE

A capacitor in an AC circuit presents an opposition to current flow called *capacitive reactance*. Unlike resistance, this opposition to current flow does not produce a power loss in the form of heat. A capacitor in the AC circuit shown in Figure 6-12 is filled to capacity by the end of the first quarter-cycle, begins emptying, and is again empty by the end of the first half-cycle. At that time the capacitor

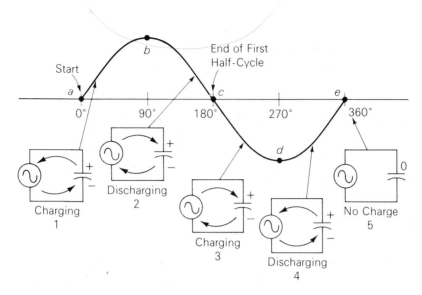

FIGURE 6-12. Charging and discharging a capacitor in an AC circuit.

begins charging in the opposite direction. At the end of three-quarters of a cycle the capacitor is fully charged in the opposite polarity and begins to discharge, reaching zero at the end of the cycle. One cycle fills and empties the capacitor twice.

If you will remember, resistance has no effect until a voltage drives a current through it. We defined 1 ohm of resistance as the amount of opposition to current flow that allows 1 amp of current to flow with an applied voltage of 1 volt. Capacitive reactance (X_C) is defined as the amount of opposition to current flow presented by a capacitor, in an AC circuit, where 1 volt will produce a current of 1 amp.

Alternating-current Ohm's Law for capacitors is shown in Figure 6-13. With X_C replacing R in DC Ohm's Law, the rules are the same.

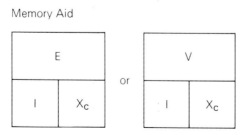

FIGURE 6-13. AC Ohm's Law for capacitors.

What Limits the Current in an AC Capacitive Circuit?

How much water could be transferred from one tank to another using a 4 liter bucket per trip? The answer is 4 liters per trip. With an 8 liter bucket, 8 liters per trip could be transferred. With 2 trips per second, 2 bucketfuls can be transferred per second. If the frequency of the trips is increased, the total volume of water transferred per second increases. With any given charging voltage, a capacitor can hold only a specific number of electrons.

> A 1 farad capacitor holds
> 1 coulomb (6.25×10^{18} electrons)
> when charged to 1 volt.

Like the bucket, a capacitor of a particular size can hold a definite quantity of electrons. A larger capacitor will permit more current transfer through the circuit with each filling or emptying. The more often the capacitor is filled and emptied, the greater the total current flow (electrons per second). Higher frequencies of alternating current result in higher total charge and discharge currents.

The current in a capacitive circuit is proportional to both the capacitance in farads and the frequency in hertz. The formula for the capacitive reactance (the AC capacitive equivalent of resistance) is:

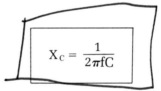

$$X_C = \frac{1}{2\pi fC}$$

The π symbol is the relationship of the diameter of a circle to its circumference. 2π(radians) is another way of saying $360°$, or one complete revolution of an AC generator that generates one cycle of alternating current ($\pi = 3.14$ approximately).

A More Convenient Version

The farad is a very large unit, much too large for most practical applications. The microfarad, or 1/1,000,000 of a farad, is a more useful unit. The formula can be modified for the purpose of expressing the value C in microfarads and making the arithmetic easier.

The modified formula is:

$$X_C = \frac{159{,}000}{fC}$$

where X_C is in ohms
f (frequency) is in hertz (cycles per second)
C is in microfarads

For the mathematician:

$$\frac{1}{2\pi} = 159 \times 10^{-3}; \quad 1 \text{ microfarad} = 10^{-6} \text{ farad}$$

$$X_C = \frac{159 \times 10^{-3} \times 10^6}{fC}$$

$$X_C = \frac{159 \times 10^3}{fC}$$

$$X_C = \frac{159{,}000}{fC}$$

Example:

Find the capacitive reactance of a 1 microfarad capacitor at 60 hertz:

$$X_C = \frac{159{,}000}{fC} = \frac{159{,}000}{60 \times 1} = 2650 \text{ ohms}$$

$$X_C = 2650 \text{ ohms}$$

CAPACITIVE REACTANCE AND RESISTANCE IN SERIES (IMPEDANCE)

In circuits where there is only resistance, the total resistance is simply the two values added together. In cases where there is both capacitance and resistance in the circuit, we encounter a new problem in calculating the total opposition to current flow. The problem comes about because the resistor's opposition to current is constant, while the capacitor offers most of its opposition toward the end of its charge cycle. Its timing is one-quarter of a cycle (90°) out of phase.

The total opposition to current flow with capacitors and resistors in the circuit is called *impedance,* and its symbol is Z.

AC Ohm's Law is summarized as follows:

With capacitance only:

$$I = \frac{V}{X_C}; \quad V = IX_C; \quad X_C = \frac{V}{I}$$

With both capacitance and resistance:

$$I = \frac{V}{Z}; \quad V = IZ; \quad X_C = \frac{V}{I}$$

Note: E = V = volts.

Phasors and Vectors

A vector is a graphical method frequently used for solving for impedance in AC circuits. A phasor is a line whose length represents the amount of resistance, capacitive reactance, or impedance of a circuit. Phasors are represented as vectors where each phasor has a direction corresponding to the amount of phase shift. Figure 6-14 illustrates the vector representation for resistance and capacitive reactance. Vector diagrams are usually plotted on graph paper. A protractor is used to measure the angles representing phase, and a ruler is used to measure line length when necessary. Line lengths can be drawn to any scale, but all lines must use the same scale. See the note in Figure 6-14.

Figure 6-15 shows how to find the impedance of a circuit when the resistance and capacitive reactance are known. The capacitive-reactance vector is shown rotated 90° clockwise (270° counterclockwise). The reactance is 40 ohms, plotted as 4 inches (10 ohms per inch). The resistance is 30 ohms, plotted as 3 inches (10 ohms per inch). The impedance is found by completing the parallelogram (dashed lines) and drawing the diagonal. The diagonal length is the impedance value. A ruler measures the length of the diagonal as 5

INDUCTANCE AND CAPACITANCE 205

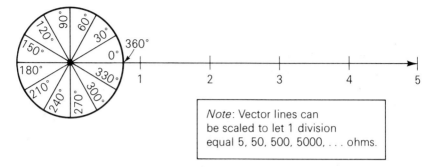

Note: Vector lines can be scaled to let 1 division equal 5, 50, 500, 5000, ... ohms.

a. The resistance vector.

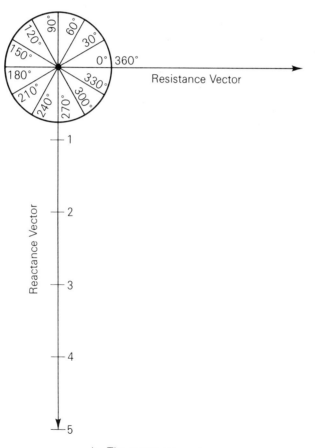

b. The reactance vector.

FIGURE 6-14. Vectors.

inches. The scale is 10 ohms per inch; 5 × 10 = 50 ohms. The impedance, Z, is 50 ohms. For the mathematically curious:

The problem can be solved by using the Pythagorean theorem for right triangles using *our* symbols.

Pythagorean theorem:

$$C^2 = a^2 + b^2$$

$$C = \sqrt{a^2 + b^2}$$

a. Altitude
b. Base
c. Hypotenuse

With our symbols:

$$Z = \sqrt{R^2 + X_C^2}$$
$$Z = \sqrt{30^2 + 40^2}$$
$$Z = \sqrt{900 + 1600}$$
$$Z = \sqrt{2500}$$
$$Z = 50 \text{ ohms}$$

The phase angle for the combined resistance and reactance is 40°* (360° − 40° = 320°) away from the 0°/360° mark. When more accurate results are required, trigonometry can be used to find the phase angle. The current in the circuit is only 40° ahead of the voltage. It would be 90° if the resistor was not in the circuit.

When an AC circuit contains only capacitance, the phase shift will be 90°, with the current leading the voltage. The addition of resistance to the circuit makes the phase shift less than 90°. The phase shift in a resistive circuit is zero degrees. Voltage and current are in phase (no phase shift) in a purely resistive circuit.

Study Problems

1. What are the amount and direction of the phase shift in an AC circuit with a capacitor (no resistance)?
2. What happens to the phase shift when resistance is added to the capacitor circuit (AC)?
3. What is a vector?

---

*40° is protractor accuracy. The true value is closer to 38.7°.

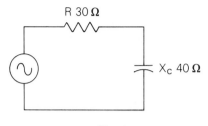

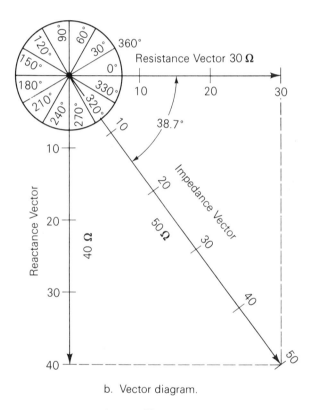

b. Vector diagram.

FIGURE 6-15. Impedance problem.

4. Using the circuit in Figure 6-16, find the impedance of the circuit for the following values of R and X_C:
   a. X_C = 4000 ohms, R = 3000 ohms
   b. X_C = 60 ohms, R = 250 ohms
   c. X_C = 1000 ohms, R = 2500 ohms
   d. X_C = 10 ohms, R = 1000 ohms
   (For this problem you will need graph paper, a ruler, and a protractor.)

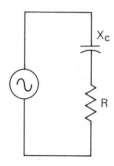

FIGURE 6-16. Circuit for Problem 4.

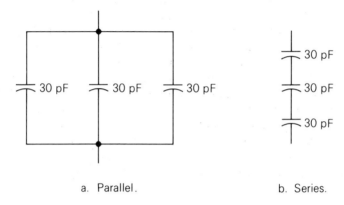

a. Parallel. b. Series.

FIGURE 6-17. Drawing for Problem 6.

5. What is the reactance of a 10 microfarad capacitor at 120 hertz?
6. What is the total capacitance in Figures 6-17a and 6-17b?

6-4 INDUCTORS

The inductor is the mirror-image counterpart of the capacitor. It stores energy in its magnetic field instead of in an electric field. The inductor "charging" behavior follows the universal-time-constant curve, but with the current lagging behind the voltage instead of leading it in time. In an AC circuit, the inductor produces a 90° phase shift, but in a direction opposite that of the capacitor.

The Bare Facts

1. In an AC circuit (remember ELI the ICE man):
    a. Inductive—voltage leads the current by 90°.
    b. Capacitive—current leads the voltage by 90°.

2. Reactance:
    a. Inductive—the current *decreases* and reactance *increases* as frequency increases.
    b. Capacitive—the current *increases* and capacitive reactance *decreases* as frequency increases.
3. Counter-voltage:
    a. An inductor presents the maximum counter-voltage at the beginning of the charge-cycle time.
    b. A capacitor presents the maximum counter-voltage at the end of the charge-cycle time.
4. Opposition:
    a. An inductor opposes changes in current.
    b. A capacitor opposes changes in voltage.
5. Energy storage:
    a. An inductor stores energy in its magnetic field.
    b. A capacitor stores energy in its electric (static) field.
6. An inductor is a coil.
7. The value of an inductor can be raised by adding an iron core. The core is generally laminated or powdered to reduce core losses.
8. Inductive action occurs only while the magnetic field is expanding or contracting.
9. The unit of inductance is the henry.
10. A 1 henry inductor produces a 1 volt counter-voltage when the current through it is changing at the rate of 1 amp per second.
11. The analog of inductance is mass (weight).
12. The time constant of a resistance-inductance circuit is equal to:

$$tc = \frac{L}{R}$$

13. At the end of 1 time constant the inductive current is 63 percent of maximum.
14. An inductor is considered to be fully "charged" by the end of 5 time constants.

210 CHAPTER 6

a. High-frequency air-core inductor.

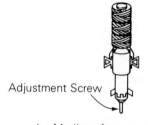
b. Medium-frequency adjustable powdered-iron-core inductor.

c. Low-frequency laminated-iron-core inductor.

FIGURE 6-18. Typical commercial inductors.

15. The formula for inductive reactance is: $X_L = 2\pi f L$ (6.28fL).

16. Impedance in an inductive circuit can be found by using a vector diagram.

An inductor is a coil, with or without a core. The unit of inductance is the *henry*.

Capacitive action occurs only while the capacitor is being charged or discharged, when voltage and current are changing. Fully charged, the capacitor stores energy in its electrostatic field. The amount of energy a capacitor can store is increased by placing an efficient dielectric in the electric field.

Inductive action occurs only while the magnetic field is in motion either expanding or collapsing, when voltage and current are changing. With the field fully expanded, the inductor stores energy in its magnetic field. The amount of energy an inductor can store is increased by placing an efficient (iron) core in the magnetic field.

The strength of a capacitor's electric field depends on the applied *voltage*.

The strength of an inductor's magnetic field depends on the *current* through it.

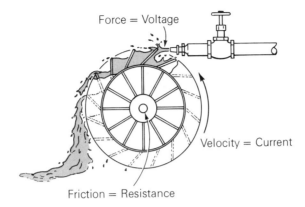

FIGURE 6-19. Rotational analog of voltage, current, and resistance.

INDUCTOR ANALOGS

Because inductive behavior is dynamic, it is time to introduce some new dynamic analogies to help visualize that behavior. Most of us have noticed that a fast-moving stream has a strong current. We have also observed that water running down a steep hill moves more swiftly than it does down a shallow hill. One of the most visual analogies we can find for our examination of inductive behavior is that of a rotating water-driven wheel. Suppose we construct the analogy with aid of Figure 6-19. In this figure we will make a translation from moving-water-stream analogs to their equivalent rotational analogs. The valve is a hydraulic resistance and is only there to illustrate a later idea. The energy of the stream leaving the nozzle forces the wheel to turn. We have converted hydraulic energy into rotational energy. Rotational force is called *torque*. The special word is used only to indicate that we are dealing with rotary motion instead of straight-line (linear) motion. The movement of the water stream (flow) is converted into wheel velocity while the water stream gives up its force and falls limply to the ground. Assume that the sides of the wheel are open mesh to eliminate the need to think about gravity operating on water trapped in the wheel cups. Gravity simply adds an additional force to complicate things, but does not actually alter the idea, so it is best to leave it out of the picture. The last analog is that of resistance, and that is present as friction in the wheel bearings.

We will be using the word *torque*, so let us take a moment to define it. Torque is any force that produces rotary motion. Figure 6-20 illustrates the definition.

FIGURE 6-20. Torque defined.

Analogs

Electrical	Hydraulic	Mechanical (rotational)
Electromotive energy	Hydraulic energy	Torque
Current	Flow	Velocity
Resistance	Friction of water against pipe and other objects	Friction of wheel bearings and other moving parts

INDUCTANCE AND MASS

Inductance resists any changes in current, just as a flywheel resists any changes in velocity. The concept of change is crucial. A flywheel stores energy and possesses a property called *inertia*. And inductance stores energy in its magnetic field and possesses a property called *reactance*. Neither flywheel nor inductor absorbs or gives up energy when the current or velocity is constant. But try to increase or decrease the current or flywheel speed and you will find an opposition to that change. That opposition is inertia.

Definitions

1. Mass is the property of objects that stores the energy of motion (kinetic energy).
2. Inductance is the property of a coil that stores electrical energy.
3. Inertia is the property of mass that resists any attempt to change its velocity.
4. Reactance is the property of inductance that resists any attempt to change its current.

The key word is *change* (in velocity or current).

Analogs

Electrical	Mechanical (rotational)
Inductance	Mass
Reactance	Inertia

If you have ever watched a freight train start from a full stop, or pushed a car, or ridden a bicycle, you're aware that movement starts slowly and increases faster near the final speed. You are also aware that in pushing a car, for example, it is hard work at first, gets a little easier, and becomes a cinch once the car is rolling at a constant speed. Jogging along behind the car, pushing at a steady speed is easy. It doesn't seem like you are pushing over 1½ tons. And, in fact, you are not. Your total effort is that required to overcome friction in bearings and so on. But run around to the front of the car while it is rolling forward and try to push it backwards. Then you will become painfully aware of the crushing force of a ton and a half of mass.* This is reactance! The stored energy is pushing back against your force.

In an inductor the situation is exactly the same. Energy stored in the field will cause the inductor to push back against any atttempt to change the current. This stored energy produces a counterforce called *counter-voltage*. If you will remember, the generator produced an output voltage only when the armature coil was moving with respect to the field, or if the field was moving with respect to the

---
*Please don't try this experiment. The word *crushing* should be taken literally.

coil. In an inductor the field expands and contracts as the current through the coil changes. This expansion or contraction of the field represents a movement of the field with respect to the coil. The inductor generates a voltage output that always opposes the applied voltage. This phenomenon is called *self-inductance,* or simply *inductance.*

Before we go any further, suppose we examine the case of opposing voltages. What happens is crucial to the understanding of inductors. For a constant voltage the current increases very gradually at first and then more and more rapidly. If there were no resistance in the circuit, the current would rise rapidly to a near-infinite value. The same thing happens when you push an automobile with a constant force. The velocity increases very gradually at first and increasingly fast as you continue pushing.

THE INDUCTIVE TIME CONSTANT

Figure 6-21 illustrates the inductance flywheel analogy. When the flywheel is at rest, the person feels the greatest counterforce while turning the flywheel. The counterforce (voltage) is at maximum when the velocity (current) is at its minimum. As the flywheel velocity (current) increases, its counterforce drops. The force (voltage)

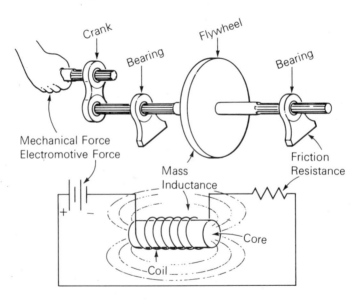

FIGURE 6-21. Inductance-flywheel analogy.

leads the velocity (current) in time, just the opposite of the behavior of a spring or capacitor.

In the case of the inductor, the lines of force in the field start near the core and expand outward. When current first starts to flow, the field tries to expand rapidly, but in doing so it generates a large counter-voltage that pushes back against the applied voltage, restricting the current flow. The counterforce is greatest when the field begins to move outward from the center of the coil, and least when the field is fully formed. As a result, the current is minimum at the beginning of the field's expansion and maximum when the field is fully expanded.

In a capacitor, no current flows when the capacitor is fully charged. In an inductor the only limit to current flow is circuit resistance. The current would be infinite if zero resistance were possible. In an *uncharged* capacitor the only limit to the charging current is circuit resistance.

In Figure 6-21 the rotating flywheel stores energy that can be made to do work until its stored energy is used up. The energy stored in the inductor's magnetic field can likewise be made to do work until its stored energy is used up. Figure 6-22a shows the velocity-force and voltage-current curve for the flywheel and inductor. Figure 6-22b shows the universal-time-constant curve for both capacitance and inductance. Notice that we have identified the upward-going curve as either inductor current *or* capacitor voltage, and the downward curve as either inductor voltage *or* capacitor current. The graph tells us a number of things. First, both devices displace current and voltage in time. In an inductor the voltage leads, while in a capacitor the current leads. The two are exact opposites, and as we will see, inductance and capacitance tend to cancel each other.

1. One inductive time constant = L/R, where L is the inductance in henrys, and R is the resistance in ohms.

2. One capacitive time constant = R × C, where C is in farads, R is in ohms, and the time constant is in seconds.

INDUCTANCE AND ALTERNATING CURRENT

An inductance in an AC circuit also produces 90° of phase shift, but with the current lagging the voltage. Figure 6-23 compares the phase shift in inductors, capacitors, and resistors. The opposition to current flow in an inductor is called *inductive reactance*, and is due to the counter-voltage. The faster the rate of field expansion (or contraction), the more turns of wire are crossed per second by the expanding

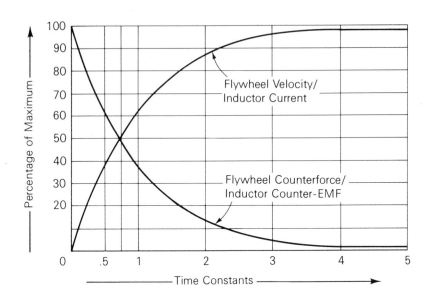

a. Velocity-force graph.

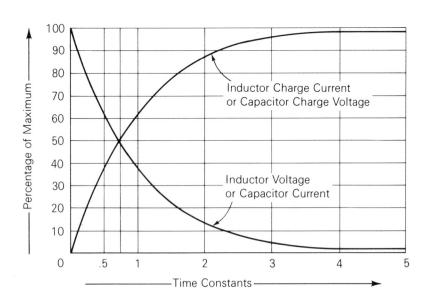

b. Universal-time-constant chart.

FIGURE 6-22. Velocity-force and universal-time-constant graphs.

INDUCTANCE AND CAPACITANCE 217

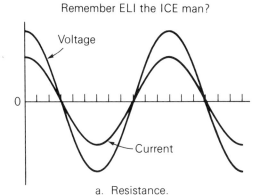
a. Resistance.

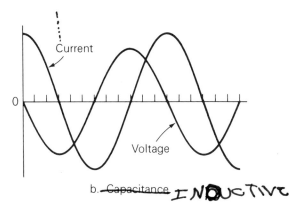
b. ~~Capacitance~~ INDUCTIVE

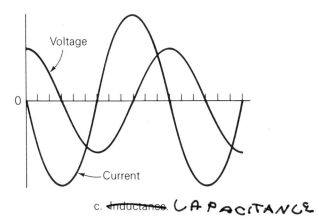
c. ~~Inductance~~ CAPACITANCE

FIGURE 6-23. Phase shift in inductors, capacitors, and resistors.

or contracting magnetic field. The more turns passed through by the field, the greater the counter-voltage, and the greater the total opposition to current flow. In a capacitor the opposition (reactance) decreases as frequency increases. And as we might expect, an inductor's opposition reactance *increases* as the frequency increases.

Capacitive Reactance

$$X_C = \frac{1}{2\pi f C}$$

Inductive Reactance

$$X_L = 2\pi f L = 6.28 \, fL$$

where X_L is in ohms
f is in hertz (cycles per second)
L is inductance in henrys

A 1 henry inductor will produce a 1 volt counter-voltage when the current through it is changing at the rate of 1 amp per second.

Inductors and Vectors

The vector diagram in Figure 6-24 indicates that the inductive-reactance and capacitive-reactance vectors point in exactly opposite directions. The two reactances cancel in the following fashion:

1. X_C is larger; X_L is subtracted from X_C. The numerical value is the total reactance and it is completely capacitive.

Example:
  $X_C = 500$ ohms, $X_L = 200$ ohms. What is the effective reactance?
  $X_{total} = X_L - X_C = 500 - 200 = 300$ ohms of capacitive reactance.
All of the inductive reactance has been canceled.

2. X_L is larger; X_C is subtracted from X_L. The result is inductive reactance.

Example:
  $X_L = 500$ ohms; $X_C = 200$ ohms. What is the effective reactance?
  300 ohms of X_L.

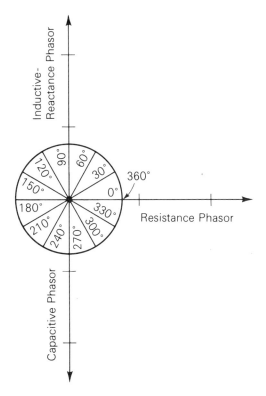

FIGURE 6-24. Vector relationships.

3. The final case is that of equal X_C and X_L. This is a very special case called the *resonant* condition. We will examine resonance in detail later in this chapter.

Impedance with Inductance, Resistance, and Capacitance

To find the impedance in a circuit composed of inductance and resistance, the procedure is identical to that used with capacitance and resistance. Figure 6-25 provides an example. When capacitance, inductance, and resistance are all present in the same circuit, it is only necessary to find the effective total reactance and plot the vector diagram with whatever X_C or X_L is left after the subtraction.

Study Problems

1. What is the amount and direction of the phase shift in an AC circuit with an inductor (no resistance)?
2. What happens to the phase shift when resistance is added to the inductive AC circuit?

220 CHAPTER 6

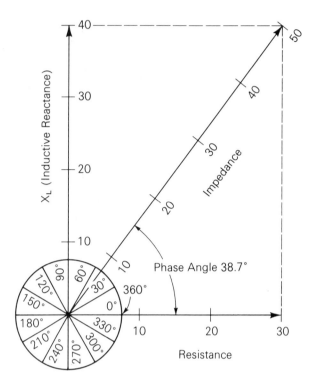

a. Vector diagram.

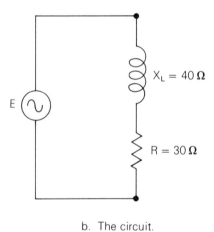

b. The circuit.

FIGURE 6-25. Finding impedance in an inductive circuit (an example).

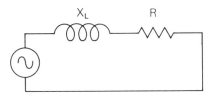

FIGURE 6-26. Circuit for Problems 3 and 4.

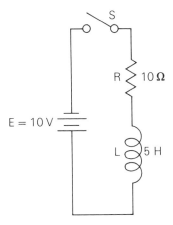

FIGURE 6-27. Circuit for Problem 5.

3. Using the circuit in Figure 6-26, find the impedance of the circuit for the following values of R and X_L:
   a. X_L = 4000 ohms, R = 3000 ohms
   b. X_L = 100 ohms, R = 250 ohms
4. If the voltage is 120 in Figure 6-26, what is the current in the circuit for Problems 3a and 3b?
   Hint: I = E/Z or I = V/Z.
5. For the circuit in Figure 6-27:
   a. What is the inductor current at the end of 2 seconds?
      (Use the universal-time-constant chart in Figure 6-22b.)
   b. What is the voltage across the resistor at 2 seconds?
6. What is the reactance of a 10 henry inductor at 120 hertz?
7. What is the total inductance of Figure 6-28?
8. What is the impedance of the circuit in Figure 6-29?

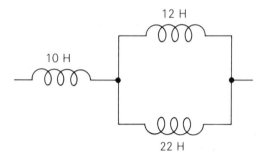

FIGURE 6-28. Circuit for Problem 7.

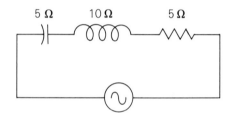

FIGURE 6-29. Circuit for Problem 8.

6-5 RESONANCE

The Bare Facts

1. A circuit is resonant when $X_C = X_L$.

2. Every series circuit that contains both capacitance and inductance is resonant at some frequency.

3. Resonant circuits are used for generating precise alternating currents electronically.

4. Resonant circuits are also used to select one frequency such as a specific TV channel out of many frequencies.

5. Resonant circuits are also used for timing devices such as watches.

6. Both electronic and mechanical watches and clocks use resonant circuits to tick off the time.

7. Resonant circuits work because of the opposite characteristics between inductors and capacitors.

8. The phase shift in a resonant circuit is zero degrees because X_C and X_L cancel each other.
9. The quality of a resonant circuit is diminished by adding resistance.
10. The quality or Q of a resonant circuit is:

$$Q = \frac{X_L}{R}$$

11. All practical series-resonant circuits are somewhat resonant over a narrow band of frequencies.
12. The higher the Q, the narrower the band of frequencies.
13. A resonant circuit will lose energy as a result of resistive losses until oscillation stops.
14. Energy must be delivered at exactly the right time to make up the losses and keep the circuit running.
15. Resonant circuits can select one frequency out of several because the energy delivered by an off-resonant frequency will not keep the circuit running.
16. The energy will be delivered at the wrong time except at the frequency of resonance.

There is a very special condition in resistor-capacitor-inductor AC circuits called *resonance*. When $X_L = X_C$, the circuit will be resonant at some frequency. This is the condition used to select one television channel or radio frequency while excluding all others. Marching soldiers break step while crossing a bridge to avoid the danger of marching at the bridge's resonant frequency. Oscillators that produce organ tones or signals for radio, television, and citizen's band transmitters use resonant circuits to determine the frequency. The crystal used to select the channel in CB radios is one of several electromechanical resonant circuits used in electronics. It is used to replace a resistor-capacitor-inductor circuit because of its precision and freedom from tampering and fewer temperature problems.

Because of the opposite characteristics of inductance and capacitance, a resonant circuit (called a *tank* circuit in electronics jargon) can be made to oscillate current back and forth in a sine-wave fashion. Because of the time difference between the inductor and

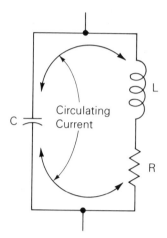

a. The tank circuit.

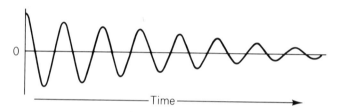

b. Decaying sine wave.

FIGURE 6-30. The resonant tank circuit.

capacitor, the capacitor will be giving up its stored energy while the field in the inductor is building. In fact, when connected as shown in Figure 6-30a, the discharging capacitor supplies the energy to build the inductor's field. When the inductor is fully "charged" its field begins to collapse, pouring its stored energy back into the capacitor. When the capacitor is fully charged, the cycle begins anew. Once started, the circulating current would continue in an AC sine-wave manner forever if it weren't for resistance in the circuit. There is always some resistance in a real circuit, and it gradually dissipates the stored energy as waste heat. The system runs down. The current through the resistor and the voltage across the circuit are a sine wave that decays as shown in Figure 6-30b. The circuit can generate a steady sine wave indefinitely if the energy lost in resistance is replaced at regular intervals.

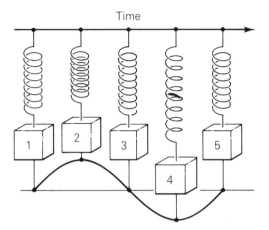

FIGURE 6-31. Weight and spring resonant circuit.

Figure 6-31 is a mechanical resonant circuit using an expansion spring and a weight. Just before Position 1, force is exerted to pull the weight down to about Position 4. When the weight is released, the energy in the spring pulls the weight up until the spring is relaxed (Position 2). The weight then falls to Position 3 and then to Position 4, stretching the spring. By Position 4, the tension of the spring stops the fall of the weight and begins a new cycle. If we were to attach a pen to the weight and move the assembly in the direction of the arrow, the pen would trace out the sine-wave curve shown underneath the weights in Figure 6-31.

A mechanical resonant system composed of a flywheel and spiral spring is shown in Figure 6-32. It is a better analogy of electronic resonant circuits because it consists of 3 independent components. The flywheel-spiral-spring system does not include earth gravity in its operation, as does the spring-weight system. A graph of flywheel velocity with time would also be a sine-wave curve like the one in Figure 6-31.

Some energy must be put into the system to start it oscillating. Once started, the flywheel will turn until the spring is tightened. Its energy has been stored in the spring and the mass slows to a halt. The spring then unwinds, driving the flywheel in the opposite direction. The flywheel is now absorbing the spring's energy. The oscillation will gradually decay as the initial starting energy is dissipated by the friction (resistance) in the bearings.

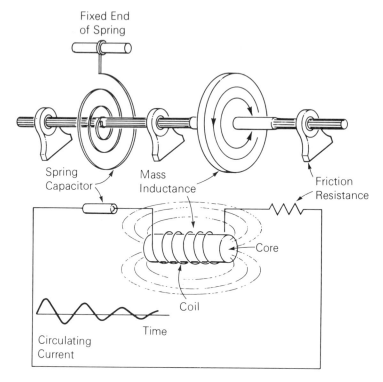

FIGURE 6-32. Mechanical resonant circuit and its electrical equivalent.

RESONANT CIRCUIT APPLICATIONS

Resonant circuits are used wherever it is necessary to generate continuous and precise vibrations at a specific frequency. They are also used to select one frequency of oscillation out of signals of many different frequencies. Television Channel 2, for example, is transmitted at a frequency different from that of Channel 4. The transmitter must generate radio waves at the precise frequency for Channel 2. The TV receiver must select the frequency of Channel 2 and reject all other frequencies (channels) when we tune to Channel 2.

Resonant circuits are also used when precise timing is required. The drawing in Figure 6-33 shows a conventional watch mechanism, a practical application of the flywheel-spring resonant circuit.

Electronic watches use resonant circuits that operate in the electronic and the mechanical world at the same time. The mineral quartz has springy properties along with mass, but it also exhibits a

INDUCTANCE AND CAPACITANCE 227

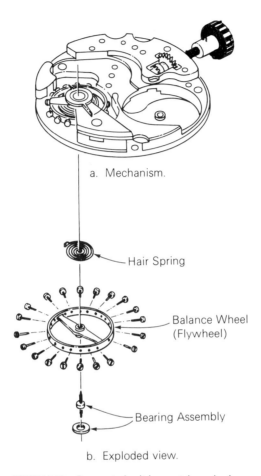

a. Mechanism.

Hair Spring

Balance Wheel (Flywheel)

Bearing Assembly

b. Exploded view.

FIGURE 6-33. Resonant circuit in a watch mechanism.

property known as the *piezoelectric effect*. If a quartz crystal is compressed slightly, it generates a small voltage between its faces. If a voltage is applied across the two faces, the crystal will compress (or expand) slightly like a spring. Because of the crystal's ability to interact with voltage, it can be connected directly into an electronic circuit as an exact replacement for a coil and capacitor. The oscillation is actually mechanical, but an electrical sine-wave voltage output is available as a direct result of the mechanical vibrations. Any resonant circuit will tend to oscillate at the resonant frequency, but if the Q (figure of merit) is low, it can easily be influenced by outside disturbances to oscillate slightly to one side or the other of the resonant frequency. The result of resistance in the circuit is a circuit that is somewhat resonant over a narrow band of frequencies instead of at a single frequency. Figure 6-34b shows the bandwidth curves for low, medium, and high-Q resonant circuits.

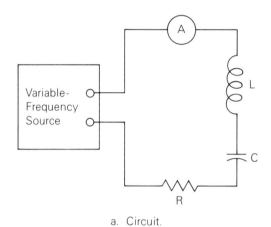

a. Circuit.

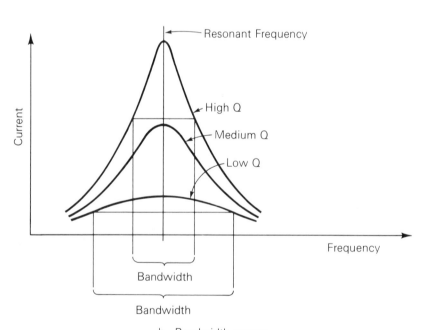

b. Bandwidth curves.

FIGURE 6-34. Bandwidth.

High Q = Low Resistance

When you are pushing someone in a playground swing, you get into the swing's natural rhythm (its resonant frequency). The wrong rhythm will slow or stop the swing instead of making it fly high. A resonant circuit selects one frequency out of a collection of others on much the same principle.

In electronic terms, a high circulating current flows in a resonant circuit when a source of sine-wave voltage is applied to it whose frequency corresponds to the resonant frequency of the circuit. Figure 6-34a shows a circuit for determining the bandwidth behavior of a resonant circuit. This circuit was used to plot the bandwidth curves in Figure 6-34b. At frequencies off resonance, the circulating current decreases rapidly as the applied frequency gets further away from the resonant frequency of the resistor-capacitor-inductor circuit.

THE RESONANT FREQUENCY AND THE EFFECTS OF RESISTANCE

Because X_L increases with frequency while X_C decreases, there must always be some frequency where $X_L = X_C$, whatever the values of C and L. This situation is illustrated in Figure 6-35.

If the resistance is kept low, the resonant circuit is a high-quality or high-Q circuit. Resistance in the circuit does not alter the resonant frequency of the circuit, but it does allow it to respond, to a degree, to

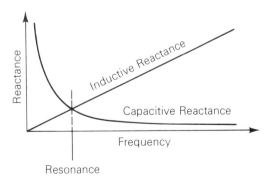

FIGURE 6-35. How inductive and capacitive reactance vary with frequency.

frequencies slightly off resonance as well as to the resonant frequency. For the mathematically interested:

---

1. The resonant frequency can be calculated by using the following equation:

$$F_o = \frac{1}{2\pi \sqrt{LC}}$$

2. Q can be calculated by:

$$Q = \frac{X_L}{R} \quad (X_L \text{ is at the resonant frequency})$$

3. Bandwidth:

$$B = \frac{\text{Resonant frequency}}{Q}$$

Study Problems

1. Describe the conditions of X_C and X_L at resonance.
2. Why must every inductance-capacitance circuit be resonant at some frequency?
3. List 3 applications for resonant circuits.
4. Define the term Q as it applies to a resonant circuit.
5. What is meant by bandwidth?
6. What element in a resonant circuit determines the bandwidth?
7. Explain how a resonant circuit can be used to tune in one radio station while rejecting all others.
8. When you pour liquid from a bottle it makes a glug-glug sound. The mass of the liquid is the inductance analog and the compression and expansion of the trapped air is the capacitance analog. Try this as an experiment and then explain why the glug-glug gets faster (higher pitch or frequency) as the bottle empties.
9. The bang-bang or squeal in household plumbing is an example of unplanned resonance. The water mass and trapped air compression form the resonant circuit. See if you can find some other everyday (planned or not) resonant systems and identify the inductance and capacitance analogs.

Chapter 7

Courtesy Smithsonian Institution

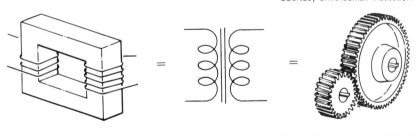

TRANSFORMERS

7-1 INTRODUCTION

As you have probably guessed, the electrical transformer is analogous to devices that translate motion. Levers, gears, pulleys, and the like are things that we can see working, and from long experience we have some understanding of the principles involved. We know that a lever can be used to move heavy objects. We have seen gears and pulleys used to change rotational speeds. We shift gear ratios in our automobile transmission to make the car go faster or to provide the extra power to climb a hill.

These simple machines accomplish the same kinds of tasks in the mechanical world that transformers do in the electrical world. There are also acoustical transformers that are important to our stereo or quad music systems. The box in which we install woofers, tweeters, and mid-range loudspeakers is a good example of an acoustical transformer. An acoustical transformer transforms the force and velocity characteristics of our loudspeakers into force and velocity characteristics that move air masses to create sound.

In the automobile, the transmission is a complex mechanical transformer system. It must transform large variations in engine speed and variations in power output at different engine speeds into an equally wide range of wheel speeds and load conditions. That seemingly simple box we call a *loudspeaker enclosure* is actually more complex in its workings than even the most complex automobile transmission. This complexity accounts for the many heated discussions among high-fi enthusiasts concerning the merits of one or another kind of enclosure. It is rare to find electrical transformer systems that even approach the complexity of an automotive transmission or a speaker box.

We will also find heat transformers used in electronics. Modern semiconductor devices tend to be temperature-sensitive. Internal heat generated by resistances in the device must be removed if we want reliable operation. In semiconductor devices, heat transformers known as *heat sinks, coolers,* or *radiators* are used to transform heat flow rates and temperature differences to ensure proper cooling. We will take a brief look at thermal transformers in this chapter.

Transformers for the different physical systems may look different, but they all share common operating principles and characteristics. Figure 7-1 illustrates some typical commercial electrical and mechanical transformer types.

Figures 7-1a and b show iron-core transformers intended for service in applications where moderate powers and low-frequency (60 hertz) operation are required. The cutaway in Figure 7-1b consists of the following parts (see numbers on the drawing):

TRANSFORMERS   233

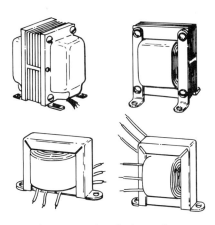

a. Iron-core electrical transformers.

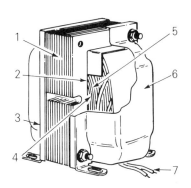

b. Cutaway view of style shown at upper left of Figure 7-1a.

c. Air-core transformer.

d. Shielded powdered-iron-core transformer.

e. Antenna transformer.

f. Iron-ceramic antenna transformer.

g. Air-core transformer.

h. Variable transformer.

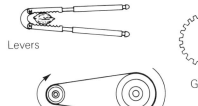

Levers

Pulleys

Gears

Mouth Piece

Acoustical transformer

i. Mechanical transformers.

FIGURE 7-1. Transformers.

1. Laminated silicon iron core
2. Insulation
3. Left-hand end shell (soft iron)
4. Insulation and anti-abrasion material between windings and core
5. Coils of copper wire
6. Right-hand end shell
7. Wires leading to the outside world

7-2 TRANSFORMERS

The Bare Facts

1. Transformers increase mechanical, electromotive, acoustical, or thermal force at the expense of current or velocity.
2. Transformers can also increase current or velocity at the expense of voltage or force.
3. Some examples of transformers are:
   Electrical
   Mechanical
     Gears
     Pulleys
     Levers
   Acoustical
     Loudspeaker enclosures
     The body of stringed musical instruments
     The horn on horn-shaped musical instruments
   Thermal
     Automobile radiators
     Air-conditioning condensers
     Transistor heat sinks
     Cooking pans
4. Most electrical transformers are very efficient. They run fairly cool because of the small amount of resistance (friction) typical of these devices.
5. Electrical transformers either step up or step down voltage.
6. All transformers are reversible. A step-up transformer can be turned around for use as a step-down transformer.

7. Electrical transformers operate on alternating current or interrupted direct current.

8. Most transformers in nonelectrical systems will operate with the equivalent of direct or alternating current.

9. Transformers step up force according to the ratio of lengths for levers, the ratio of the number of teeth in primary and secondary gears, and the ratio of the number of turns on primary and secondary windings in an electrical transformer.

10. The power output of a transformer never exceeds the input power.

11. Any gain in voltage is exactly offset by a decrease in current. Any gain in force is exactly offset by a decrease in velocity.

12. The power in the primary of a transformer equals the power in the secondary.

WHAT DOES A TRANSFORMER DO?

A transformer steps up voltage and steps down current, steps up current and steps down voltage, or provides isolation for safety.

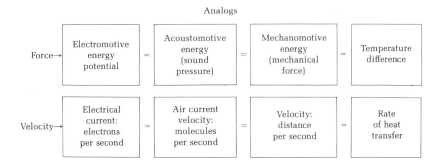

In an electrical, mechanical, or any other type of system a transformer increases force at the cost of a proportional decrease in velocity. A transformer may also be "turned around" to increase velocity at the expense of force. The man moving the heavy safe in Figure 7-2 illustrates the sort of simple lever that we will use as an analog of the electrical transformer. We have all had almost daily experience with a variety of levers. When we examined work and power in Chapter 3, we found that work involves both force and distance.

FIGURE 7-2. Man and lever.

Work = Force × Distance

The man in Figure 7-2 can use the lever to transform the distance he moves his arm into the greater force needed to move the heavy safe. He has no problem in moving his arm through a greater distance, but it may be impossible for him to make his muscles produce enough force to do the job without the lever.

In our system of mechanical/electrical analogs, mechanical force is the analog of electromotive force (voltage), and electrical current is the analog of mechanical velocity. In our discussion of levers we have talked about force and distance, but we have no electrical analog for distance. Suppose we move the longer segment of the lever in Figure 7-3 from Point A to Point B in 1 second. If the distance from A to B is 2 meters (m), the velocity of this end of the lever is 2 meters per second. The short end of the lever moves from Point C to Point D in the same 1 second period. The velocity at the short end of the lever is only 1 meter per second, just half the velocity of the longer end. We have transformed velocity into force. The force at the short end is twice that at the long end, while the velocity (as well as the distance) at the long end is twice as great as the velocity of the short end. Because the ratio of velocities is always the same as the ratio of distances, we can rewrite the work relationship in terms of velocity. When we are talking about force and distance, we call it *work;* when we talk about force and velocity, we call it *power.*

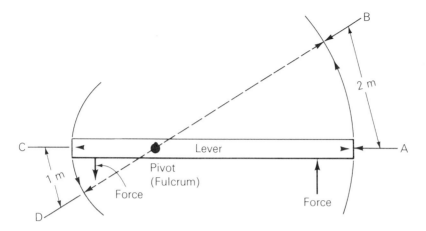

FIGURE 7-3. The distance/velocity relationship.

$$\text{Work} = \text{Force} \times \text{Distance}$$

and

$$\text{Power} = \text{Force} \times \text{Velocity}$$

or, in electrical terms,

$$\text{Power} = \text{Voltage} \times \text{Current}$$

ENERGY CONSERVATION

Not long after the invention of the wheel, clever inventors began an attempt to defy a basic law of nature, as we now know it. These efforts have yielded a long line of so-called perpetual-motion machines in which the idea is to get something for nothing by creating energy. Every perpetual-motion machine has either been operated from some hidden energy source or failed to function at all.

The German physicist Hermann Von Helmholtz (1821–1894) stated that energy can be changed from one form into another, but it cannot be created or destroyed. This law is referred to as the *law of conservation of energy*. It is easy to mistake an energy transformation for energy creation. Some of the language used in discussing amplifying devices seems to suggest getting something for nothing, and it can be misleading. Anything that appears to be energy creation is actually some form of energy conversion.
Conservation:

$$\boxed{\text{Input work} = \text{Output work}}$$

and

$$\boxed{\text{Input power} = \text{Output power}}$$

In any real device we expect to get less power out than was put in. There is always some friction or resistance, and some energy is converted to heat, which is lost to the surroundings. We can *never* expect to get more out than we put in. The transformers in all systems are usually quite efficient; output powers of 80 to 99 percent of the input power are common.

Transformer Efficiency

Like everything else in this imperfect world, transformers are less than 100 percent efficient. The wire used in the windings wastes some power due to ordinary resistance. In addition, localized currents are induced in the iron transformer core, just as though the core was another winding. Cores are made of thin sheets insulated from each other and laminated into a stack, or made of small grains of iron insulated with a glue-like binder. Such methods minimize these wasteful core currents, called *eddy* currents. Other cores are made of ceramic materials that have excellent magnetic properties, but are not electrical conductors.

Transformer efficiency:

$$\boxed{\text{Efficiency} = \frac{\text{Power out}}{\text{Power in}} \times 100 \text{ percent}}$$

Whenever we have conditions where the power source can deliver enough power to the load, but with the wrong proportions of force and velocity, we can use a transformer to deliver more force through less distance to the load. If the load requires less force but a greater velocity or distance than the power source produces, we can use a transformer to reduce the force and increase the velocity (or distance).

For example, an automobile engine develops considerably less power at low RPM (revolutions per minute). However, low speed is exactly where we need the greatest force at the wheels. The greatest force is required at the wheels when we are just putting the car in motion. The solution to the problem is to run the engine at a higher RPM and to transform that excess speed (RPM) into the increased force we need at the wheels. For this task we use a mechanical transformer called a *transmission*. On the other hand, if we wish to make the wheels turn faster than the engine is able to turn, we can transform some of the engine force into greater wheel speed. Modern automatic transmissions actually sense both speed and force requirements for a given situation and automatically select the proper gear ratio.

If we want to operate a television picture tube or a neon sign, we will need several thousand volts to do the job. The household outlet delivers only about 120 volts. A voltage step-up transformer is the answer. In other cases the 120 volts of the household outlet could be excessive. Semiconductor devices, transistors, and integrated circuits are generally not capable of safe 120 volt operation.

A resistor can be used in some applications to reduce the voltage, but resistors are power wasters and not suitable for certain applications. The transformer wastes almost no power and produces very little heat in its operation. When voltage is to be increased, a transformer is essential.

If we move 2 grams through a distance of 2 centimeters, the work done is 2 grams × 2 centimeters, or 4 gram-centimeters. If we move 4 grams through a distance of 1 centimeter, the work done is 4 grams × 1 centimeter, or 4 gram-centimeters. If the work continues for a period of time, we are talking about power. In electrical terms, 2 volts × 2 amps = 4 watts, and 4 volts at 1 amp = 4 watts. In all cases the product of voltage and current (velocity) at the input of the device must be equal to the product of voltage and current at the output.

In the previous example we have a mechanical advantage of 4 to 1, or a voltage step-up for the electrical transformer. All the levers in Figure 7-4 have a mechanical advantage of 4 to 1. Whatever force is applied at the A end will be magnified 4 times at the B end. The B

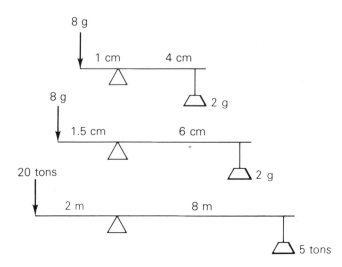

FIGURE 7-4. Illustration of force step-up and velocity step-down ratio. The ratio is the same for all of these mechanical transformers.

end of the levers will travel one-fourth the distance at one-fourth the velocity of the A end. The true lengths of the levers are not important; it is only the ratios that matter. (The abbreviations g and cm used in Figure 7-4 mean grams and centimeters, respectively.)

In transformers of all kinds, actual voltage values, lengths, numbers of turns, and so on are unimportant. Transformer behavior is governed by ratios of these quantities, not by their actual values. A 1 volt input to the transformer in Figure 7-5a will produce an output voltage of 2 volts. An input voltage of 8 volts will produce an output voltage of 16 volts, and so on. The lever in Figure 7-5b also has a ratio of 1 to 2. A force of 1 pound or 1 gram or 1 of any other unit of force on the input will produce an output force of 2 of the same unit.

While the voltage is being doubled in the output coil, the current in the output is being reduced to half that in the input coil. In the case of the mechanical transformer in Figure 7-5b, the output velocity is reduced to half the velocity of the input side as the output force is doubled.

The transformer in our example (Figure 7-5a) is called a *step-up* transformer, indicating a voltage increase from input to output. In transformer terminology the input is called the *primary*, and the output is the *secondary*. The transformer, like the lever, is reversible and can be used to step down voltage by exchanging the input and output. Step-down transformers are as common as step-up transformers. Levers are also used to step down force, usually for the sake of increasing distance or velocity. The gear, which is simply a

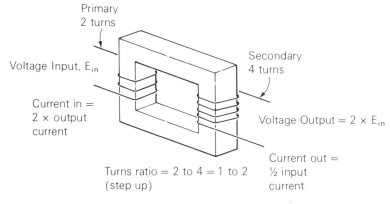

a. The electrical transformer.

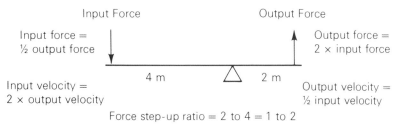

b. The mechanical transformer.

FIGURE 7-5. Transformers and ratios.

"round" lever with teeth, is also used to increase or decrease force and to increase or decrease velocity.

Here we must mention a difference in the nature of electromagnetic and mechanical coupling. In mechanical coupling a constant velocity can be coupled from input to output. In electromagnetic coupling the current (velocity) must be constantly changing for coupling to take place. A constant velocity is analogous to direct current. Most mechanical transformers work just as well with both constant velocity (direct current) and variable velocity (alternating current). An electrical transformer will not operate with a *steady* direct current (constant velocity).

In electronics textbooks the statement is often made that transformers cannot be used with direct current. The statement is true but incomplete. We can make a transformer work if we periodically interrupt the direct current and allow the magnetic field to alternately expand and collapse. The field must be changing for inductive action to take place, but it doesn't matter how we go about keeping it changing.

There is a well-known prank in which the electronics initiate is sent to fetch a direct-current transformer. Having read the "statement" in the previous paragraph somewhere, the pranksters always expect the novice to return empty-handed. In reality, he can return with any available transformer, along with instructions to simply interrupt the primary current flow periodically.

As a matter of fact, there are a good many applications where *interrupted* direct current is transformed to higher or lower voltages. An outstanding application is that of the automobile ignition system shown in Figure 7-6. The object is to get a spark across the spark-plug gap to ignite the mixture of fuel and air. An adequate spark under the temperature, pressure, and other conditions within the engine cylinder requires 20,000 volts or more. The rest of the automotive electrical systems operate on 12 volts DC. The only obvious solution is some kind of transformer. In the simplified system in Figure 7-6, a cam turned by the engine periodically opens and closes a switch called *breaker points* in automotive terminology. The interrupted 12 volts from the battery is transformed up to the 20,000 or so volts required to develop a spark across the spark-plug gap.

If we apply a direct current to the primary of a transformer, we will get a secondary voltage only until the current stops changing in the primary. The period of transformer action depends on the fact that the inductive reactance of the coil tends to oppose the current and

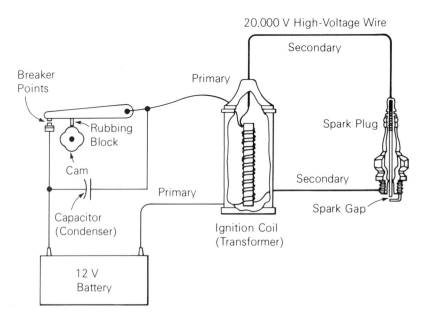

FIGURE 7-6. Transformer used with direct current in an automobile ignition system.

delay the complete formation of the magnetic field. Like a teeter-totter when it hits bottom, the transforming action ceases as soon as the magnetic field can expand no further.

In the majority of applications where transformers are used, transformer input and output voltages are sine-wave alternating current. The transformer action is smooth and continuous. This action is more easily visualized if we replace the simple lever analog with its rotational equivalent, pairs of gears.

In case you have forgotten, the word *force* is replaced by the word *torque* whenever rotary motion is discussed. There is no essential difference between the two, but torque tells us that force is being applied in the special way of rotating machinery. Figure 7-7 illustrates the torque concept, and Figure 7-8 shows a transformer and its gear analog.

It is usual to specify gear ratios in terms of the number of teeth in the gears. A gear with 10 teeth driving a gear with 100 would provide a 10 to 100, or a 1 to 10 ratio. The torque would be increased 10 times. A convenient memory aid is: teeth = turns. Caution is needed, though, if you are studying gears in a book on mechanics or physics, because they are often discussed in terms of rotational-speed (velocity) increase instead of torque step-ups as we have emphasized it here. Both transformers in Figure 7-8, electrical and mechanical, increase voltage or mechanical force in exchange for a decrease in current or velocity. In this instance, the forces are voltage and torque. Again, both transformers can be reversed for voltage or torque step-down.

a. Torque defined.

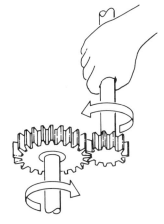
b. Torque transformer (step up).

FIGURE 7-7. Torque.

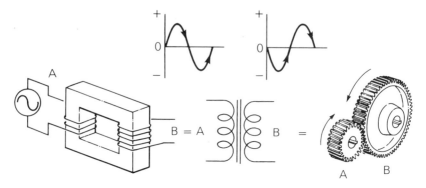

FIGURE 7-8. Electrical transformer and its gear analog.

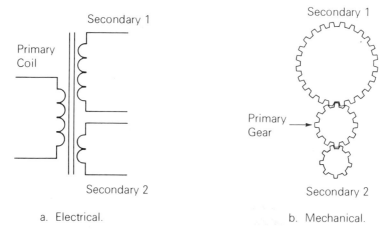

a. Electrical. b. Mechanical.

FIGURE 7-9. Multiple-winding transformer and its gear analog.

Multiple-Winding Transformers

Transformers with several windings to provide several different voltages are common in practical systems. When a number of different voltages are required, the most economical approach uses several windings on a common core. The most expensive part of iron-core transformers is the fabrication of the core; a few extra turns of wire add little to the cost. Air-core coils are usually wound on some nonmagnetic material to provide mechanical support for the coils. Even in a transformer with a cardboard or plastic core, the support structure is often the most expensive part of the transformer. Figure 7-9 shows a multiple-winding transformer and its gear analog. For the mathematically inclined, here are some important equations:

Electrical:

$\dfrac{V_{in}}{V_{out}} = \dfrac{N_{primary}}{N_{secondary}}$	V_{in} = Input voltage V_{out} = Output voltage $N_{primary}$ = Number of turns in primary coil $N_{secondary}$ = Number of turns in secondary coil

Gears:

$\dfrac{T_{in}}{T_{out}} = \dfrac{N_{primary}}{N_{secondary}}$	T_{in} = Input torque T_{out} = Output torque $N_{primary}$ = Number of teeth in primary gear $N_{secondary}$ = Number of teeth in secondary gear

For those who are not comfortable with mathematical shorthand, here is an easy way to calculate turns, ratios, and voltages:

> Divide the number of turns in any winding by the voltage of the winding.

$$\text{Turns per volt} = \text{Turns} \div \text{Volts}$$

Example:

A multiple-winding transformer is intended for use with ordinary 120 volts, 60 hertz household power. The 120 volt winding has 600 turns. How many turns does each of the following windings have?

1. 10 volt
2. 24 volt
3. 500 volt

Solution: Find the number of turns per volt for any winding where both turns and volts are known.

246 CHAPTER 7

$$\text{Turns per volt} = 600 \text{ turns} \div 120 \text{ volts}$$
$$600/120 = 5 \text{ turns per volt}$$

Now, for each winding it takes 5 turns for each volt. We must multiply volts × turns per volt:

1. 10 volt winding
   10 volts × 5 turns per volt
   10 × 5 = 50 turns

2. 24 volt winding
   24 volts × 5 turns per volt
   24 × 5 = 120 turns

3. 500 volt winding
   500 volts × 5 turns per volt
   500 × 5 = 2500 turns

Let us suppose that we are going to add a winding for a geodetic bodangle. As everyone knows, a bodangle is a mythological electronics device that requires 17 volts to operate it. How may turns must our extra winding have?

Solution: If each volt requires 5 turns and we need 17 volts, then: 17 volts × 5 turns

$$17 \times 5 = 85 \text{ turns}$$

Suppose that this transformer has a winding that consists of 60 turns, but the information printed on the transformer is blurred so we can't read what voltage this particular winding is supposed to produce. Can we calculate it? We know that it takes 5 turns to provide 1 volt, so we simply find out how many groups of 5 turns there are.

$$\frac{60 \text{ turns}}{5 \text{ turns per volt}} = 12 \text{ volts}$$

Problem:
Given the transformer in Figure 7-10, list the voltage and the number of turns for each winding.

THE REFLECTED LOAD CONCEPT

So far we have only implied that power was being transformed. It is time to look at loaded transformers with a definite power to be delivered. Let's look at some details of the operation of the circuit shown in Figure 7-11. The primary voltage is 1 volt. That 1 volt is stepped

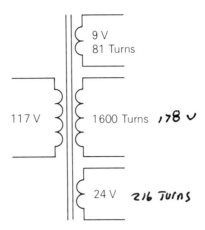

FIGURE 7-10. Transformer problem.

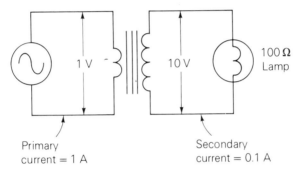

a. Loaded transformer.

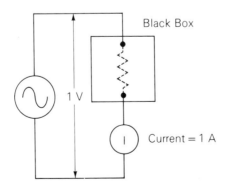

b. Reflected load.

FIGURE 7-11. Reflected loads.

up to 10 volts in the secondary, where it forces 0.1 amp of current through the 100 ohm lamp. We find that current by using Ohm's Law:

$$\text{Current} = \frac{\text{Voltage}}{\text{Resistance}} = \frac{10 \text{ volts}}{100 \text{ ohms}} = \frac{1}{10} \text{ or } 0.1 \text{ amp}$$

Because the voltage in the secondary is 10 times that of the primary, the current in the primary *must* be 10 times that of the secondary. Ten times 0.1 amp is 1 amp. In the chapter on basic electricity we found that resistance has reality only in terms of the question, "How much current will a given voltage force through a circuit?"

In figure 7-11b we can assume that there is a resistance of:

$$\frac{1 \text{ volt}}{1 \text{ amp}} = 1 \text{ ohm in the box}$$

The generator "sees" a resistance of 1 ohm, and it is of no importance whatsoever what kind of hardware is in the box. The generator in Figure 7-11a finds conditions identical to those for the generator in Figure 7-11b. Both generators "conclude" that their load is a 1 ohm resistor. Being a bit smarter than the generator, *we* know that the real load, the lamp, has 100 ohms of resistance and we are curious about what is happening. Obviously, the transformer has not only transformed voltage and current values, but it has also caused a resistance transformation. The 100 ohm resistance of the lamp in the transformer secondary has been transformed down to 1 ohm of effective resistance in the primary.

Because the transformer is usually used with alternating current, it is more accurate to talk about impedance transformation instead of resistance transformation. The transformation ratio for the load impedance in Figure 7-11 is 1 to 100, not 1 to 10 as it is for the voltage. It works out that:

Impedance ratio = Turns ratio × Turns ratio
(Impedance ratio = Turns ratio squared)

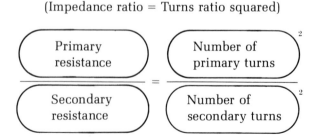

$$\frac{\text{Primary resistance}}{\text{Secondary resistance}} = \frac{(\text{Number of primary turns})^2}{(\text{Number of secondary turns})^2}$$

In mathematical shorthand:

$$\boxed{\frac{R_p}{R_s} = \frac{N_p^2}{N_s^2}}$$

One of the common uses for transformers is that of transforming resistances (impedances). In order to transfer the maximum amount of power from one device to another, *the output resistance of the source must equal the input resistance of the load.* In our high-fidelity music system, for example, we want to deliver the maximum possible acoustical power to the air. The amplifier may contain an electrical transformer to match the amplifier output impedance to the loudspeaker. An acoustical transformer, the speaker box, matches the acoustical impedance of the loudspeaker to the acoustical impedance of free air molecules in a room.

At present there seems to be a primitive urge to match impedances as a matter of routine. If we must transfer the maximum power from one device to another, it is important to match the impedance of the source or generator to the impedance of the load. All we need do is select an appropriate transformer, and connect it between source and load. However, delivering maximum power to the load is inappropriate in a great many instances. A tourniquet to stop bleeding in an arm or leg may be exactly the right thing. It would be advisable to think twice, though, before putting a tourniquet around the victim's neck to stop the bleeding from a head wound.

Impedance matching is important where it is appropriate, and can be self-defeating in other situations. Impedance matching is important only when we must transfer as much power (voltage × current) as possible from one device to another. This is a special case because more often we are concerned with providing a specific voltage (or torque) to a load, or conversely, we are interested in greater current (or velocity) for the load.

Every device that produces a voltage contains some internal resistance. That resistance is simply a fact that we must accept. The load, whether it's a loudspeaker, a lamp, or anything else, also has an internal resistance, and its value is not normally under our control. When an automobile is just moving away from the curb, we need all the torque we can get. But to make the wheels turn at the speed required to move the car at 50 miles per hour while the car is barely moving would only result in spinning wheels and excessive tire wear. On the high end of the speed range we need a lot of velocity (rapidly turning wheels), but very little torque. At some cruising speed we would have the largest torque-velocity product, and the maximum power delivered to the wheels. Impedance matching is appropriate only for the particular cruising speed. For the other conditions in the example, impedance matching is counterproductive. The transformer is capable of 3 coupling modes:

1. It may transform voltage (or torque) up to a higher value, with a corresponding reduction in current (velocity).
2. It may transform current (velocity) up to a higher value, with a corresponding reduction in voltage (torque or force).
3. It may transform the impedance of the load (up or down) to match the impedances of load and source. This is the condition for delivering maximum power to the load.

Each of the 3 modes has its own particular applications, and no one mode is more important than the others. In this section we will be concerned with the matched-impedance mode, but we cannot forget the others. Keep in mind that we are always stepping voltage up and current down, or the opposite, even when our main concern is impedance matching.

If the internal resistance of the source of the voltage and the internal resistance of the load are equal in value, we have the optimum conditions for the transfer of power. Given a particular source and a particular load device, it would be a happy accident if their internal resistances should turn out to be equal.

Because of the transformer's ability to transform impedances, we can use it as a matching device by selecting the proper number of primary and secondary turns. Commercial units are available that specify primary and secondary impedances; we need only select the correct one for each particular application. If we have a 60 watt power amplifier in a music system, for example, we want to convert as much as possible of that 60 watts into acoustical power. We might use an electrical transformer to match the output impedance of the amplifier to the voice-coil impedance of the speaker. The next problem would be to match the acoustical resistance (impedance) of the speaker cone to the impedance of free air. We would use an acoustical transformer in the form of a loudspeaker enclosure. The electrical-impedance transformer can be quite scientifically designed. Acoustical transformer design is the result of scientific experimentation and trial and error. Some advertising copy almost suggests that a trace of the magic arts is included in the design process. The perfect loudspeaker enclosure is yet to be designed. When and if it is designed, it will probably prove to be very bulky and expensive.

TRANSFORMER PHASING

Transformers can be connected together as shown in Figure 7-12b. However they must be connected so that the two output voltages aid each other instead of opposing. Figure 7-12b shows the two transformers connected out of phase. When one sine wave is at its positive peak, the other is at its maximum negative value. Like the gears in Figure 7-12a, the two voltages (force) are fighting each other. The result is zero volts to the load and no current flow through it. The cure is simple. It requires only that the two leads on *one* transformer be exchanged. The process of getting the transformers connected so the two voltages add instead of subtract is called *phasing*. If the two output voltages are not equal, the effective voltage will be the difference between the two if they are connected in opposition, and the sum of the two if they are connected as aiding voltages.

When two (or more) loudspeakers are connected to a music system, acoustical forces can tend to cancel if they are pushing air masses in opposition. Complete opposition resulting in no sound at all is extremely rare, but can be approached with special equipment. Sound does not consist of a single simple sine wave, but a multitude of sine waves of varying frequencies and phases. Earphones have been developed for high-noise environments in which out-of-phase noise can be introduced into the phones to cancel the environmental noise getting to the ears directly.

In stereo systems, some right-channel music material is mixed into the left channel and some left-channel material is mixed into the

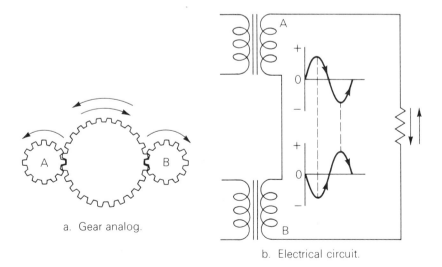

a. Gear analog.

b. Electrical circuit.

FIGURE 7-12. Transformers out of phase.

right channel. A 100 percent separation would create a dead zone in the center. Improperly phased speakers can cancel out much of the cross-channel music, causing that same kind of hole in the sound pattern. The effective acoustical power is also reduced. Swapping the two leads on one speaker cures the problem. Quadraphonic systems are subject to the same kind of problem, and proper phasing is required for best results.

7-3 THE ISOLATION TRANSFORMER

A transformer with equal input and output voltages might seem to be a do-nothing device, but it is a common item. As the name implies, its function is to isolate. The wiring in a house or factory has one side of the power line at the same potential as the fireproof conduit that encases the wires. Touching one side of the power line while in contact with a water pipe, the conduit, or some other grounded conductor can result in a severe or even fatal shock. If an isolation transformer is used, both sides of the power line must be touched at the same time to present a shock hazard. This is a far less likely accident than contacting one wire and some grounded object.

We often find different resistance values in conduits, fittings, and so on, and the ground wire. This frequently results in a small potential difference between the plug connection and the conduit, which are both supposed to be at ground potential. This small voltage can often cause problems in sensitive electronic equipment. The transformer's magnetic coupling effectively eliminates this problem. All power-line-operated transformers function as isolation transformers, but the name *isolation transformer* is reserved for the one-to-one-ratio transformer.

7-4 THE AUTOTRANSFORMER

The autotransformer has nothing to do with cars. It is the one transformer type that does not provide the normal transformer isolation. The isolation is bypassed by connecting one primary and one secondary wire together internally. The autotransformer with one primary and one secondary has only 3 connecting wires. Autotransformers are normally restricted to certain special applications. Variable transformers are usually autotransformers, as are some television high-voltage transformers.

7-5 TRANSFORMER FAILURES AND TESTING

Transformers are very reliable components. The principal failures in transformers are open windings, which can be detected by the ohmmeter, and shorted turns due to insulation breakdown. Shorted turns cannot be detected by the ohmmeter. Transformers that handle some power will generally overheat when shorted turns are present. Since overheating can also be caused by shorts in the circuit powered by the transformer, external circuits must be disconnected to be sure that the short is in the transformer. Sometimes a circuit short will make the transformer run so hot that the insulation will break down, causing the transformer to short internally. It is always wise to suspect that component failure may have been caused by something outside the component itself.

Shorted turns in transformers that do not handle much power will not result in overheating, and more sophisticated tests are needed. Some transformers have impedance ratings given in ohms, but this is not the DC resistance measured by the ohmmeter. A winding with an impedance of several thousand ohms may have a DC resistance of only a few ohms, as measured by the ohmmeter. The ohmmeter can be used to detect an open circuit in an inductor or transformer, but it cannot detect shorted turns or measure impedance or inductance.

7-6 HEAT TRANSFORMERS

In heat systems we are often interested in transferring as much heat power as possible. The radiator in an automobile is there for the purpose of transferring as much heat as possible from the engine into the air. The radiator serves as a thermal transformer. In heat systems, the analog of potential difference (voltage) is temperature difference. The analog of current and velocity is the *rate of heat transfer.*

Thermal resistance, like electrical resistance, varies among different materials. Some materials are good heat conductors and some are not. We insulate our homes with materials that are poor thermal conductors—that is, good thermal insulators. Metals are good conductors for both heat and electricity, but mica, for example, is a good conductor of heat and a very poor electrical conductor. In electronics we find that many devices require that we hold the temperature down if we are to have predictable electronic results. The transformer for this task is called a *heat sink.*

Figure 7-13 shows a very common and long-used thermal transformer. The frying pan is heated over a relatively small area by a very hot flame. The rest of the pan starts out at room temperature, but gradually the heat spreads until the entire pan is far above room temperature. The frying pan is an impedance-matching transformer because it transfers maximum heat power to the food being cooked. Cooking pans vary widely in their efficiency and effectiveness. Different foods represent different thermal impedances, so the imped-

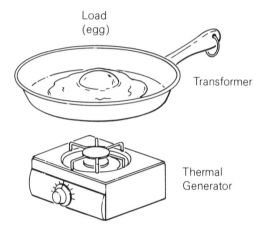

FIGURE 7-13. Primitive heat transformer.

ance match is only approximate. The perfect thermal transformer would be at the same temperature all over. Thermal resistance restricts the heat flow and wastes heat power. Aluminum pans are often used to provide a lower thermal resistance, and copper bottoms are often added because of copper's better thermal conductivity.

Thermal transformers such as automobile radiators and transistor coolers can be more scientifically designed. In an automobile radiator, the thermal impedance of hot water is matched to the thermal impedance of free air molecules. The two impedances to be matched are fairly constant, and a good match is possible. We are apt to see a good many examples of inadequate impedance matching on any hot day in the form of disabled cars by the side of the road. Like the near-perfect acoustical transformer, a near-perfect thermal transformer for a car would tend to be large and expensive.

Heat sinks for cooling transistors are large-area, air-cooled radiators with high thermal conductivity. Two versions are shown in Figure 7-14.

FIGURE 7-14. Transistor heat sinks.

7-7 ANTENNAS

We might guess that the antenna connected to our television receiver and the antenna at the transmitter are radiation transformers. This idea is speculative because even though we can build highly effective antennas, we know little about electromagnetic energy. We can't be sure what is really happening. Yet the idea that antennas are radiation transformers is supported by quite a bit of evidence. It is also appealing as a further indication of consistency in nature. Figure 7-15 shows some common antenna forms.

FIGURE 7-15. Antennas.

Study Problems

1. Given the bicycle that began this chapter, answer the following questions:
   a. From the bicycle rider's point of view, is the bicycle a torque step-up or torque step-down transformer?
   b. Use a ruler to roughly determine the ratio of the input to output torque.
2. Under what circumstances can an electrical transformer be used with direct current?
3. How does the input power of a transformer compare with the output power?
4. When torque is increased by a gear system, what quantity is reduced? In what proportion is it reduced?
5. When voltage is stepped up in an electrical transformer, what quantity is reduced?
6. Why are transformer cores laminated?
7. Explain how reflected impedance takes place in an electrical transformer.
8. How is an autotransformer different from the ordinary transformer?
9. What is the purpose of an isolation transformer?
10. Are most variable transformers also isolation transformers?
11. Are most other transformers that are intended to be connected to 110 volt power lines isolation transformers?
12. What is meant by the term *transformer phasing?*
13. Why is speaker phasing necessary in a stereo music system?
14. A mechanical transformer increases _____ at the expense of velocity.
15. An electrical transformer increases voltage at the expense of _____.
16. When an electrical transformer has more turns in the primary than the secondary, is it called a step-up or a step-down transformer?
17. In Question 16, what is being stepped up or down?
18. If voltage is stepped up, what is stepped down?
19. Which of your senses can detect a transformer's good or poor efficiency? How can that particular sense provide you with the information?
20. Are all transformers reversible? What about an acoustical horn? (*Hint:* think of a shipboard application of the acoustical horn.)
21. In order to step up voltage when a direct-current input is involved, what must be done to the current flow?

22. What determines the ratio of input to output voltage in an electrical transformer?
23. Define work in terms of force and distance.
24. Define power in electrical terms.
25. Define efficiency for electrical transformers.
26. Define torque.
27. What is the electrical equivalent of torque?

A graphic summary of the contents of this chapter follows.

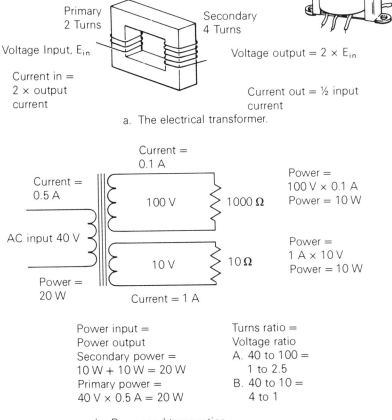

FIGURE 7-16. Summary.

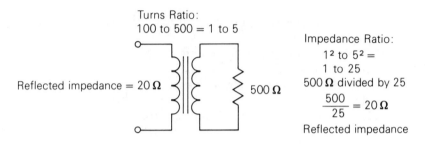

c. Reflected impedance (for impedance matching).

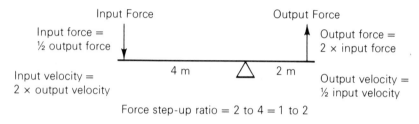

d. The mechanical transformer (the lever).

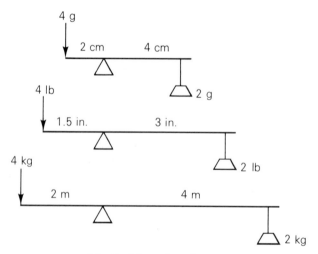

e. Identical force transformers.

FIGURE 7-16. (continued)

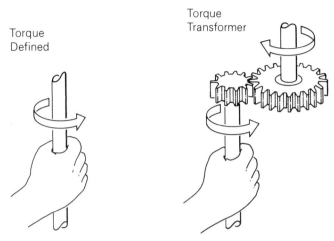

f. Mechanical transformers (gears).

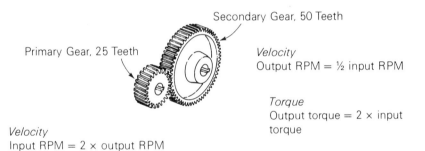

Primary Gear, 25 Teeth

Secondary Gear, 50 Teeth

Velocity
Output RPM = ½ input RPM

Torque
Output torque = 2 × input torque

Velocity
Input RPM = 2 × output RPM

Torque
Input torque = ½ output torque

g. Ratios.

FIGURE 7-16. (continued)

Chapter 8

A Digital Mechanical Amplifier, 1750 (Courtesy Smithsonian Institution)

INTRODUCTION TO AMPLIFIERS

8-1 INTRODUCTION

The Bare Facts

1. Amplification is the control of a larger power by a smaller power.

2. The output of a linear amplifier is a *replica* of the input signal. In other kinds of amplifiers a low-power input signal controls a larger-power output signal that is not a replica of the input signal.

8-2 WHAT IS AMPLIFICATION?

Amplification is *always* the control of a larger power (voltage or current) by a smaller power (voltage or current). In all electronic amplifying devices some electrical mechanism varies the flow of a stream of electrons supplied by some power source, such as a battery. Any device in which a larger power (voltage or current) is controlled by a smaller power (voltage or current) is defined as an amplifier.

Because of its ability to control power, the common light switch is an amplifier. You apply very little force over a very small distance to turn on several hundred watts of power. Two 100 watt light bulbs equal a little more than ¼ horsepower. If it were not for power amplification in the switch, the effort required to turn on the light would be equal to that of lifting a 137 pound weight a foot off the ground in 1 second. The difference is that you supply only *control* power and not the power actually required to run the lights.

To carry the idea a little further, suppose you intend to signal a friend by turning the lights on and off in Morse code. When you move the switch on and off in a specific pattern, the friend receives a 200 watt communication beam that carries exactly the same information that you are putting into the switch with probably a watt or so of mechanical effort. An important point here is that your friend receives the same *information* that you are putting into the switch, but he does not actually receive any of your input effort. Your total effort is expended in moving the switch and none of that energy reaches the remote observer. He receives an exact copy of your coded finger "movements," but the energy he receives is supplied by some remote power plant. The power is there and you simply control it.

ANALOGS

All amplifier devices and circuits have certain properties in common. We will discuss these properties with the aid of an electromechanical and a fluid analog. Take note here that these concepts are difficult to understand in terms of electronic circuits because of their dynamic nature. The analog can help make what is happening in electronic circuits easier to understand.

An important property of every electronic amplifier is feedback. It is too important an idea to be left out, but it is almost impossible to understand feedback without the use of some analogy. Usually mathematical analogies are used; however feedback equations are some of the most difficult to be found in electronics.

Our analogies may be inferior to mathematics from a quantitative standpoint, and may have a Rube Goldberg character, but they are adequate for understanding what feedback is and how it operates.

8-3 AMPLIFIER CHARACTERISTICS

The Bare Facts

1. Amplifier gain characteristics are determined partly by properties of the amplifying device and partly by external resistors and other factors.

2. All amplifiers produce power gain. They may also produce current gain or voltage gain, or both.

3. External components can be added to emphasize voltage gain at the expense of current gain or vice versa.

4. Gain is defined as the controlled quantity (output) divided by the control quantity (input).

5. An on-off type of amplifier control is called *digital*.

6. Continuous control is called *analog*.

7. Power gain = output power/input power
   Voltage gain = output voltage/input voltage

8. *"Input"* generally refers to the signal or control input, not to the power input.

9. Current gain = output current/input current

10. It is possible to view an amplifier as a transformer with *apparent* efficiencies of several hundred percent.

11. Efficiency appears to be greater than 100 percent because the amplifier is actually controlling power derived from a source that is totally independent of the input power.

12. Positive feedback is regenerative and often unstable.

13. Amplifiers may be inverting or noninverting.

14. An inverting amplifier produces maximum output (voltage, current, or power) when the input (voltage, current, or power) is at its minimum.

15. A noninverting amplifier produces maximum output when the input is maximum.

16. Cut-off in an amplifier is a condition in which no current flows in the output circuit.

17. Saturation in an amplifier is a condition in which a change in the input cannot produce a further increase in the output voltage or current.

All amplifiers (with a rare exception) produce power gain. Most of them also provide voltage gain, current gain, or both. Generally, we are particularly interested in either voltage or current gain, so we design the amplifier to get the most we can of that particular kind. An amplifier can step up voltage or current. If we have a voltage gain of 10, an input voltage of 1 volt will produce 10 volts out. If we put 2 volts in we will get 20 volts out, and so on. In general an increase in voltage results in a corresponding reduction in available current gain.

If it seems that either the author has become confused and is still talking about transformers, or there is actually some relationship, you are on the right track. There is a very real similarity between the concepts of transformer behavior and amplifier device behavior. Both can be made to step the "voltage" up (or down) and to step "current" up or down. The key difference is that an amplifying device, viewed as a transformer, can appear to be several thousand percent efficient without getting something for nothing. The transformer cannot. The efficiency appears to be greater than 100 percent because the amplifier is simply controlling power derived from a source that is totally independent of the input power. The power *must* be provided or we are indeed trying to get something for nothing, and that can't happen.

There are also other differences that must not be overlooked. The amplifying devices we will be discussing later are not electromagnetic devices, and do not suffer from core limitations and other problems that belong to electromagnetic devices. As you might suspect, there are electromagnetic amplifiers, but their modern applications are very limited. However, the relay, an electromagnetic-mechanical amplifier, is a very common device in current technology.

Historically, the first commercial amplifier was a transformer-like magnetic device used by the Bell Telephone people to amplify signals along telephone lines. Vacuum tubes soon replaced them, however. During World War II, German shipboard gun control systems used transformer-like magnetic amplifiers that proved less vulnerable to battle conditions than the tube-type amplifiers used by the Allies. Magnetic amplifiers currently find limited application in large industrial control systems, but they are being replaced by more effective semiconductor devices.

THE RELAY AS AN AMPLIFIER

The relay was the first amplifier to play a significant part in practical communications. When the distance between telegraph stations was long, the resistance of the wire would reduce the signal strength to the point where it could not operate the receiving sounder. The solution to the problem was an electromagnet with switch contacts attached. The electromagnetic switch was placed between stations at intervals along the line. The weakening telegraph signals energized the coil in response to the telegrapher's key. The magnet coil pulled down an armature that closed and opened switch contacts, connecting and disconnecting a battery in response to the telegrapher's code transmissions. In this way a full voltage copy of the telegrapher's original signal was *relayed* to the next telegraph station or to the next relay station. Figure 8-1 illustrates the telegraph system with a relay.

AMPLIFIER SPECIFICATIONS

The relay is an on-off device and is therefore classified as a digital amplifier. Figure 8-2 demonstrates the meaning of voltage, current, and power gain in a relay circuit. The relay operating characteristics in the example are taken from the specifications sheet for a commercially available relay. The numbers in the example are real, not just made up.

The relay specifications:

>Coil (input) voltage 6 volts
>Coil (input) current 0.012 amp
>Coil resistance 500 ohms
>Contact current (maximum) 2 amps
>Contact voltage (maximum) 220 volts

The output circuit in Figure 8-2 consists of a 100 watt lamp, a 100 volt power source, and the relay contacts. The 100 watt lamp is the device being controlled and is called the *load*. The relay contacts are

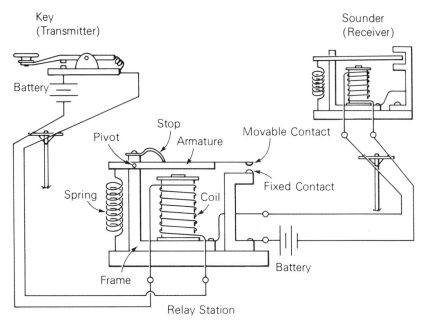

FIGURE 8-1. Telegraph system with relay station.

rated at 220 volts, 2 amps, so it can easily handle the 100 volt, 1 amp load we are using. In amplifier terms, the output circuit in the example has the following characteristics:

> Output power = 100 watts (100 volts × 1 amp)
> Output voltage = 100 volts
> Output current = 1 amp

It is the requirements of the load we wish to control that usually determine the output circuit characteristics in an amplifier. The input characteristics are determined by the requirements of the relay coil in this example. They are:

> Input voltage = 6 volts
> Input current = 0.012 amp (12 mA)
> Input power = input voltage × input current
> Input power = 6 volts × 0.012 amp
> Input power = 0.072 watt

Now let's look at the three gain figures. The gain numbers are simply ratios. They have no units such as volts, amps, or watts attached.

266 CHAPTER 8

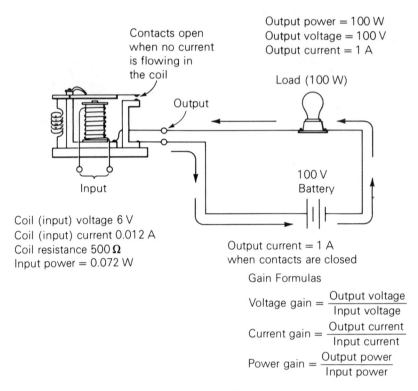

a. Noninverting amplifier.

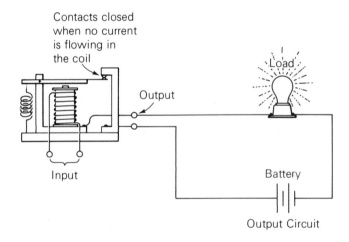

b. Inverting amplifier.

FIGURE 8-2. The relay as an amplifier.

INTRODUCTION TO AMPLIFIERS 267

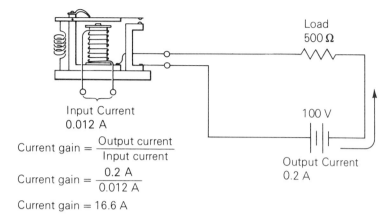

Input Current
0.012 A

Current gain = $\dfrac{\text{Output current}}{\text{Input current}}$

Current gain = $\dfrac{0.2\text{ A}}{0.012\text{ A}}$

Current gain = 16.6 A

Output Current
0.2 A

c. Relay amplifier with a 500 Ω load resistance.

FIGURE 8-2. (continued)

Voltage Gain

$$\text{Voltage gain} = \dfrac{\text{Output voltage}}{\text{Input voltage}}$$

For our example:

$$\text{Voltage gain} = \dfrac{100\text{ volts}}{6\text{ volts}} = 16.6$$

The voltage gain is 16.6.

The voltage gain number tells us that the relay amplifier in the example can control an output voltage that is 16.6 times greater than the input voltage required to control it.

Current Gain

$$\text{Current gain} = \dfrac{\text{Output current}}{\text{Input current}}$$

For our example:

$$\text{Current gain} = \dfrac{1\text{ amp}}{0.012\text{ amp}} = 83.3$$

Power Gain

$$\text{Power gain} = \dfrac{\text{Output power}}{\text{Input power}}$$

For our example:

$$\text{Power gain} = \frac{100 \text{ watts}}{0.072 \text{ watt}} = 1388.8$$

Maximum Ratings

Every amplifier has basic ratings that cannot be exceeded without damaging the device. In the relay example: The maximum output current is 2 amps because that is as much current as the contacts can handle without being damaged. The maximum output voltage is limited to 220 volts according to the relay specifications sheet. The limitation is probably due to contact arcing problems or to insulation breakdown at higher voltages. According to the manufacturer's specification sheet, the maximum input voltage is 6 volts. Higher input currents will cause the coil to run hot. The relay coil could burn out quickly or deteriorate slowly depending on the amount of excess current. In either case we can expect a premature relay failure if we exceed the current or voltage rating.

HOW THE LOAD AFFECTS AMPLIFIER GAIN

The voltage, current, and power gain in an amplifier are primarily controlled by the load value. The maximum gain figures possible depend on the characteristics of the individual amplifier. The load resistance will usually determine the actual gain up to the amplifier's built-in limits. The lamp used as a load in the example (Figure 8-2) is a 100 watt lamp. We can do a little calculation and discover that the lamp has a resistance of 100 ohms. Suppose we replace the lamp with a 500 ohm resistor to act as the load. With a 100 volt power supply (battery) and a 500 ohm resistor for the load we can calculate the output current, using Ohm's Law.

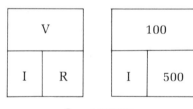

I = 100/500

The output current is 0.2 amp. The relay amplifier current gain is: Output current/input current; with the new output current value of 0.2 amp, the current gain is:

$$\frac{0.2 \text{ amp}}{0.012 \text{ amp}} = 16.6$$

The current gain with the 100 watt lamp for a load was 83.3. See Figure 8-2c. Here it is important to understand that this lower current gain figure does not mean that the amplifier is less satisfactory than the previous higher gain version. In both cases the input circuit can turn the load *on* properly and in each case the load gets all the current it needs. The amplifier with the 500 ohm load has less current gain than the amplifier with a 100 ohm load, but a gain of 16.6 is all that is needed to do the job with the 500 ohm load.

INPUT RESISTANCE

The manufacturer's specifications sheet for the relay in the example also lists the coil resistance as 500 ohms. The input circuit resistance is another important specification for any amplifier. If we were using an AC input signal to drive the coil we would have to talk about input impedance. The relay coil is inductive, and the input current would decrease as the frequency increased. The particular relay in the example is a DC relay and would not work properly with AC. However, AC relays are available that have a shaded core to prevent the relay from chattering as the AC input voltage varies. A relay coil also has some capacitance because the individual turns of wire are insulated and are wound close together. The capacitance is called *distributed capacitance.* Any coil has some distributed capacitance. In electronic amplifiers the input impedance is nearly always capacitive with little or no inductance. The relay and magnetic amplifier are the only common amplifiers that are highly inductive.

INVERTING AND NONINVERTING AMPLIFIERS

The relay amplifier in Figure 8-2a closes the contacts and lights the light when current flows in the input coil. When the input is on, the output is also on. This is called a *noninverting amplifier.*

The relay amplifier circuit in Figure 8-2b is called an *inverting amplifier* because the output is on when current is not flowing in the relay coil. When the input coil is on, the output lamp is off. When the input coil is off, the output lamp is on. A majority of electronic amplifiers are of the inverting type.

AN ANALOG RELAY AMPLIFIER

The digital on-off relay amplifier is practical and common in electrical and electronic circuits. The analog amplifier that we will be examining next is not common or very practical. We will use it only to illustrate the difference between digital and analog amplifiers and to introduce the concept of bias. The analog relay amplifier could be made to work, but to make it work well would be difficult. Also, there are so many superior electronic amplifying devices to choose from that there is really no reason to manufacture such a device.

In Figure 8-3a, a variable resistor takes the place of the contacts attached to the relay armature. When current flows through the coil, the armature is pulled down toward the magnet pole. As the armature moves, the resistance in the load circuit decreases and the lamp gets brighter.

As the coil current increases, the armature moves a greater distance and the lamp brightness increases further. The lamp brightness is

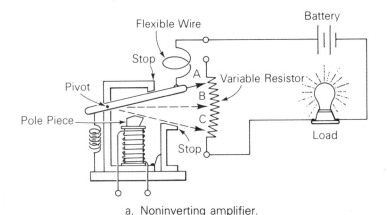

a. Noninverting amplifier.

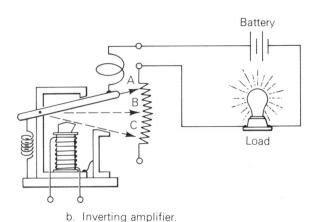

b. Inverting amplifier.

FIGURE 8-3. Analog relay amplifiers.

directly proportional to the relay coil current. By changing the coil current we can change the brightness of the lamp over a range from dark to full brightness. The amplifier is noninverting.

In Figure 8-3b the variable resistor is connected so that the resistance increases as the coil current increases. The light gets dimmer as coil current increases. This amplifier is an inverting amplifier.

GAIN IN ANALOG AMPLIFIERS

At this point we need to alter our previous gain formulas slightly to allow for the difference between digital and analog amplifiers. We must include in our formulas the fact that an analog amplifier has a range of *on* output voltages (or currents), instead of only one *on* voltage. What becomes important in analog amplifiers is how much *change* in output voltage is caused by a smaller *change* in input voltage. The formula for voltage gain in an analog amplifier reads:

$$\text{Voltage gain} = \frac{\text{The change in output voltage}}{\text{The change in input voltage}}$$

(The Greek letter Δ (delta) is used as shorthand for the words "a change in.")

The Gain Formulas for Analog Amplifiers

1. Voltage gain = Δ output voltage/Δ input voltage
2. Current gain = Δ output current/Δ input current
3. Power gain = Δ output power/Δ input power

BIAS

Bias in an amplifier is analogous to the idle adjustment on your car. The idle is set for 500 revolutions or so per minute when the car is not moving. The engine must be kept running if you are to get a fast response when the traffic signal changes to green. The idle adjustment does the same thing as the accelerator under your foot. The only differences between the idle screw and the driver's foot pedal are the idle adjustment's limited control range and the fact that the idle screw is set and forgotten (at least until the next tune-up). The driver's accelerator pedal is analogous to an amplifier's signal input. Yes, the automobile engine is an amplifier; the small amount of power delivered by your foot to the accelerator controls a much larger power delivered to the wheels.

When you need to back out of the driveway you shift into a reversing gear because the engine cannot reverse its direction of rotation. With the exception of the *operational amplifier*, nearly all electronic amplifiers can amplify only the current that flows in one direction.

They are not reversible. This is a problem because most analog information is alternating current and the current flow reverses in alternating current.

The solution to the problem is to bias the amplifier so that it has an output voltage (or current) midway between its full on and full off condition. This bias level is called class A, and is the most common bias condition. Other bias levels are used for special applications, but we won't get into them now.

For the purpose of this example we have added the one-way electronic valve called a diode in Figure 8-4 because the relay is normally reversible. The diode is a real device that we will study shortly. For now, all you need to know is that a diode allows current to flow only in one direction. This makes the relay nonreversible like most electronic amplifiers.

In Figure 8-4 the bias battery delivers enough current to the relay coil to keep the wiper on the variable resistor midway between points A and C. The bias adjustment resistor provides a way to adjust the bias to the desired level. The relay moves the variable resistor wiper from point A to point C with a 12 milliamp input current. In that case, 6 mA of bias should pull the wiper to point B.

Now that we have set the bias, let's see what it does for us. Without the bias the diode would not let the negative half of the sine wave

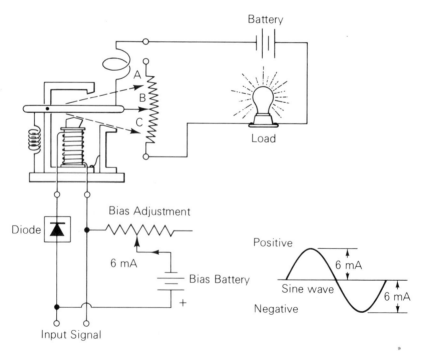

FIGURE 8-4. Relay amplifier with a bias circuit.

signal through, and that half would cause no action at the output. We could amplify only the positive half of the sine wave signal. This is a very severe distortion of the signal and would be intolerable in sound, for example.

During the positive half of the sine wave signal the bias and signal currents add because they are the same polarity. At the positive peak of the signal the 6 mA signal adds to the 6 mA bias current to move the wiper from point B to point C. When the positive half of the signal goes back to zero, the wiper is back at point B.

During the negative half of the sine wave signal the wiper moves from point B toward point A. The negative signal current subtracts from the positive bias current. At the peak of the negative half of the sine wave, the negative 6 mA of the signal current subtracts from the 6 mA bias current. This produces an input current of zero. The wiper on the amplifier variable resistor has moved to point A.

The input current has varied from +6 mA to +12 mA, back to +6 mA and down to zero, following the input signal. The input current was always positive; it never passed through zero to negative. No current reversal occurred.

In many cases only the shape of the wave is important, and it doesn't matter that the current doesn't reverse. In applications in which current reversal is important, it is easily accomplished with a transformer or resistor capacitor circuit. In many cases it is possible to get the bias voltage from the same power source that provides power for the load. An example is shown in Figure 8-5.

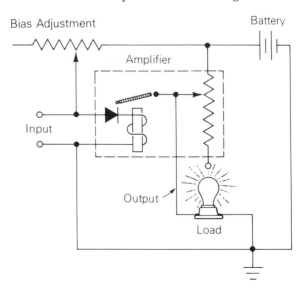

FIGURE 8-5. Bias current is often taken from the same power supply that provides power for the load.

Study Problems

1. Define voltage gain.
2. Define current gain.
3. Define power gain.
4. What is bias?
5. Are most amplifiers capable of amplifying alternating current?
6. Define amplification.
7. What is the difference between digital and analog amplifiers?
8. Define inverting amplifier.
9. Define noninverting amplifier.
10. Is the input impedance of most electronic amplifiers a combination of resistance and inductance or a combination of capacitance and resistance?

8-4 FEEDBACK

The Bare Facts

1. When a part of the output signal of an amplifier is fed back to the amplifier input, it is called *feedback*.
2. There are two kinds of feedback, positive and negative.
3. Positive feedback tends to cause amplifier instability.
4. Positive feedback exaggerates noise, hum, and distortion that might be present in an amplifier.
5. Positive feedback is generally avoided in amplifiers except for amplifiers that are meant to oscillate.
6. An oscillating amplifier (or simply oscillator) is used to generate audio tones, radio signals and so on.
7. An oscillator normally has no input. Its signal begins as electrical noise and a tuned circuit filters out all but the desired frequency.
8. The howl so often heard from public address systems is a common example of the oscillation resulting from positive feedback.
9. Negative feedback tends to cancel noise, distortion, hum, and so on that an amplifier produces.
10. Negative feedback is often used to improve amplifier performance.
11. A loss of some amplifier gain is always the price paid for the benefits of negative feedback.

12. Feedback always alters the input resistance of an amplifier.

13. In negative feedback, the feedback signal works against the input signal.

14. In positive feedback, the feedback signal aids the input signal.

AN ANALOGY

In order to broaden our understanding of amplifiers a bit more and to introduce the concept of feedback, let's look at a fluid analogy. A valve can be an amplifier if a small input can control a larger fluid flow. The hydraulic valve (amplifier) shown in Figure 8-6 can control a large flow when a smaller flow is applied to the input-signal fluid stream. Energy is supplied by the gasoline engine to drive the pump. The pump forces water around the circuit, driving the turbine wheel. We assume that the shaft of the turbine is connected to a machine that does some kind of useful work.

A similar hydraulic drive system was proposed for automobiles, but apparently was too expensive. In the automotive system, the needle valve in Figure 8-6 would be moved mechanically by the accelerator pedal that increases or decreases the car's speed. The turbine would turn the wheels.

The problems of pollution and fuel efficiency might invite another look at the hydraulic system some day. In a fluid-control system

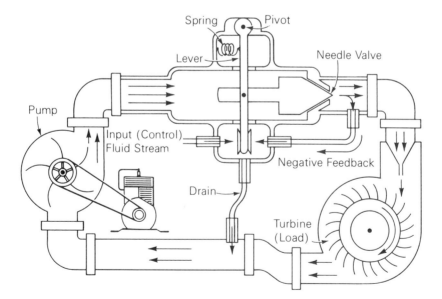

FIGURE 8-6. Inverting hydraulic amplifier with negative feedback.

something like the one in Figure 8-6, the engine could run at a constant speed. The system could, at least theoretically, be designed always to operate the engine at the speed at which it was most efficient and least polluting.

Feedback is one of the most common and complex elements in nature, technology, and social intercourse. Nearly every interaction we encounter involves either negative or positive feedback. Feedback always involves action. It is never passive. Positive feedback is regenerative like an avalanche or a minor disagreement between nations that builds up into war. Controlled positive feedback is used in oscillators, to be discussed later.

Negative feedback is calming and stabilizing. In amplifiers, negative feedback can reduce the distortion due to amplifier imperfections, cancel undesirable temperature drift, and minimize the effects of variations in component values due to manufacturing imperfections. Positive feedback can exaggerate these same problems, or, when properly controlled, it can produce the oscillations that generate radio waves.

Some feedback is inherent in any amplifying device, but it is generally necessary to add more external feedback to make the device practical. Suppose we examine feedback in the hydraulic amplifier analog in an effort to get some feel for its properties before we examine it in electronic amplifiers.

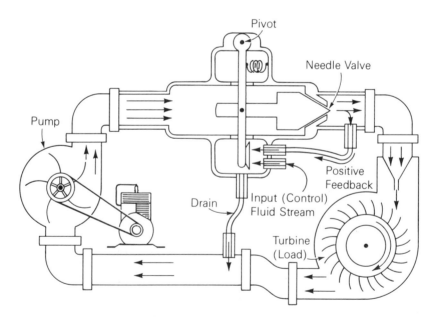

FIGURE 8-7. Noninverting hydraulic amplifier with positive feedback.

FEEDBACK IN AMPLIFIERS

When we connect a part of the output signal back to the input of an amplifier we have added feedback to the amplifier. There are two kinds of feedback, negative and positive.

Negative feedback tends to oppose and cancel part of the input signal. In electrical terms the negative feedback voltage or current opposes and thus subtracts from the input voltage or current.

Positive feedback is connected in a way that *adds* to the signal input voltage.

Look at the fluid amplifier analogy in Figures 8-6 and 8-7. Note how the negative feedback stream pushes against the input (control) while a positive feedback stream pushes in the same direction as the input stream. In this analogy:

1. Stream force is the analog of voltage.
2. The spring-loaded lever arm is the analog of the coil in the relay amplifier.
3. The needle valve is the analog of the variable resistor attached to the armature of the relay amplifier.
4. The turbine is the analog of the lamp or resistor in the relay amplifier.
5. The pump is the analog of the battery that powers the load in the relay example.
6. The gasoline engine is the energy source. In a battery the energy source is chemical and cannot be separated from the battery itself.
7. Pipes and tubes are the analog of connecting wires.

Negative Feedback

Negative feedback tends to oppose any *change*. It is a defender of the status quo. In the fluid amplifier in Figure 8-6, fluid from the output circuit is fed back to the input control circuit. The sample of the output fluid pushes against the lever in a direction that tries to open the needle valve as the signal tries to close it. In most feedback cases the input signal exceeds the feedback input. It takes a greater input signal force to close the valve when feedback is present because the negative feedback stream produces a force that pushes against the signal stream.

Notice that the spring in Figure 8-6 tends to hold the needle valve open just as the feedback stream does. But there is an important difference. The spring is static. It just sits there and pulls with a nearly constant force, like the bias voltage in the relay amplifier example. On the other hand, the force of the feedback stream depends on the partial opening and closing of the needle valve. The feedback stream is dynamic and changing, and its dynamic character is what makes it so important. Its dynamic behavior is also what makes it difficult to understand and what makes feedback mathematics so horrendous.

Positive Feedback

Positive feedback tends to further increase any increase in the amplifier output. You have probably encountered positive feedback in public address systems. The howl begins at low volume and builds up, often in spite of the operator's efforts to stop it. Positive feedback tends to get out of hand if not controlled.

The hydraulic amplifier in Figure 8-7 illustrates positive feedback. A small opening of the needle valve by the control "signal" stream drives fluid back through the feedback line to open the valve a little more. As increased fluid flows in the output circuit, additional positive feedback fluid drives the valve open still wider. If not checked, the positive feedback can drive the needle valve all the way open with only enough *input* (control) signal to start the regenerative action. The input control signal would then (probably) cease to have any further control over the valve action or output flow. Positive feedback tends to build up rapidly and override the input signal control.

Study Problems

1. Which kind of feedback tends to be unstable?
2. A change in output conditions should occur when there is a change in what quantity?
3. What effect does feedback have on input resistance in an amplifier?
4. How is voltage gain usually controlled in an amplifier?
5. Give some everyday examples of positive and negative feedback. Spend a couple of days observing before you answer. A silly argument that builds up to a shouting match is an example of one kind of feedback (which kind?).
6. A couple are arguing in a marriage counselor's office. The counselor intercedes and the couple calms down. What kind of feedback has the counselor introduced into the system?

INPUT RESISTANCE

Both positive and negative feedback alter the normal amplifier input resistance. Either kind of feedback can lower or increase the input resistance. Whether the feedback increases or decreases the input resistance depends on the kind of feedback and how it is connected. There are several possible connections.

In the fluid amplifier in Figure 8-6 you can see that the input stream must exert extra force to move the control lever a given distance because the feedback stream is pushing against it. The lever offers increased resistance to movement because of the feedback stream. In this case, negative feedback has caused an *increase* in input resistance.

In the fluid amplifier with positive feedback (Figure 8-7), the feedback stream helps the input stream push the control lever. The input stream finds less resistance to its force because of the feedback stream's added force. In electronic amplifiers it is electromotive force (voltages) instead of fluid forces that aid or oppose each other.

BACK TO THE RELAY AMPLIFIER

Let us go back to the electrical world and examine feedback circuits in a relay amplifier. A look at relay amplifier feedback will make feedback in the next section on electronic amplifiers easier to understand.

In the circuits in Figure 8-8a and b, a variable resistor adjusts the amount of feedback. You will notice only one difference between this circuit and the relay amplifier bias circuit in Figure 8-5. In the feedback circuit the bias connection is made to the amplifier output instead of directly to the battery. The battery provides a constant current bias while the feedback connection provides a bias current that varies as the output voltage varies. The amplifier with negative feedback (Figure 8-8a) has a fairly large bias current when the signal input current is small. At first this looks like it might be a case of positive feedback because of the initial bias current provided. However, feedback must be considered in terms of changing conditions not static conditions. The feedback current *decreases* as the input current *increases*, and in that sense the feedback current is subtractive. When the feedback is subtractive it is negative feedback.

In the positive feedback circuit of Figure 8-8b, the feedback current is initially a low value, but increases as input current increases. The feedback current adds to the input current and is therefore positive feedback.

Study Problems

1. In negative feedback amplifiers does the feedback signal add to or subtract from the input signal?
2. Answer question 1 for positive feedback.

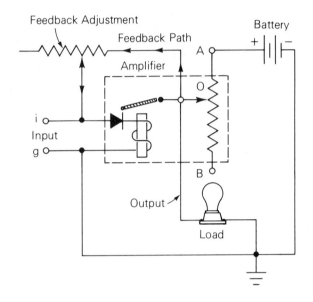

a. Relay amplifier with negative feedback.

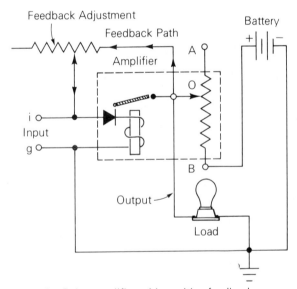

b. Relay amplifier with positive feedback.

FIGURE 8-8. Relay amplifiers with feedback.

INTRODUCTION TO AMPLIFIERS 281

3. Is feedback static or dynamic?
4. Does feedback have any effect on an amplifier's input resistance?
5. Which kind of feedback tends to make an amplifier unstable?
6. Why might one want to add negative feedback to an amplifier?
7. How does bias differ from feedback?

8-5 THE OPERATIONAL AMPLIFIER

The Bare Facts

1. The operational amplifier (op-amp) is the nearest thing to a universal amplifier that the technology has to offer. It is a complete functional amplifier system in a tiny package.

2. Modern op-amps are no more expensive than a few transistors, but are a complete packaged amplifier system.

3. The voltage gain of an op-amp is set by 2 external resistors. (One of them is a negative-feedback resistor.)

4. In the inverting-amplifier configuration, the feedback lowers the amplifier's input resistance.

5. In the noninverting-amplifier configuration, the negative feedback increases the input resistance.

6. The voltage follower has a very high input resistance due to the large amount of negative feedback.

7. The voltage follower is used as an impedance transformer or current step-up transformer. Transformer action is one-directional from input to output.

8. The voltage-follower configuration has a voltage gain of unity (1).

CHARACTERISTICS OF OPERATIONAL-AMPLIFIERS

The following are basic characteristics of the op-amp:

1. High input impedance (resistance).

2. Low output impedance (resistance).

3. Very high open-loop (without feedback) voltage gain (100,000 or more).

4. Provision for external feedback to set the voltage gain to any desired value.

5. Very good stability.

6. Capable of frequency response from DC to some high frequency.

7. One 180° (−) input and one zero-degree (+) input.

The op-amp is a nearly universal amplifier. It derives its name from its original use in analog computers. Even in the early days, it was recognized as an ideal amplifier for almost every purpose. The price and bulk were far from ideal, however. Op-amps were priced at several hundred dollars each. As a result, their use was severely limited. Modern integrated-circuit op-amps are available at a very low cost.

The op-amps available in integrated-circuit form may contain 20 or more transistors and a number of resistors; yet the complete functional circuit costs little more than 4 or 5 transistors. It is doubtful that op-amps will entirely replace individual components, but they are certain to take over many more of the tasks where individual transistors have been used in the past.

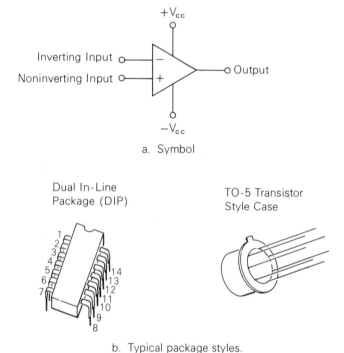

FIGURE 8-9. The operational amplifier.

The op-amp is symbolized by the shape in Figure 8-9a. We will look at the kind of circuitry inside the unit later. We will also examine practical uses in other chapters.

Figure 8-10 shows the 3 most basic op-amp circuits. The circuit in Figure 8-11a is an inverting amplifier. The output is the reverse of the input. For a sine wave the phase is inverted, making the output signal 180° out of phase with the input. This inversion is common to most amplifier devices. The reason will become apparent when we study

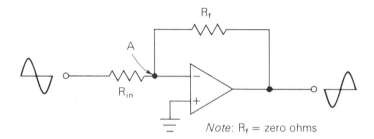

a. Inverting amplifier.

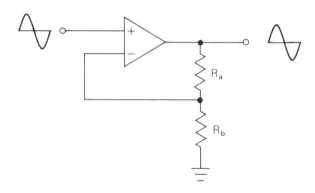

b. Noninverting amplifier.

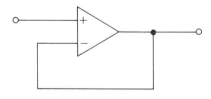

c. Transformer.

FIGURE 8-10. Operational-amplifier circuits.

transistors. The voltage gain is the simple ratio of 2 external components, R_{in} and R_f:

$$\text{Voltage gain} = -\frac{R_f}{R_{in}}$$

The minus sign simply indicates a signal inversion. R_f is an external negative-feedback resistor. The feedback in this case lowers the input resistance at Point A drastically. The input resistance as seen by the input signal is equal to R_{in}.

The circuit in Figure 8-10b is a noninverting amplifier. The voltage gain is:

$$\text{Voltage gain} = \frac{R_a}{R_b}$$

For the mathematician:

$$\text{Voltage gain} = 1 + \frac{R_a}{R_b}, \text{ but for higher gains there is a negligible difference.}$$

In this circuit the negative feedback is applied to the other input (−), and the amplifier input resistance is increased by the feedback. How much it is increased depends on the amount of feedback through R_a.

The circuit in Figure 8-10c is similar to that of 8-10b, except that there is 100 percent feedback. The feedback raises the input resistance drastically. Theoretically the input resistance is infinite with 100 percent feedback, but in practice the best we can hope is that it is very high. The circuit is called a *voltage follower* and is primarily used as an impedance transformer or current step-up transformer. The 100 percent feedback reduces the voltage gain to 1 (unity) or slightly less.

The operational amplifier is a multistage integrated-circuit amplifier with open-loop (without negative feedback) voltage gains of 50,000 to 1 million. Its open-loop input resistance is high and its output resistance is low. It always features 2 inputs: an inverting input and a noninverting input. The op-amp generally requires 2 batteries or power supplies. The two power supplies allow both positive and negative output voltages. The op-amp can amplify AC.

8-6 SEMICONDUCTORS

The Bare Facts

1. Semiconductors have a conductivity between insulators and metals.
2. The conduction in semiconductor materials is highly temperature-dependent.
3. The most frequently used semiconductor in modern technology is silicon.
4. An intrinsic semiconductor is a highly purified semiconductor crystal.
5. Silicon has 4 valence electrons.
6. Silicon is doped with 3 or 5 valent atoms to reduce its temperature sensitivity and increase its conductivity.
7. Three valent dopants form a silicon crystal called *P-type*.
8. Five valent dopant atoms form a silicon crystal called *N-type*.
9. A P-type crystal has no free electrons (at absolute zero), and the electrons that make up current flow must come from outside the crystal.
10. An N-type crystal has many free electrons contributed by the dopant atoms.
11. Vacancies in the crystal where electrons are missing are called *holes*. Electrons flow from hole to hole.
12. N-type dopant atoms are called *donors* because they contribute free electrons.
13. P-type dopant atoms are called *acceptors* because they provide holes to accept electrons from an external source.
14. The simplest practical semiconductor device is a junction diode.
15. The diode is a continuous crystal, part of which is P-type and part N-type.

16. The point of transition in the crystal from P-type to N-type is called a *junction*.
17. An area around the junction that is free of both holes and electrons is called the *depletion zone*.
18. The diode is a one-way electronic valve. Current will flow only in one direction through the device.
19. Forward bias is the conducting condition in a junction diode.
20. Reverse bias is the nonconducting condition.

Because all modern amplifier devices are semiconductor-based, let's look at the properties of semiconductors.

At absolute zero (−273°C), silicon is a near-perfect insulator. Above absolute zero, the higher the temperature, the better silicon becomes as a conductor. Within the range of earth surface temperatures, where humans can exist, semiconductors range from poor to good conductors. This change in conductivity with temperature is a drastic phenomenon. In silicon, each 10° rise in temperature triples its conductivity.

If a block of silicon is hooked up and heated as shown in Figure 8-11, the current in the circuit will rapidly increase as the temperature of the block increases. The current also increases if the voltage is increased. But increasing the voltage increases the current only proportionally, obeying Ohm's Law, while increasing the temperature increases the current at a much faster rate.

The temperature, over which we have but little control, exerts a

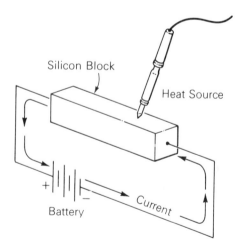

FIGURE 8-11. Heat and conduction in silicon.

greater effect on the current than does the voltage, which is what we can control in electronics. To build useful semiconductor devices it is necessary to control this temperature problem by making the current more voltage-dependent than temperature-dependent. This is accomplished by adding certain impurities to the semiconductor material to make free electrons available in large amounts. These impurity-produced electrons are not temperature-dependent, but they are voltage-dependent.

SILICON CRYSTALS

The elements silicon and germanium are the two most important semiconductors in modern technology. Silicon has gradually replaced germanium as processing methods for silicon have matured. Compounds such as gallium arsenide are used in special applications such as light-emitting diodes (LEDs) and photoelectric cells, but have not been much used in amplifying devices.

Germanium and silicon each have 4 valence electrons in their outer orbits. When there are many atoms of silicon or germanium in their crystalline form, electron-sharing occurs. The result of this sharing behavior is that the formerly free valence electrons become tightly bound in their particular places in the crystal. Figure 8-12 shows a diagram of this covalent bond. An electron associated with

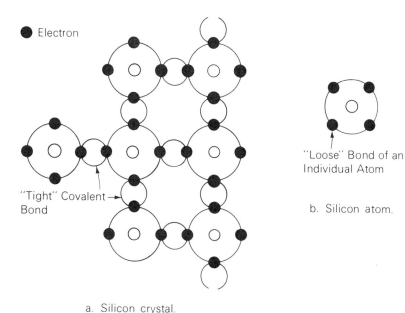

a. Silicon crystal.

FIGURE 8-12. The covalent bond.

an individual silicon atom would require but little energy to free it from the atom. When a large number of atoms form covalent bonds, they bind the electron to a pair of atoms so tightly that it takes about 14 times as much energy to break the electron free.

In semiconductors these covalent bonds are still so weak that many of them are broken at any temperature above absolute zero, with the number increasing as the temperature rises. Whenever a covalent bond is broken by heat energy, an electron becomes free to be moved by an electric field if one is present. When the electron leaves its home in the crystal, it leaves behind a *dangling bond*. This dangling bond has an attraction for any electron passing by, and will capture it if it comes near. The dangling bond is the positive charge of one of the protons in the atom that has lost an electron. The dangling bond (hole, in semiconductor talk) is tightly anchored to its spot in the crystal. Nevertheless, it is convenient to pretend that it is as mobile as an electron.

Although we cannot easily visualize anchored holes moving under the influence of an electric field, we can easily visualize changing concentrations of holes, as electrons fill more holes in one area than in another. Figure 8-13 shows a crystal in which B is an electron whose covalent bond has been broken. The dangling bond A is the hole.

CONDUCTION IN SILICON

When there is no electric field, free electrons are fairly evenly distributed throughout the block of silicon. As electrons are freed by thermal energy, they are also being captured by holes, and each capture results in the destruction of a recently created electron-hole

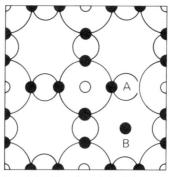

A = Dangling bond (hole)
B = Freed electron

FIGURE 8-13. Bond broken by heat energy.

pair. When an electron is captured by a hole, the hole no longer exists and the electron is no longer a *free* electron. Electron-hole pairs are constantly being created and destroyed in about equal numbers.

When an electric potential is applied across the block, electrons are attracted to the positive end of the block. Electron-hole pairs are being formed throughout the block, but because of the electric field there is a heavy concentration of electrons at the positive end and fewer electrons at the negative end. Those holes at the negative end of the block are not capturing electrons, because there are almost none available to capture. They have been attracted to the positive end of the crystal. At the negative end of the block there is an excess of holes. At the positive end of the block there is an abundance of electrons, and holes are filled almost as soon as heat energy breaks them loose.

We now have a block of semiconductor material with a high concentration of electrons at one end and a high concentration of holes at the other. Electrons have moved to the positive end of the block, while holes *appear to have moved* to the negative end. In a pure silicon crystal, all available carriers—holes and electrons—are the result of heat-ruptured covalent bonds. Conduction can be controlled by providing controlled amounts of electrons (or holes) that do not depend on heat for their freedom.

N-Type Silicon

If an atom of some impurity with 5 valence electrons is introduced into the crystal, there is 1 unattached electron introduced into the crystal. Figure 8-14 is a sketch that shows this situation. The atom to the right of center is called a *donor* atom because it donates 1 free

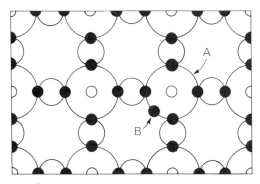

A = Donor atom
B = Donated electron

FIGURE 8-14. The N-type silicon crystal and donor atoms.

electron to the system. In addition to having 5 valence electrons, the donor atoms must fit into the crystal structure easily. This restricts available donors to 3 elements: arsenic, phosphorus, and antimony. One impurity atom is added to each one hundred thousand to 1 or 2 million atoms of silicon. The process of adding impurity atoms is called *doping*.

With one donor atom per million silicon atoms, a 1 cubic centimeter block would have 10^{16} (10,000,000,000,000,000) free electrons contributed by the donors. In addition to the donated electrons, there are about 10^{10} electrons that have been freed because of heat-ruptured bonds. The majority of the electrons available come from the donors and are free at all temperatures much above absolute zero. At room temperature there are 10^{16} donor electrons and about one-millionth of that number that are the result of broken bonds. Donor electrons are in the vast majority, and their number is essentially constant at all usual temperatures. The addition of donor atoms creates a less temperature-sensitive semiconductor.

Because there are more electrons than holes, and because the electrons carry a negative charge, a semiconductor doped with donor-type atoms is called *N-type* (N for negative). The N-type semiconductor contains only free electrons. At normal earth surface temperatures, heat-ruptured bonds produce some electron-hole pairs. The thermally generated electrons simply join the donor electrons. The holes "move" in a direction opposite to the direction taken by the electrons.

P-Type Silicon

In order to make most practical semiconductor devices, we need another type of semiconductor, called the *P-type*. It has only free holes (at absolute zero) and no free electrons. P-type semiconductors are also made by doping the pure silicon. In P-type silicon, however, the doping atoms have only 3 valence electrons. These doping atoms are called *acceptors* because each one contributes a dangling bond (or hole). These holes can accept electrons supplied from outside the block. Figure 8-15 shows the P-type crystal and acceptor atoms. The P-type dangling bonds (holes) created by the acceptor atoms can absorb electrons when they are supplied by an external power supply. P-type silicon has no free electrons, and electrons for current flow must come from somewhere outside the P-type crystal.

Figure 8-16 is a much simplified description of conduction and nonconduction states for pure silicon, silicon doped with donor impurities, and silicon doped with acceptor impurities. P-type dopants are aluminum, boron, gallium, and indium.

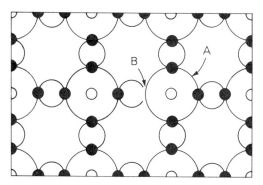

A = Acceptor atom
B = Impurity-created hole

FIGURE 8-15. P-type silicon crystal and acceptor-created hole.

8-7 THE PN JUNCTION DIODE

The junction diode is made by joining a block of P-type semiconductor with a block of N-type semiconductor in a continuous crystalline structure. The manufacturing process starts by diffusing donor and acceptor impurities into a block of pure silicon. The diffusion is accomplished at temperatures near 1000°C; impurities melt into the pure silicon like butter into hot toast. P-type (acceptor) impurities are introduced into one end of the block, while N-type (donor) impurities are introduced into the other end. Within the block there is a more or less abrupt transition from P-type to N-type material. This transition is called a *junction*. Figure 8-17 is a sketch of a junction diode. The junction diode is essentially a one-way electrical valve, often called a *rectifier*.

THE DEPLETION ZONE

When the crystal is formed, electrons begin to diffuse across the junction into holes, forming a transition or depletion zone that contains neither free electrons nor holes. The process is self-arresting and ceases after the formation of a very narrow transition zone. In the sketch in Figure 8-17, as the transition zone grows wider, electrons must cross a widening gap where electrons and holes are few and far between. The amount of recombination decreases rapidly as the distance away from the junction increases.

The N and P regions are electrically neutral in the beginning, but as the crystal is formed (as it cools), electrons diffuse across the

292 CHAPTER 8

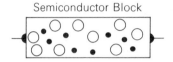

Semiconductor Block

- • Electron
- ○ Hole (dangling bond)
- ⊙ A Hole that has captured an electron

Conditions:
1. All carriers, electrons, and holes are heat-generated.
2. Electrons and holes are generated in pairs, and are distributed evenly through the semiconductor block.

a. Pure silicon with no EMF applied.

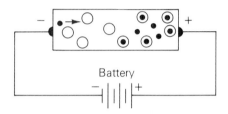

Conditions:
1. All carriers, electrons, and holes are heat-generated.
2. Electrons and holes are equal in number because they are always formed in pairs.
3. Free electrons are attracted to the positive end of the block, filling nearly all the holes in the positive end.
4. Electrons in the positive end of the block that have not been captured by holes are returned to the positive end of the battery.
5. Electrons enter the negative end of the block from the battery and move from hole to hole toward the positive end of the block.
6. The amount of current is a function of temperature, and increases exponentially as the temperature increases.

b. Pure silicon with EMF applied.

Conditions:
1. N-type silicon has only free electrons, no holes.
2. These free electrons are free at most ordinary temperatures, and conduction is not very dependent on temperature when a potential is applied.

c. N-type silicon with no EMF applied.

FIGURE 8-16. Conduction in silicon.

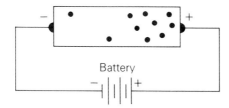

Conditions:
1. Electrons are attracted to the positive end of the block. Electrons in the block are supplied by five-valent impurity atoms.
2. Conduction is not much influenced by temperature.
3. A few holes, approximately one for each million electrons, exist as the result of heat-ruptured covalent bonds.

        d. N-type silicon with EMF applied.

Conditions:
1. The P-type semiconductor has only holes, no electrons (at absolute zero). Holes are provided by trivalent impurity atoms.
2. Some electrons exist in the free state owing to heat-ruptured covalent bonds, but the number is very small compared to the number of holes supplied by the donor atoms.

        e. P-type silicon with no EMF applied.

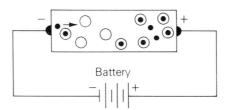

Conditions:
1. All the electrons involved in conduction must be supplied from outside the block, in this case by the battery.
2. Electrons injected into the block move freely in and out of available holes.
3. The holes are supplied by trivalent dopant atoms. Conductivity is not a function of temperature, except the conductivity owing to the small number of additional carriers produced by heat-ruptured covalent bonds.

        f. P-type silicon with EMF applied.

FIGURE 8-16. (continued)

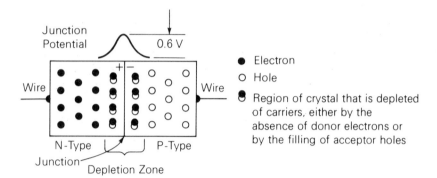

FIGURE 8-17. The junction diode.

junction and are pulled into holes on the other side. The N side loses some of its free electrons to holes on the P side, leaving the N side slightly positive. This makes the zone next to the junction on the P side slightly negative because of the electrons that have come from the N side. In silicon, a potential barrier equal to approximately 0.6 volt is formed. Further movement across the junction is prevented until an external voltage greater than the 0.6 volt barrier voltage is applied across the junction. This junction voltage is approximately 0.6 volt for all silicon junctions.

The important thing to remember about the depletion zone is that it sets a minimum threshold potential. Current cannot flow through a silicon junction until a potential of approximately 0.6 volt is applied across it.

THE REVERSE-BIASED JUNCTION

The reverse-bias condition is the off, or nonconducting condition. Figure 8-18 shows the reverse-bias battery polarity and the hole and electron distributions in a reverse-biased junction.

THE FORWARD-BIASED JUNCTION

The forward-bias condition is the conducting condition. Figure 8-19 shows the power-supply (battery) polarity and the distribution of electrons and holes in the forward-bias connection. The negative field at the N end of the device repels free electrons toward the junction. The positive field at the P end of the block repels holes toward the junction. The arrows within the block show the direction of carrier movements within the block. The arrows outside the block show the direction of electron movement. If the battery voltage is higher than the junction potential (0.6 volt), electrons are forced

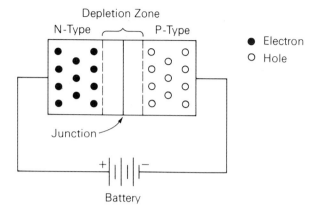

FIGURE 8-18. The reverse-biased junction.

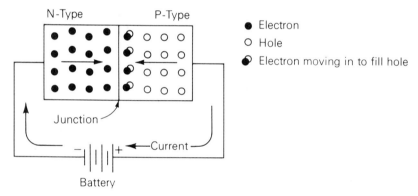

FIGURE 8-19. The forward-biased junction.

through the transition zone across the junction and into the P region. An electron crossing the junction into the P region is captured by a hole. The defecting electron has left the N region one electron short. The N crystal, which was electrically neutral, now becomes one unit positive because the crystal now has one more proton than it has electrons. This positive charge draws an electron from the negative terminal of the battery. One electron has left the battery. At the P side of the block, the electron has been drawn into a hole. Once the hole is filled, it is no longer a hole. One hole has been lost while an electron has been gained by the P crystal. The P side of the crystal now has an excess electron and consequently a unit negative charge. The positive terminal of the battery draws an electron out of the block to restore the charge balance in the P end of the crystal. One electron has left the negative battery terminal, and one electron has returned to the positive battery terminal. If we can visualize hundreds of millions of electrons crossing the junction, we have a rough picture of forward-biased junction action. Forward bias is the conducting condition.

JUNCTION CAPACITANCE

A capacitor consists of 2 conductive plates separated by an insulating material (dielectric). A reverse-biased diode meets these requirements. It has 2 areas that are rich in carriers (conductors) and that are separated by the depletion zone. This depletion zone has no available carriers and is therefore an insulator. The dielectric constant of a silicon depletion zone is approximately 12. The amount of capacitance in a junction diode is small enough so that it becomes important only at high frequencies. Special junction diodes are manufactured in such a way as to provide fairly high capacitances. These diodes are used as voltage-variable capacitors in oscillator resonant tank circuits and other applications, and are called *varactors*. The capacitance of a reverse-biased junction is varied by varying the bias voltage. Increasing the voltage increases the width of the depletion zone, decreasing the capacitance.

Study Problems

1. Define semiconductor.
2. Name the principal factor that governs the amount of current flow through a pure semiconductor material.
3. Define intrinsic (pure) semiconductor.
4. List the P-type dopant elements.
5. List the N-type dopant elements.
6. Define electron-hole pair.
7. Define dangling bond.
8. What is the majority carrier in an N-type crystal?
9. List the way in which semiconductors can be made less temperature-dependent.
10. Define doping.
11. Define N-type semiconductor.
12. Define P-type semiconductor.
13. Explain how a reverse-biased junction can behave as a capacitor.
14. Define valence electrons.
15. Define covalent bonds.
16. Define hole.

17. Define donor.
18. Define acceptor.
19. Make a drawing showing carrier movement when a silicon junction is newly formed.
20. What is the depletion-zone potential in a silicon junction device?
21. What is the depletion zone?
22. How does voltage affect the amount of current through a forward-biased junction-diode circuit?

Chapter 9

Reproduced with permission of A.T. & T. Co.

ELECTRONIC AMPLIFYING DEVICES AND CIRCUITS

9-1 INTRODUCTION

Now that we have learned what amplifiers do, it is time to look at some other basic devices that make electronic miracles possible. In the last chapter we studied semiconductors because nearly all modern amplifier devices are made of semiconductor materials. In this chapter will see how amplifying devices work, and examine their special characteristics. In following chapters we will see how the devices fit into electronic circuits and systems. We will begin with one of the simplest amplifier devices in our arsenal, the junction-field-effect transistor.

9-2 THE JUNCTION-FIELD-EFFECT TRANSISTOR (J-FET)

The Bare Facts

1. The junction-field-effect transistor (J-FET) consists of a thin channel of N-type (or P-type) material and a pair of PN junctions with the channel forming the N part of both junctions.

2. The power input element is called the *source*. The control element is the *gate*, and the output element is the *drain*.

3. Current through the N-type channel is varied by increasing or decreasing the reverse bias on the two junctions.

4. The off condition of the J-FET is called *pinch-off*.

5. The amplifier figure of merit for the J-FET is called *mutual conductance*.

The J-FET in Figure 9-1 consists of a thin filament of N-type silicon with a slab of P-type material on each side, forming the 2 junctions. The N-type material forms a channel through which source-to-drain current can flow. The 2 P-type slabs are wired together to form a control element called the *gate*. Figure 9-1a shows the channel in its nearly wide-open condition. In this condition the only depletion zone is that formed when the junctions are manufactured. Current flows freely through the channel. In Figure 9-1b, several volts of reverse bias are applied to both junctions. The depletion zone spreads into the channel, squeezing it shut and turning off the source-drain current. Remember, a depletion zone is an area swept clear of both electrons and holes—an insulating area. The condition is called *pinch-off*.

Between the two extremes of the nearly wide-open and completely pinched-off channel lie a wide range of channel widths and resulting channel currents, illustrated in Figure 9-1c. It is within this control

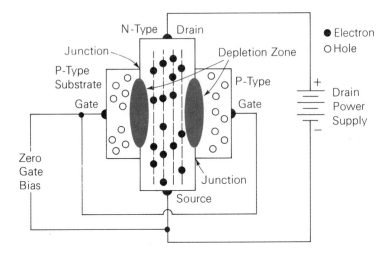

a. J-FET with zero volts gate bias and an open channel.

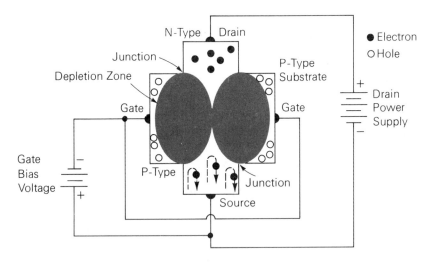

b. J-FET with channel biased to pinch-off.

FIGURE 9-1. The J-FET.

range that the J-FET normally operates. It is also within this range that amplification takes place. The amplification figure of merit in the J-FET is called *mutual conductance* (or *transconductance*), and is the dynamic ratio of drain current to gate voltage:

$$\text{Mutual conductance} = \frac{\Delta \text{ Drain current}}{\Delta \text{ Gate voltage}}$$

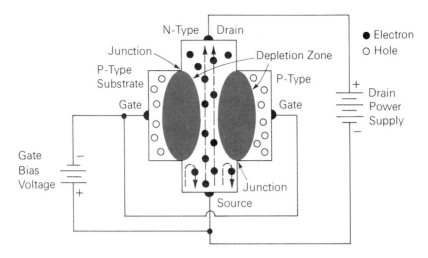

c. J-FET with gate bias that leaves the channel partly open.

FIGURE 9-1. (continued)

In mathematical shorthand:

$$G_m = \frac{\Delta I_D}{\Delta E_g}$$

Note: Δ means "a change in," and G_m is mutual conductance. The notation I/E represents Ohm's Law upside down (conductance), and the unit is the mho. The G_m numerical range is from 100 to 6000 micromhos for the J-FET (micromho = 10^{-6} mho).

9-3 THE METAL-OXIDE FIELD-EFFECT TRANSISTOR (MOS-FET) (DEPLETION MODE)

The Bare Facts

1. The MOS-FET consists of a substrate, a channel, and an insulated metal plate.

2. The metal plate serves as the control element (gate).

3. Control is accomplished by placing a negative charge on the gate that drives electrons out of the channel into the substrate. As electrons are pushed out of the channel, channel current decreases.

4. The MOS-FET figure of merit is mutual conductance.

The MOS-FET operates on the same basic principle as the J-FET, except that the control element (gate) that pinches the channel is a metal plate insulated from the channel by a very thin layer of silicon dioxide (glass). In Figure 9-2a the gate is in the zero-bias condition and electrons flow freely through the N-type silicon channel. In Figure 9-2b several volts of negative bias voltage are applied to the gate, driving the free electrons into the P-type substrate. The electrons are captured by holes in the substrate, making them unavailable for channel conduction. The source-to-drain current is cut off.

Between these two extremes, full conduction and cut-off, is a range where some electrons remain in the channel and others are driven into the substrate by the gate bias. When there are many electrons in the channel, a relatively high source-to-drain current flows. A higher negative gate voltage drives more electrons into the substrate, reducing the channel current proportionately. The partially conducting channel condition is shown in Figure 9-2c. The amplifying figure of merit for the MOS-FET also is called *mutual conductance* (G_m).

$$G_m = \frac{\Delta \text{ Drain current}}{\Delta \text{ Gate voltage}}$$

The range is about 100 to 6000 micromhos.

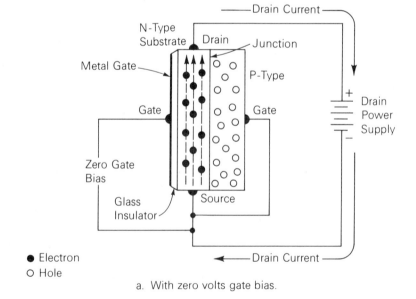

a. With zero volts gate bias.

FIGURE 9-2. The MOS-FET.

ELECTRONIC AMPLIFYING DEVICES AND CIRCUITS   303

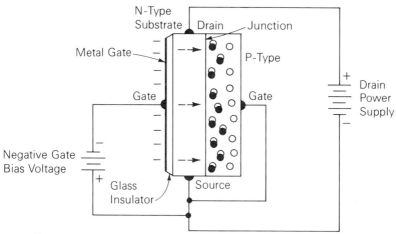

b. MOS-FET biased to channel cutoff.

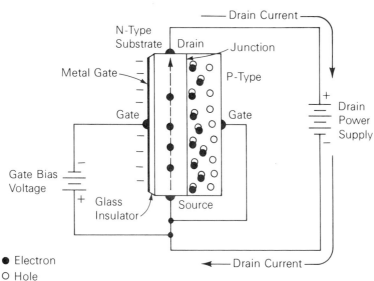

c. MOS-FET with partially conducting channel.

FIGURE 9-2. (continued)

THE ENHANCEMENT-MODE MOS-FET

There is another kind of MOS-FET that is the mainstay of digital integrated circuits. The enhancement-mode MOS-FET is normally off and requires a gate voltage to turn it on. Figure 9-3 shows how it is constructed. Unlike depletion-mode devices, the enhancement MOS-FET has no built-in channel.

Electrons are drawn up from the P-type substrate by a positive charge on the gate to form an electron bridge from source to drain.

One of the characteristics of enhancement-mode MOS-FETs that makes it particularly useful in digital circuitry is its critical turn-on voltage. This voltage can be controlled during manufacture over a range of less than 1 volt to about 5 volts. This threshold voltage helps keep electrical noise from accidentally turning the MOS-FET on.

9-4 VACUUM TUBES

The Bare Facts

1. The vacuum tube produces a stream of electrons by boiling them off a metal or oxide element called a *cathode*.

2. The plate in a vacuum tube collects the electrons that manage to travel the distance from the cathode.

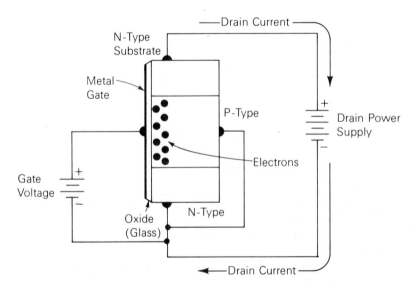

FIGURE 9-3. Enhancement MOS-FET.

3. The grid is the control element and prevents some or many electrons from reaching the plate, depending on the amount of negative voltage applied to it.

The vacuum tube was king of the electronics world until the introduction of the transistor in the early 1950s. It is still found in older equipment and in some special applications in newer equipment. The simplest version of the vacuum tube is the diode, which provides no amplification but serves as a one-way valve for electrons. The triode tube is a modification of the vacuum diode and is capable of amplification.

Figure 9-4 shows the elements of the vacuum diode. The *heater* (also called the *filament*) heats an oxide-coated cathode that literally boils off a copious supply of electrons. The plate serves to scavenge the electrons as shown in Figure 9-5.

THE DIODE VACUUM TUBE

The diode is a 2 element device. The heater is considered to be a part of the cathode assembly. The heater is always there, but is generally not indicated in schematic drawings as a part of the basic assembly. The filament (heater) circuit is normally shown separately to avoid complicating the drawing and making it difficult to follow. We will therefore omit the heater from future drawings.

In Figure 9-4, electrons are simply being boiled off the cathode. Nothing else is happening. In Figure 9-5 the plate is made positive (+) with respect to the cathode. The negatively charged electrons are drawn to the positive plate. As electrons leave the cathode, the

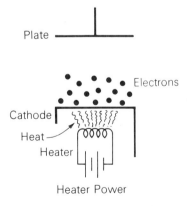

FIGURE 9-4. The vacuum-tube diode.

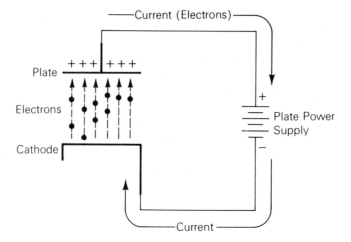

FIGURE 9-5. The conducting vacuum diode.

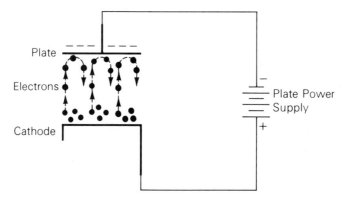

FIGURE 9-6. The nonconducting vacuum diode.

cathode becomes deficient in electrons and draws electrons from the plate power-supply battery to replenish the cathode supply. Electrons scavenged by the plate are returned to the positive plate of the battery. This is the conducting direction of the tube.

When the plate is negative with respect to the cathode, it repels electrons and no current flows. The plate is *cold*, and produces no electrons for current in the reverse direction. The nonconducting condition is shown in Figure 9-6.

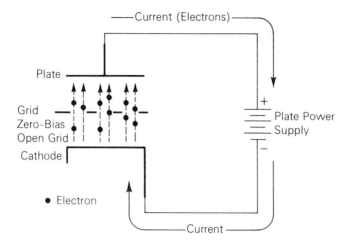

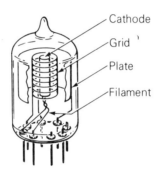

FIGURE 9-7. The triode tube with "open" grid.

THE TRIODE VACUUM TUBE

The diode can be made to amplify by adding a third element called the *grid*. The grid is an element that controls the amount of current reaching the plate from the cathode. The grid is the equivalent of the J-FET gate. Figure 9-7 shows a schematic of the triode and a cutaway drawing of a commercial triode tube. In this figure there is no potential applied to the grid, so the electrons pass through it unopposed. The tube is in a full conducting condition with zero volts on the grid. It behaves just like a diode.

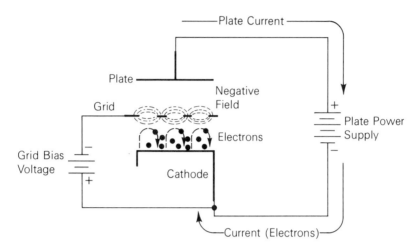

FIGURE 9-8. The triode biased to cut-off.

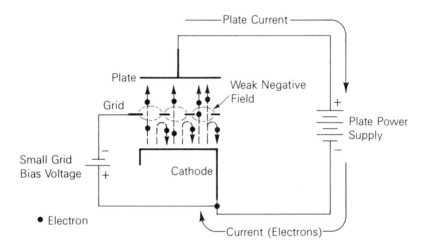

FIGURE 9-9. The triode partially turned on.

The voltage applied to the grid is called a *bias* voltage. A high negative bias voltage will stop all plate current flow, a condition (shown in Figure 9-8) known as *cut-off*. The strong negative field on the grid repels electrons back toward the cathode, preventing any electrons from reaching the plate and cutting off the plate current. Cut-off in a tube is the equivalent of pinch-off in a J-FET.

When the negative field is somewhere between zero and strong enough to cut off the plate current, some electrons (the faster-moving ones) get through the grid. The grid voltage can be varied between these extremes, providing a valve-like action where plate current is proportional (inversely) to the negative bias voltage. Figure 9-9 illustrates this partially on condition.

Study Problems

1. Explain how the control mechanism in a J-FET operates.
2. How is channel conductivity control accomplished in a MOS-FET?
3. Explain how the "channel" current is controlled in a vacuum tube.
4. What is cut-off?
5. What is pinch-off?
6. What is the function of the substrate in a MOS-FET?
7. Can you find the capacitor in a MOS-FET, a J-FET, and a vacuum tube? Explain.

9-5 THE BIPOLAR (JUNCTION) TRANSISTOR

The Bare Facts

1. Unlike depletion MOS-FETs and J-FETs, the bipolar "channel" is normally closed and a control voltage is required to open it.
2. MOS-FET and J-FET devices do not draw any appreciable gate current. The bipolar transistor requires base (the gate equivalent) current to make it operate.
3. The amplifier figure of merit for bipolar transistors is beta (β) = collector current divided by base current.
4. In the case of bipolar transistors, the following have similar functions:
   a. Gate, grid, and base
   b. Source, cathode, and emitter
   c. Drain, plate, and collector.
5. The 3 layers of the bipolar transistor may be arranged as either NPN or PNP.
6. The base region is very thin and lightly doped.
7. Emitter and collector are much larger and more heavily doped.
8. In normal operation, the base-emitter junction is forward-biased and the collector-base junction is reverse-biased.
9. The collector current is always beta times the base current.
10. Typical values of beta range from 50 to 250.

11. Because of temperature sensitivity and broad manufacturing tolerances, transistor specifications (parameters) are far from reliable.
12. Temperature variations and manufacturing tolerances are easily corrected by using negative feedback.
13. Voltage gain is achieved by adding collector load resistance.
14. Bias is a voltage or current that establishes an idling (quiescent) source-drain or emitter-collector current.
15. Bias is necessary to allow both an increase and decrease in output current to accommodate AC signals.
16. The actual bias point is carefully selected by the circuit designer. Once it is selected, it must be maintained very close to that value.
17. The designer-selected bias point tends to vary with temperature and from transistor to transistor (of the same type number) because of large manufacturing tolerances.
18. All of these variations can be stabilized through the use of negative feedback.
19. In addition to stabilizing the quiescent operating point, negative feedback also controls voltage gain and input resistance.
20. A capacitor can be used to eliminate signal feedback without affecting the DC feedback required for stability.
21. A transistor circuit must have a stability-factor value of 10 or lower if the circuit is to be acceptably stable.

The control mechanisms of the devices we have examined so far all work in a similar way. Each device (except the enhancement MOSFET) conducts fully with no voltage applied to the control element (gate or grid). Each controls the current (drain or plate) with voltage that repels electrons to control channel current flow. The amplifying figure of merit for each device is its mutual conductance (G_m).

Field-effect transistors

$$G_m = \frac{\text{Drain current}}{\text{Gate voltage}} = \frac{\Delta I_D}{\Delta E_g}$$

Vacuum tubes

$$G_m = \frac{\text{Plate current}}{\text{Grid voltage}} = \frac{\Delta I_p}{\Delta E_g}$$

The bipolar or junction transistor uses a current-control mechanism that is quite different from devices we have already studied. The bipolar transistor was discovered accidentally in 1948. Only after Doctors Shockley, Brattain, and Bardeen found they had a functional device on their hands at Bell Telephone Laboratories was this new mechanism explained. Although the control mechanism is different, the bipolar transistor shares fundamental amplifier properties with FETs and vacuum tubes.

TRANSISTOR CONSTRUCTION

The transistor consists of 2 junctions in a continuous crystal. The configuration may be PNP or NPN (see Figure 9-10). A contact is made to each crystal area, making it a 3-terminal device. One end block is called the emitter, the middle block is called the base, and the other end block is called the collector. The base region of the crystal is very thin, 1 micron or less. Twenty-five microns equal about 0.001 inch. This thin base region is essential to normal transistor operation, as we shall see shortly.

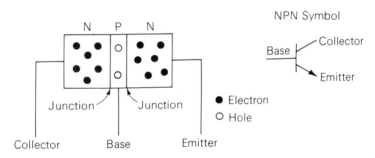

a. Symbolic representation.

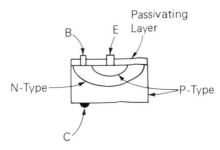

b. Actual construction.

FIGURE 9-10. The structure of the transistor.

TRANSISTOR AMPLIFICATION

For the following discussion, refer to Figure 9-11 and keep the following facts in mind:

1. The base region is very thin and has few available holes (in the example the base region is P-type).
2. The base is lightly doped, reducing the number of available holes still further.
3. Assume that the base-emitter bias is less than the junction voltage of the base-emitter junction. The base-emitter junction is therefore reverse-biased, and there is no base-emitter current.
4. The collector-base junction is reverse-biased by, let's say, 10 volts. Thus, there is no collector-base current flowing. There is also no emitter-to-collector current flowing because both junctions are reverse-biased.

So far the transistor has been in the off condition. Now let us bring the emitter-base bias voltage up above the junction voltage (0.6 volt), forward-biasing the junction and starting base-emitter current flow. A great many electrons are drawn through the emitter block toward

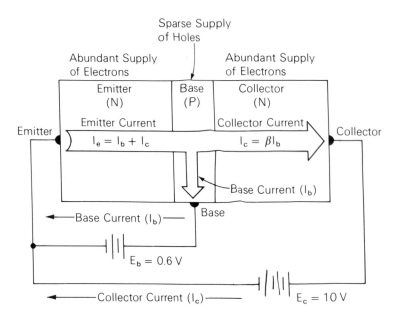

FIGURE 9-11. Transistor amplification.

ELECTRONIC AMPLIFYING DEVICES AND CIRCUITS   313

the junction. These electrons arrive at the junction to find holes in the base region, combine with them, and result in base current as in any forward-biased junction. When electrons are drawn into the base region, there are only enough holes in the base to capture one percent of them or less. The rest of the electrons cannot go back to the emitter because the field is pushing them away. These uncombined electrons gather near the collector-base junction, where they are attracted by the positive field of the collector battery. Uncaptured electrons in the base are drawn into the collector region, causing an emitter-to-collector current flow.

Electrons leaving the emitter block take two paths. One percent or less combine with holes in the base to form base-emitter current, while most of them move through the collector block to form an emitter-collector current. The large emitter-collector current is dependent on the flow of the much smaller emitter-base current. If the base-emitter voltage is increased slightly, more electrons are drawn into the base region—far more than the base can use. The result is a small increase in base current and a much larger increase in collector current.

REVERSE-POLARITY DEVICES

J-FETs, MOS-FETs, and bipolar transisitors are all available for reverse-polarity operation. These devices operate on the same principle as their opposite-polarity counterpart, but all polarities are reversed. The channel in J-FETs is changed to a P-type silicon and the gate becomes N-type. In MOS-FETs the substrate is changed to an N-type while the channel becomes P-type. The bipolar transistor is changed to a PNP configuration. The symbols for these devices and their complements are given in Figure 9-12. There is no reverse-polarity version of the vacuum tube.

BIPOLAR-TRANSISTOR PARAMETERS (SPECIFICATIONS)

Here we will take a brief look at device parameters. Manufacturer-specified device parameters are not very reliable, but with enough negative feedback, it doesn't matter whether they are or not.

In addition to the transistor's figure of merit (β, or beta) there are several other important parameters.

Currents

I_c: The collector current (*Note:* The collector and emitter currents are approximately equal.)
I_b: The base current
I_e: The emitter current ($I_e = I_c = \beta I_b$)

314 CHAPTER 9

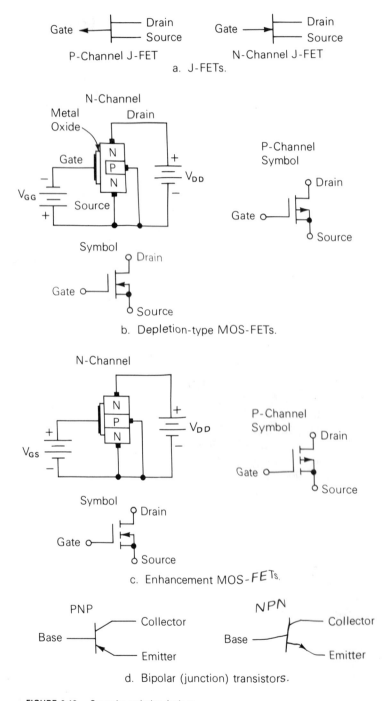

FIGURE 9-12. Opposite-polarity devices.

I_c(max): The manufacturer's absolute maximum steady-state collector current. Normal operation is at less than 40 percent of this figure.

E_c(max): The manufacturer's absolute maximum rating for the reverse-bias voltage across the collector-base junction. Even an instantaneous overvoltage can destroy the transistor.

Current gains

Beta (β): The current gain of the transistor. Range: 10 to 200.

$$\beta = \frac{I_c}{I_b}, \text{ a simple ratio, no dimension}$$

I_g: The current gain of the complete circuit. (Note that I_g is always less than β.)

Junction resistances

a. *Base-emitter junction*

R'_e: The forward-biased base-emitter junction resistance.

$$R'_e = \frac{25 \text{ (millivolts)}}{I_c \text{ (in milliamps)}} \text{ The Shockley relationship.}$$

R'_e is in ohms. Range: 100 ohms to 5000 ohms.

The Shockley relationship is the result of a great many measurements, and is no more than a fair approximation for any particular case. However, it is the best we can do in predicting the base-emitter junction resistance. This resistance is important because it is a built-in negative-feedback resistor and will play a critical role in determining circuit parameters in many practical circuits. In spite of the unreliable nature of this parameter, it is probably the only important one in practical circuits (except for absolute maximum ratings).

b. *Collector-base junction.* The collector-base junction is always reverse-biased and always has a resistance of 1 to 2 megohms.

Leakage current

I_{co}: The reverse-bias leakage current of the collector-base junction.

Because the leakage current results from thermally produced carriers, the temperature is normally stated as a part of the parameter. The emitter is not connected. The symbol I_{cbo} is also used for I_{co}. I_{co} is of little importance in a practical circuit. Techniques used to stabilize other temperature variations more than take care of any potential leakage-current problem.

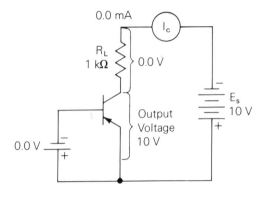

a. With zero volts input.

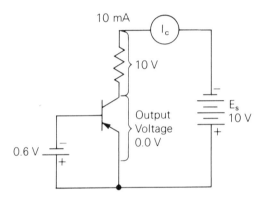

b. With 0.6 V input.

FIGURE 9-13. Voltage gain.

Voltage Gain

The transistor can be used as a voltage amplifier because of the large allowable* (and usually necessary) collector supply voltage. The base-emitter circuit requires a range of only a few tenths of a volt to vary the base-emitter junction current from zero to maximum. The collector-base junction may be capable of handling a voltage of from 10 to 50 volts (sometimes more). In Figures 9-13a and b, the transistor has a collector supply voltage of 10 volts (a typical value) and a collector load resistor of 1 kilohm. (The abbreviation for kilohm is $k\Omega$, and mA means milliamperes.)

---

*Compared to the base-emitter voltage.

In Figure 9-13a there is a base-emitter bias voltage of zero volts. This means that there is no base-emitter current, and therefore no collector current. Under these conditions there is no voltage drop across the collector load resistor (R_L). (Assume a second fixed bias voltage that is within a millivolt of overcoming the barrier potential.)

$$\boxed{E = IR} \qquad E = 1{,}000 \text{ ohms} \times 0.010 \text{ amp} = 10 \text{ volts}$$

If the entire supply voltage is dropped across the collector load resistor, the voltage drop across the transistor, and consequently the output voltage, must be zero volts. The voltage gain is equal to the change in output voltage divided by the change in input voltage.

$$\boxed{\text{Voltage gain} = \frac{\Delta \text{ output voltage}}{\Delta \text{ input voltage}} = \frac{\Delta E_o}{\Delta E_{in}}}$$

In this example the change (Δ) in input voltage is 0.6 ΔE_{in} = 0.6 volt. The voltage gain of the circuit is:

$$V_g = \frac{\Delta E_o}{\Delta E_{in}} = \frac{10 \text{ volts}}{0.6 \text{ volt}} = 16.66 \text{ volts}$$

Practical voltage gains range from 2 to several hundred, depending on the circuit design. Figure 9-14 shows some typical bipolar transistor package styles.

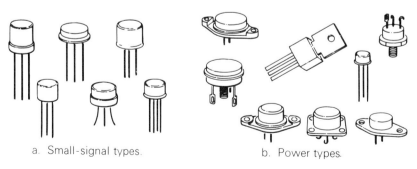

a. Small-signal types.   b. Power types.

FIGURE 9-14. Typical transistor case styles.

Electronic devices are biased to an idling current that best suits the particular application. Figure 9-15 illustrates the need for bias. The J-FET in Figure 9-15a is conducting all of the current it can at zero bias. The positive part of the input signal cannot widen the channel to increase that current, so the positive part of the wave form is not replicated as source-drain current variations, and is lost in the output. The negative part of the input signal reduces the width of the channel and the channel current in proportion to its voltage. This part of the input signal appears as output signal.

In Figure 9-15b, the bias squeezes the channel somewhat to allow the signal to expand the channel on the positive half-cycle of the signal. The bias does not close the channel completely, or the negative half of the signal would be lost in the output. By proper selection of the bias voltage, which controls the channel width, the signal can be made to expand and contract the channel in a symmetrical fashion.

Once the bias point has been selected, the quiescent (no-signal) collector (or drain) current must stay fixed at that value. The selection of the proper bias point is a fairly complex problem because none of a transistor's principal characteristics is constant. For example, the mutual conductance in FETs and beta in bipolar transistors both vary widely with different bias levels. If a circuit is to operate as the designer intended, the bias level must stay constant in spite of the effect of temperature variations on the operating (quiescent) current, and in spite of very broad manufacturing tolerances in transistors. These tolerances could be tightened, but only at extreme cost. There is really no need for tighter tolerances because the techniques used to stabilize against temperature variations automatically make the circuits tolerant of wide variations in transistor parameters.

Negative feedback is the principal tool used to provide bias stability (quiescent-point stability) and to compensate for large manufacturing tolerances. Negative feedback is added to the transistor circuit by adding external feedback resistors.

Table 9-1 compares the characteristics of the various amplifier devices.

PRACTICAL BIPOLAR TRANSISTOR CIRCUITS

Any practical transistor circuit requires proper bias and negative feedback to stabilize the operating point established by that bias. As soon as enough negative feedback is added to a transistor circuit, the feedback controls nearly all of the circuit characteristics. The characteristics of the devices themselves become of little importance, as long as they meet certain minimum requirements. In particular, the

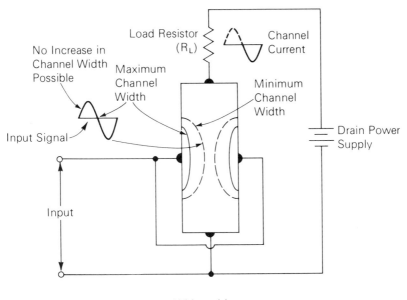

a. Without bias.

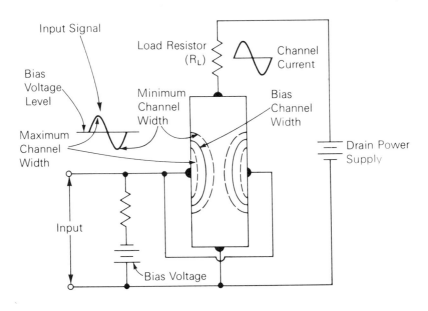

b. With bias.

FIGURE 9-15. J-FET with and without bias.

TABLE 9-1 AMPLIFIER DEVICE SUMMARY

Device	J-FET	MOS-FET	Vacuum Tube	Bipolar Transistor
Control element	Gate	Gate	Grid	Base
Direction of controlled-current flow	From source to drain	From source to drain	From cathode to plate	From emitter to collector
Figure of merit	Mutual conductance $G_m = \dfrac{\Delta \text{ drain current}}{\Delta \text{ gate voltage}}$	Mutual conductance $G_m = \dfrac{\Delta \text{ drain current}}{\Delta \text{ gate voltage}}$	Mutual conductance $G_m = \dfrac{\Delta \text{ plate current}}{\Delta \text{ grid voltage}}$	Beta $\beta = \dfrac{\text{Collector current}}{\text{Base current}}$
Units for figure of merit	Micromho	Micromho	Micromho	No unit. Simple ratio.

device must have a high enough figure of merit (mutual conductance, or beta).

Suppose we look at typical practical bipolar circuits, their characteristics, and what each circuit component does. There are 3 bias/stability circuits in common use with bipolar transistors:

1. Base bias with collector feedback
2. Base bias with emitter feedback
3. Emitter bias with emitter feedback

The circuit in Figure 9-16a is a transistor amplifier with base bias and collector feedback. The circuit is capable of voltage gains of up to 200. A fraction of the output signal is fed back to the input through the feedback resistor $R_{b1}a\ R_{b1}b$. Bias current is also supplied through the feedback resistor, but if the feedback resistor has a small enough value for adequate stability, it will deliver far too much bias current. The bias current is actually controlled by R_{b2}, which bleeds off the excess current. R_{b2} is called the *bias* resistor because it sets the transistor operating point by determining how much base-emitter bias current flows.

Stability Factor

The stability factor is directly related to the amount of feedback. The stability factor for any bipolar transistor circuit must be 10 or smaller for the circuit to be stable enough to be practical. The stability for the circuit in Figure 9-16a is:

ELECTRONIC AMPLIFYING DEVICES AND CIRCUITS 321

$$S = \frac{R_{b1}a + R_{b1}b}{R_L}$$

Example:

Let $R_L = 5{,}000$ ohms and $R_{b1}a + R_{b1}b = 50{,}000$ ohms.

Is the circuit stable enough to be practical?

$$S = \frac{50{,}000}{5{,}000} = 10$$

The answer is yes, the circuit is stable enough to be practical. This means that the circuit will not vary too much with temperature and that transistor replacement will not be a problem. The circuit should also be fairly tolerant of transistors that are not exact replacements.

Input Resistance and Current Gain

The feedback resistors $R_{b1}a$ and $R_{b1}b$ lower the input resistance to a very low value, which reduces the circuit current gain to the value of S.

$$\text{Current gain} = S = \frac{R_{b1}}{R_L}$$

The price of stability in most bipolar transistor circuits is drastically reduced current gain. This is one important reason for the emphasis on voltage gain. Stability must be had at any cost, and reduced current gain is that cost. In this circuit the current gain is reduced by feedback. In other circuits, other means must be used. The fact that the feedback drastically lowers the input resistance makes the circuit unattractive unless signal feedback can be eliminated without eliminating the DC feedback necessary for stability.

To raise the signal input resistance (impedance), the feedback resistor is split in two and a large capacitor (C_{bp}) is used to bleed off the AC signal feedback. The capacitive reactance to the AC signal is very low and the signal feedback takes a shortcut through it, back to the negative terminal of the battery, and does not reach the input. The DC feedback is not affected by the capacitor.

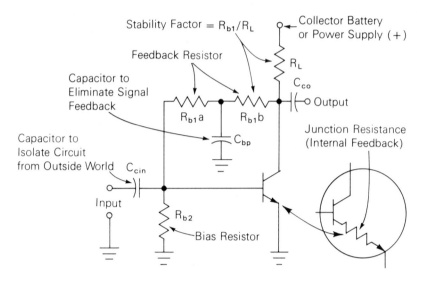

a. Base bias with collector feedback.

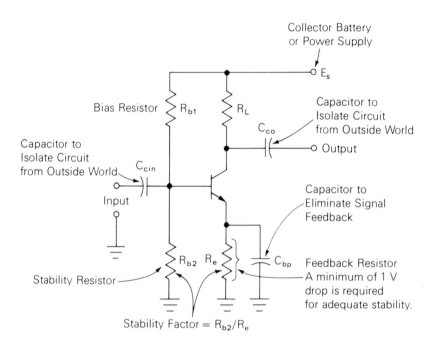

b. Base bias with emitter feedback.

FIGURE 9-16. Practical bipolar transistor-amplifier circuits.

ELECTRONIC AMPLIFYING DEVICES AND CIRCUITS 323

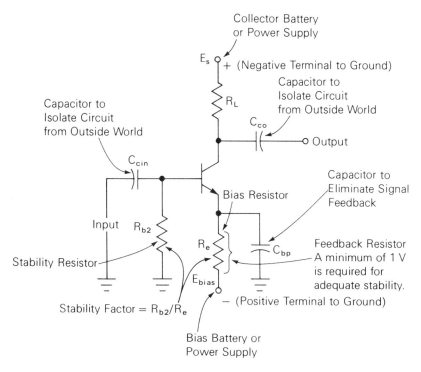

c. Emitter bias with emitter feedback.

FIGURE 9-16. (continued)

Voltage Gain

The voltage gain in any bipolar transistor circuit is controlled by the emitter feedback. In this case there is no external emitter feedback resistor, but there is always some internal emitter feedback due to the emitter junction resistance (R'_e). This resistance is typically 25 to 50 ohms. The voltage gain is always the ratio of R_L to the emitter feedback resistance. If we assume a junction (feedback) resistance of 50 ohms for the circuit in Figure 9-16a, the voltage gain is:

$$\text{Voltage gain} = \frac{R_L}{R'_e} = \frac{5000 \text{ ohms}}{50 \text{ ohms}}$$

$$\text{Voltage gain} = 100$$

The input capacitor C_{cin} prevents any device that might be connected to the input from upsetting the bias. The output capacitor C_{co} isolates the transistor's collector circuit from the outside world for direct current.

Base Bias with Emitter Feedback

In Figure 9-16b an external emitter resistor is added to provide the negative feedback. This resistor is always much larger than the internal emitter junction resistance, typically 1000 to 2000 ohms. The feedback resistor (R_e) must have at least a 1 volt drop across it to ensure adequate feedback.

The stability factor is:

$$S = \frac{R_{b2}}{R_e}$$

And again S must have a value of 10 or less for a practical circuit.

The bias current is derived directly from the collector supply battery through R_{b1}. R_{b2} is the stability resistor.

The voltage gain in this circuit is:

$$\text{Voltage gain} = \frac{R_L}{R_e}$$

The capacitor C_{bp} is added in most real circuits to eliminate the emitter feedback for the signal. This increases the AC voltage gain without upsetting the DC stability. The maximum voltage gain with the capacitor omitted is less than 10, but the voltage gain can be increased to 100 or more by putting the capacitor in.

Emitter Bias with Emitter Feedback

The circuit in Figure 9-16c is identical to that of Figure 9-16b, except that the bias resistor R_{b1} has been removed and bias current is supplied through the feedback resistor by a separate battery or power supply. Voltage-gain and bias-stability rules are the same as those for Figure 9-16b.

Study Problems

1. If we distinguish between J-FET and bipolar transistors by calling one an *enhancement-mode* device and the other a *depletion-mode* device, which would be called what? Explain. (*Hint:* What is the "channel" conduction with zero volts bias?)

2. Match the items in the following two columns (more than one answer is possible):
   1. Base   a. Grid
   2. Collector   b. Source
   3. Emitter   c. Plate
                    d. Cathode
                    e. Drain
                    f. Gate
3. What is the bipolar-transistor figure of merit called? Define it.
4. Make a statement about base thickness and doping level.
5. For normal transistor operation, match the items in the following columns:
   1. Base-emitter junction   a. Forward biased
   2. Collector-base junction b. Zero biased
                               c. Reverse biased
6. A certain transistor has a beta of 50. One milliamp of base current is flowing. How much collector current is flowing?
7. Why do manufacturers make very little effort to tighten the tolerances in transistor parameters?
8. What technique is used to control quiescent-operating-point variations due to temperature?
9. Define bias.
10. Define quiescent operating point.
11. Why is bias necessary? When?
12. What controls voltage gain and input resistance?
13. Why is feedback often eliminated for the signal, but maintained for DC stability? (two cases)
14. What is the maximum allowable stability-factor number?

9-6 THE PRACTICAL J-FET AMPLIFIER

Figure 9-17 shows a practical J-FET amplifier stage. R_s provides negative feedback for DC stability and C_s bypasses signal feedback. The voltage gain of the circuit with capacitor C_s connected is: $V_g = G_m R_L$ where G_m is the figure of merit for the 2N3086. (Remember that MΩ is the abbreviation for megohm.)

9-7 THE DIFFERENTIAL AMPLIFIER

The differential amplifier is the nearest thing to a universal amplifier stage yet built. It is the amplifier type used in operational amplifiers. The differential amplifier is the most stable and versatile of available circuits. It features unusually good stability and good voltage gain from DC to several megahertz. There are 2 available inputs; one is

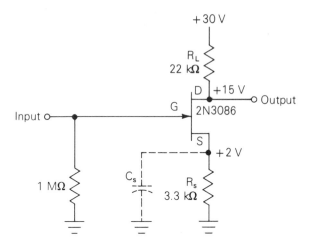

FIGURE 9-17. J-FET amplifier.

inverted 180° at the output, and one is without inversion. The basic circuit is shown in Figure 9-18.

The circuit in Figure 9-18 has 2 inputs—one 180° inverting input, and one noninverting input. The 2 inputs can be used separately with the unused input grounded, or both inputs can be used to amplify the algebraic difference between the 2 signal voltages. (Therefore the name *differential amplifier*.) There are also 2 possible outputs, either of which can be used separately or in special cases, both at the same time. Because of the balanced circuit, temperature drift is canceled in the emitter resistor.

Transistors must be matched very closely if the circuit is to live up to the performance expected of it, but this is easily done in an integrated circuit. Several stages of the differential-amplifier circuit are cascaded to get the desired voltage gain in an operational amplifier. The total gain in any string of amplifier stages is equal to the product of the voltage gains of all the individual stages.

9-8 SWITCHING AMPLIFIERS

The Bare Facts

1. Switching amplifiers are on-or-off devices with no intermediate levels.

2. Switching amplifiers provide gain like any other amplifier.

3. The thyristor family is the most important group of moderate-power to very-high-power switching amplifiers.

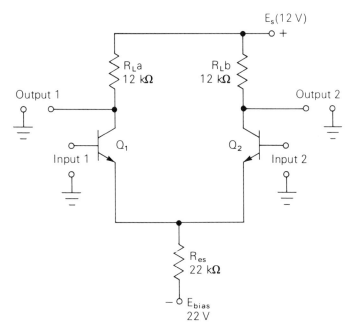

FIGURE 9-18. The differential amplifier.

4. Switching amplifiers can be used to simply turn things on or off, or to provide a good approximation of linear (analog) operation in certain cases.

5. The thyristor family is a group of 4-layer junction diodes.

6. The chief member of the family is the thyristor itself.

7. The thyristor is also called a *silicon-controlled rectifier* (SCR).

8. SCRs are available with ratings of 50 to 1200 volts, and 0.5 to 10,000 amps.

9. The triac is a pair of integrated back-to-back SCR devices in a single case for use with 60 hertz AC.

10. The SCR and the triac are true switching amplifiers featuring large power gains.

11. The Shockley diode is a low-power thyristor-family device used to turn on SCRs.

12. There is also a bidirectional version of the Shockley diode for turning on triacs.

13. Shockley-type diodes are switches, but they do not provide amplification.

14. The unijunction transistor (UJT) is a special device that is often used to turn on thyristors.

15. The unijunction transistor is a switching amplifier.

16. The UJT is *not* a member of the thyristor family.

In this section we will examine a group of amplifying devices that operate in a strictly on-or-off mode. These devices are different from the kinds of digital devices used in computers and logic systems. All but one of the devices in this section are members of the thyristor family and are used where a moderate to very large amount of power is controlled. They are used in lamp dimmers, motor speed-control systems, and a variety of other power-control applications.

THYRISTORS

The thyristor* family is a group of 4-layer PNPN silicon-junction semiconductor devices. They are used to switch high power levels without mechanical switching and its problems of contact wear and arcing. There are also lower-power-level thyristor devices used for triggering the higher-power thyristors and for timing and similar applications. These 4-layer devices operate in a strictly on-or-off mode, and cannot be used for linear (analog) operation in the usual sense.

However, they can *simulate* linear (analog) control in many industrial and consumer control applications. The household light dimmer is an example. As we turn the knob from dim to bright, the control *appears* to be continuous, and from the observer's point of view it could well be analog (if not always 100 percent linear). But the dimmer is distinctly digital, in the sense that the power delivered to the lamp is *always* either fully on or fully off, with no intermediate values.

The internal-combustion engine is a rough analog of the situation. Although the engine rotates at what seems to be a constant speed (RPM) it is actually turned by a series of separate (almost digital) explosions in the cylinders. A heavy flywheel is attached to the engine crankshaft to smooth out the firing pulses. Because of the mechanical (flywheel) inertia, the crankshaft cannot stop or start instantly. The individual cylinder explosions are "averaged" into a fairly constant engine speed. Lamp filaments take time to cool down and electric motors take time to change speed. This makes them good candidates for the thyristor on-off kind of control to vary brightness or speed.

---

* The thyristor is also known as a silicon-controlled rectifier.

It might be well to point out that the power-line frequency of 60 hertz was chosen partly because the human eye does not respond to changes in brightness that occur that rapidly. Some early power lines delivered AC power at 25 or 50 hertz. In both cases a lamp seemed to flicker as the alternating current went through its cycle. The flicker in a 50 hertz system is annoying but tolerable. At 25 hertz the flicker is downright irritating.

The ratio of on time to off time in a switching amplifier is called the *duty cycle*. By varying the duty cycle, motor speed, lamp brightness, oven temperatures, spot-weld heat, and so on, can be controlled in what seems to be a continuous analog fashion. The thyristor has a chance to cool down during the off time. Thyristors are also excellent replacements for mechanical relays and high-power switches. The thyristor is quite inexpensive, very efficient, and has a high power gain. The high power gain allows the SCR to be controlled by small signals from transducers, photocells, and low-power transistor control circuits.

The thyristor, like ordinary junction diodes, is a one-way electronic valve. Current can flow through it in only one direction. Two or more thyristors can be connected to control both halves of an AC sine wave; however, a bidirectional thyristor called a *triac* is available for this purpose. In the following pages we will examine the most important members of the thyristor family.

Theory of Operation

The crucial point to remember about the thyristor family is that they all operate in an avalanche mode. The avalanche conduction mode is initiated by leakage carriers within the depletion zone of a reverse-biased semiconductor junction. Before we discuss the avalanche mechanism, let's review reverse-bias leakage. Remember that thermally generated electrons always leave a "hole" behind. In semiconductor terminology a thermally freed electron and the hole it leaves behind are called an *electron-hole pair*.

Figure 9-19 shows a reverse-biased diode with electron-hole pairs being thermally generated within the depletion zone. The hole and the electron labeled A are forward-biased and represent leakage current. This leakage current is the only current in the reverse-bias direction, and is so small that it can be ignored in ordinary diodes. However, in thyristors, where there is considerable current gain, leakage currents become important. Let's take a look at the reverse-bias part of the junction-diode conduction curve in Figure 9-20. Notice that increasing the reverse bias voltage increases the reverse current only slightly until a critical voltage called the *zener* point is reached. At

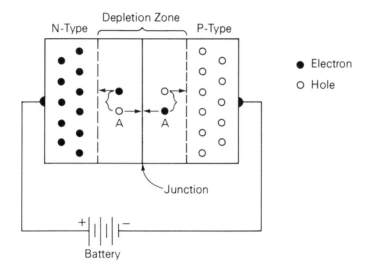

FIGURE 9-19. Reverse-bias leakage current.

the zener knee of the curve the current suddenly rises from microamperes or nanoamperes to milliamperes or amperes. There are two principal mechanisms involved in production of this current—field emission, and ionization by collision.

As we increase the voltage in the reverse-bias direction, the depletion zone becomes wider and wider. At some critical point the field is large enough that an electron can gain enough momentum to strip electrons from atoms it collides with. The collision-dislodged electrons pick up speed until they collide with other atoms, freeing still more electrons. Figure 9-21 is a sketch of this process, which is called an *avalanche*. Unless it is controlled, it will continue building up like an avalanche until the device is destroyed.

Members of the thyristor family are designed to operate in a limited avalanche mode. Operation in a limited avalanche mode is possible as long as the junction temperature is held below a certain level. To hold the junction temperature down, the current through the device is restricted by placing a current-limiting resistance in the circuit and by removing the heat that is generated. The current-limiting resistance is normally the resistance of the device being controlled by the thyristor.

The Four-Layer Diode

The simplest member of the thyristor family is the 4-layer diode. The rest of the family consists of modifications of the 4-layer diode. The 4-layer diode is generally a low-power device used to turn on

ELECTRONIC AMPLIFYING DEVICES AND CIRCUITS   331

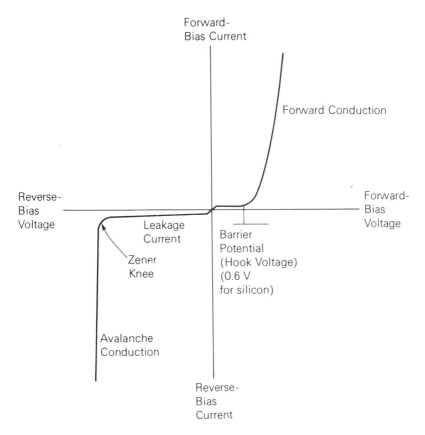

FIGURE 9-20. Silicon-diode conduction curve.

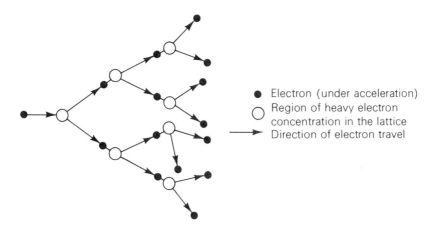

FIGURE 9-21. The avalanche mechanism.

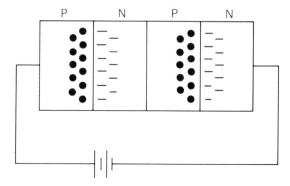

a. Forward-bias condition.

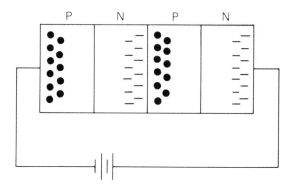

b. Reverse-bias condition.

FIGURE 9-22. The four-layer diode.

higher-power thyristors. In its low-power form it is often called the *Shockley* diode, after its inventor. Figure 9-22 shows the 4-layer diode construction and its forward- and reverse-bias conditions.

Figure 9-22a shows the forward-biased condition. The 2 outer junctions are forward-biased and the center junction is reverse-biased. If we assume that the zener-knee voltage for each junction is 100 volts, we would have to increase the battery voltage to 100 volts to drive the center junction over the zener knee, and into avalanche before conduction can take place.

In Figure 9-22b both outer junctions are reverse-biased, while the center junction is forward-biased. If each of the outer junctions requires 100 volts to drive it over the zener knee, we would have to increase the battery voltage to 200 volts before avalanche conduction could begin in the 2 outer junctions. Six-tenths of a volt is enough to make the center, forward-biased junction begin conducting. To start

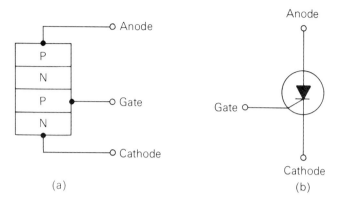

FIGURE 9-23. The SCR and its symbol.

the 4-layer diode conducting would require slightly more than 100 volts in the forward-biased direction and slightly more than 200 volts in the reverse-biased direction. Below 100 volts in the forward-bias direction, and 200 volts in the reverse direction, the device is in a nonconducting state.

Once the avalanche is initiated it will be self-sustaining, with device current limited only by the load resistance. Current will continue to flow until the circuit is opened or until the supply voltage is reduced to near zero volts. The avalanche will continue, once started, until the number of carriers (current) is reduced below some critical point where the avalanche process dies out. This critical current is called the *holding* current.

The difference between the turn-on voltage and the turn-off voltage is called *hysteresis*. Hysteresis is an important characteristic in all 4-layer devices. Ordinary diodes (and zener diodes) in the avalanche mode have negligible hysteresis.

The Silicon-Controlled Rectifier

The silicon-controlled rectifier (SCR) is the chief member of the thyristor family and is the only member that uses the name *thyristor*.

The SCR has a built-in triggering gate that provides a convenient way of turning the 4-layer diode on. Once the SCR has been turned on, the gate cannot turn the device off. To turn the SCR off, the anode-cathode current must be reduced to a value below the holding current.

Figure 9-23 shows the SCR structure and its symbol.

Figure 9-24 shows some typical SCR case styles.

As the name *avalanche* implies, the turn-on mechanism in an SCR is an unstable one, and can easily be started. In the case of the SCR, a

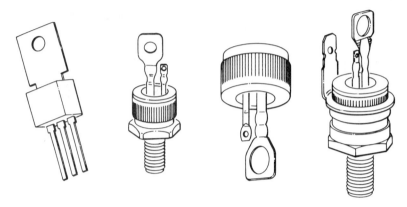

FIGURE 9-24. Some common SCR case styles.

positive potential on the gate draws electrons toward the gate terminal and also into the depletion zone of the center junction. These few electrons cause enough disturbance to initiate the avalanche. Once the avalanche has started, the gate loses all control and cannot significantly increase or decrease the cathode-anode conduction.

TRIACS

The triac consists of 2 back-to-back SCR devices integrated into a silicon chip. The construction is more complicated than the SCR but it provides bidirectional control. The triac can be used to control an alternating current using both halves of the AC cycle. Bidirectional control can be accomplished with SCRs, but it requires 2 of them and often involves fairly complex gate-control circuitry.

The triac is intended for AC operation and can be gated by a signal of either polarity, although gate current and voltage vary among the 4 gating conduction modes.

UNIJUNCTION TRANSISTORS

The unijunction transistor (UJT) is not a member of the thyristor family, nor is it a bipolar or FET transistor. The UJT is one of the most popular SCR triggering devices. Figure 9-25 shows the simplified structure, the symbol, and an analog of the UJT.

This device consists of a bar of N-type silicon similar to Figure 9-25b. A small PN junction is formed partway down the N-type bar. The resistance of the N-type bar forms a voltage divider that reverse-biases the junction by some fraction of the voltage between Base 2 (B_2) and Base 1 (B_1). The ratio of R_{b1} and R_{b2} in the analog (Figure 9-25c) is called the *intrinsic standoff ratio*.

ELECTRONIC-AMPLIFYING DEVICES AND CIRCUITS 335

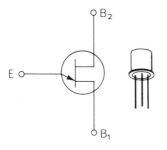
a. Symbol and typical case style.

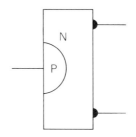
b. Simplified structure.

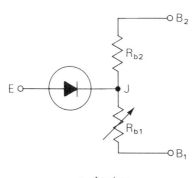
c. Analog.

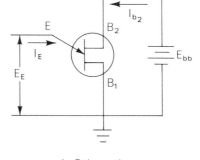
d. Schematic.

FIGURE 9-25. The unijunction transistor.

For the following explanation, examine Figures 9-25b and c. When the UJT has a voltage source connected between Base 1 and Base 2 and the emitter is at zero volts, a very small current flows. If a voltage E_E is applied between the emitter and B_1, the junction is reverse-biased until E_E is raised to a value high enough to overcome the diode reverse bias provided by the voltage divider R_{b1}, R_{b2}.

When the junction becomes forward-biased, holes are injected into the N-type bar and current flows from emitter to Base 1. This injection of carriers reduces the resistance R_{b1}. When R_{b1} is reduced, the voltage-divider ratio is altered and the diode becomes more heavily forward-biased. The process becomes regenerative.

The reduction in the resistance R_{b1} also reduces the resistance between B_1 and B_2, causing the current between Base 1 and Base 2 to increase. Because E_{bb} is comparatively high (typically, 30 volts), the interbase current is large compared to the emitter current. Because this current is larger than the emitter current, the UJT has considerable current gain. We will examine some applications of the UJT in the next chapter.

Study Problems

1. Make a sketch of the SCR structure.
2. On your sketch of the SCR structure, show the distribution of holes and electrons for (a) the forward-bias condition, and (b) the reverse-bias condition.
3. Describe the avalanche mechanism in a four-layer diode.
4. What are the two important differences between the SCR and the Shockley 4-layer diode?
5. Once a four-layer device is turned on, what is required to turn it off?
6. Can an SCR be turned off by applying a large gate pulse?
7. How does the triac differ from the SCR?

Chapter 10

Painting by Kelly Williams

AMPLIFIER SYSTEMS

10-1 INTRODUCTION

In this chapter we will examine some complete amplifier, oscillator, and radio systems. We will also look at some important system characteristics including distortion and noise problems, and bandwidth requirements. In our discussion of bandwidth we will use the term *octave*, so perhaps this place is as good as any to define the term.

Each time a frequency doubles, it has changed by 1 octave. 440 hertz is a concert A on a piano. 880 Hz would be A an octave higher. 1760 hertz would be an A 2 octaves above concert A. The piano covers a range of A_0 (27.5 hertz), to C_8 (4,186 hertz), a little over 7 octaves.

Most of our inquiry in this chapter will involve linear (analog) amplifiers. We will also look at the most important of the switching amplifiers that are becoming so popular. These switching amplifiers provide us with speed control in home and home-shop appliances, and control over the lighting in our homes. The last two chapters will cover the very special digital switching amplifiers that we associate with computers.

10-2 AMPLIFIER CLASSIFICATIONS

The Bare Facts

1. Amplifiers fall into the following categories:
   a. Audio amplifiers
   b. Radio and intermediate-frequency amplifiers
   c. Video amplifiers
   d. Switching amplifiers (digital, on-off)
   e. Instrumentation amplifiers

2. Classifications a, b, and c are called *linear* or *analog* amplifiers.

3. Switching amplifiers are also called *digital*, depending on type and application.

4. An octave is a doubling in frequency. The span from 2000 hertz to 4000 hertz is 1 octave.

5. Audio-frequency amplifiers operate in the range of 20 to 20,000 hertz (cycles per second). The audio span (bandwidth) is approximately 8 octaves.

6. Video amplifiers cover a frequency range of about 12 octaves, from 20 hertz to 5 megahertz. Some special video amplifiers have an even greater bandwidth.

7. Radio-frequency and intermediate-frequency amplifiers are *tuned* amplifiers with a bandwidth of generally less than one-half an octave. The operating center frequency can range from 25,000 hertz to 1,000,000,000 hertz or greater.

8. Instrumentation amplifiers generally must have a bandwidth that starts at zero hertz (direct current), and goes up to a frequency adequate for the particular application.

The first 3 amplifiers listed above are based roughly on the frequency at which they are intended to operate. Audio covers from about 20 to 20,000 hertz (cycles per second). Radio-frequency amplifiers generally operate above 1 megahertz (1 million cycles per second). Intermediate-frequency amplifiers operate over about the same range. The difference between radio-frequency and intermediate-frequency amplifiers is that radio-frequency (RF) amplifiers are operator-turnable, whereas intermediate-frequency (IF) amplifiers are factory-tuned to a set frequency. When you change stations on your FM tuner or change channels on your TV set you are *tuning* an RF amplifier.

More important than the actual frequency at which the first 3 amplifier classes work is their bandwidth. An audio amplifier that could faithfully reproduce all the notes on a piano would have to have a bandwidth of over 7 octaves (20 hertz to 20,000 hertz). The term *bandpass* is a synonym for *bandwidth*, and is more descriptive but not quite as common as bandwidth.

A video amplifier is used to amplify picture signals for television. In a picture, large dark areas represent the low-frequency end of the bandwidth, while the higher frequencies determine the amount of fine detail. The home television receiver has video amplifiers with a bandwidth of about 4 megahertz. Studio equipment and some industrial television systems might need up to 12 megahertz of bandwidth to reproduce enough fine picture detail.

Radio-frequency (RF) and intermediate-frequency (IF) amplifiers serve to select one narrow band of frequencies out of a large spectrum of frequencies. In TV the tuner contains an RF amplifier that has a bandpass just wide enough to let Channel 4 (for example) get through, but not Channel 3 or 5. Figure 10-1 compares the bandwidth of radio-frequency, audio-frequency, and video amplifiers. The RF bandpass curve is not to scale. If it was, it would be little more than a wide vertical line. (Also, remember that MHz is the abbreviation for megahertz.)

Switching amplifiers cannot provide a partly on condition. They turn completely on or completely off with no intermediate conduction state. They can, however, turn on and off rapidly and in many

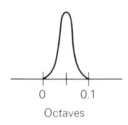

a. Radio-frequency amplifiers.

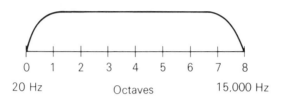

b. Audio-frequency amplifiers.

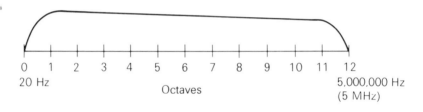

c. Video amplifiers.

FIGURE 10-1. Bandwidths.

instances can *simulate* intermediate conditions. Switching amplifiers are more efficient than conventional (linear) amplifiers when high power levels must be controlled.

AUDIO-FREQUENCY AMPLIFIERS

The Bare Facts

1. 15,000 hertz is the typical upper frequency that an adult can hear.
2. The highest frequency an audio amplifier should be able to handle is a very subjective thing.

3. Distortion consists of any sounds that appear in the loudspeaker that were not in the original music or speech.
4. Distortion changes the original wave shape in some fashion.
5. Distortion creates new vibrations that were not a part of the original music or speech.
6. The basic wave shape is a sine wave.
7. All complex wave shapes are composed of many sine waves.
8. The sine waves in a complex sound (such as that of a musical instrument) are related to a fundamental tone (or frequency) by simple multiples, 2 times the fundamental, 3 times the fundamental, and so on.
9. These related frequencies are called *harmonics* or *overtones*.
10. Any wave shape that is not a perfect sine wave contains harmonics.
11. The harmonic content is a prime factor in making each kind of musical instrument sound different from all other kinds.

Audio amplifiers are used for sound reproduction and recording. They are used in applications such as amplifying the vibrations of guitar strings and the output of organ and music synthesizer tone generators. Other applications may also come to mind.

Audio frequencies generally cover a range of from 20 hertz to 20 kilohertz, although many high-quality amplifiers cover a somewhat wider range. Many audio designers and users consider it necessary to exceed an upper frequency of 50 kilohertz. This makes it possible to clearly reproduce such subtleties as finger noise on guitar strings and bow-string noise in violins.

The average 25-year-old can seldom hear frequencies above 15 kilohertz. However, it is generally agreed that an amplifier with a frequency capability of 20 kilohertz or more sounds better than one with a 15 kilohertz limit. How much better and in what ways it is better are still subjects of debate. In the early 1950s, listening tests showed that listeners did not like the sound of speech or music reproduced by amplifiers with frequency responses much greater than about 18 kilohertz. The disagreement between current opinion and those tests is possibly due to the higher distortion figures in amplifiers of the time. It may also have been a case of conditioning, because one rarely had an opportunity to hear a wide-range amplifier in those days. Hearing is a very subjective phenomenon, and in spite of all the research there are a great many unknowns.

DISTORTION

Distortion in an audio amplifier is any sound that appears in the output of the system that was not contained in the original music or speech. The amplifier either alters the sound in some way or produces some electronic vibrations of its own that it adds to the original sound. Some noise is produced by any amplifier, and even though it is a form of distortion we will examine it separately because of its very special nature.

Distortion alters the sound by causing some change in the wave shape of the electrical signal as it passes through the amplifier. The basic wave shape is a sine wave. All complex wave shapes are composed of many sine waves of different (related) frequencies and amplitudes. Figure 10-2 shows some common kinds of distorted sine waves that occur in audio amplifiers.

Any change in the wave shape tends to generate new frequencies that did not exist in the original sine-wave signal. These signals are called *harmonics* and are simple multiples of the original sine-wave frequency: 1, 2, 3, 4, 5 . . . times the original frequency. A pure sine wave is the *only* wave shape that contains only a single frequency. Any other wave shape must contain the base or *fundamental* frequency along with some harmonics.

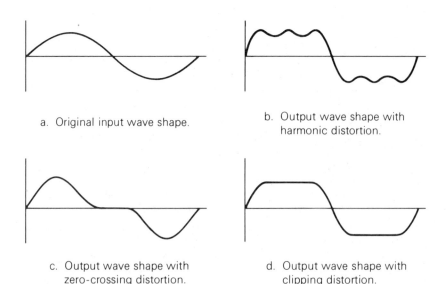

a. Original input wave shape.

b. Output wave shape with harmonic distortion.

c. Output wave shape with zero-crossing distortion.

d. Output wave shape with clipping distortion.

FIGURE 10-2. Common kinds of distortion in amplifiers.

We must look at the situation in two ways:

1. Any waves shape that is not a sine wave consists of a fundamental sine wave and a number of harmonically related sine waves.
2. Any time a sine wave is altered in shape, new harmonically related sine waves are added.

Figure 10-3 demonstrates how added harmonics alter a sine wave and the other side of the coin, the harmonic content of an altered sine wave.

The concept of wave shape and harmonics is a "which came first, the chicken or the egg?" proposition. If you will compare the resultant in Figure 10-3b with the zero-crossing distortion in Figure 10-2c, you will notice that they are alike. The amplifier, in producing zero-crossing distortion, has generated a second-harmonic signal. Zero-crossing distortion is also called *second-harmonic* distortion.

A comparison of the amplifier distortion in Figure 10-2b with the resultant wave shape in Figure 10-3e shows that the amplifier distortion is an odd-harmonic distortion. This kind of information can often tell the engineer or technician just what section of the amplifier system is producing the distortion.

A comparison of Figures 10-2d and 10-3f indicates that clipping distortion comes very close to producing a perfect square wave. An amplifier deliberately designed to produce clipping distortion is often used to convert a sine wave into a square wave for certain applications. It is largely due to the harmonic content and the consequent wave shape that each musical instrument has its own unique sound and is readily identifiable by the ear.

There are other factors too that make each musical instrument unique to the ear. These are almost as important as the harmonic content. For example, each instrument generates its own unique noises, and if they are missing, the sound is not quite right for that particular instrument. The music and speech wave shapes that we amplify are so very complex that no convenient way has yet been found for measuring the distortion an amplifier produces in music reproduction. Such measurements can be made with current technology, but the process is closer to a research project than a routine laboratory measurement. A sine wave is most frequently used for measurement purposes. Figure 10-4 shows the typical wave shape and harmonic content of a few common musical instruments.

Intermodulation Distortion

When an amplifier causes the significant distortion of a sine wave, a form of distortion called *intermodulation* distortion can occur if

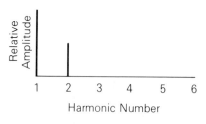

a. Harmonic spectrum diagram.

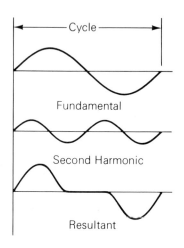

b. Result of adding the second harmonic.

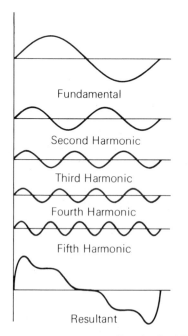

c. Result of adding additional harmonics.

FIGURE 10-3. Sine waves and complex wave shapes.

AMPLIFIER SYSTEMS 345

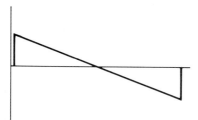

d. Result when an infinite number of harmonics are added.

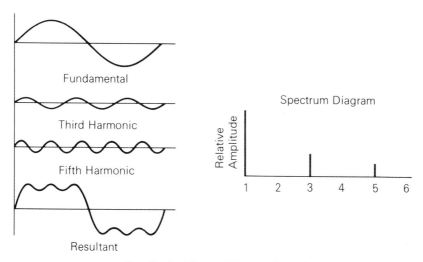

e. Result of adding odd harmonics only.

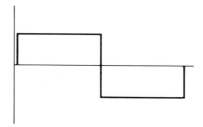

f. A perfect square wave, resulting from an infinite number of odd harmonics.

FIGURE 10-3. (continued)

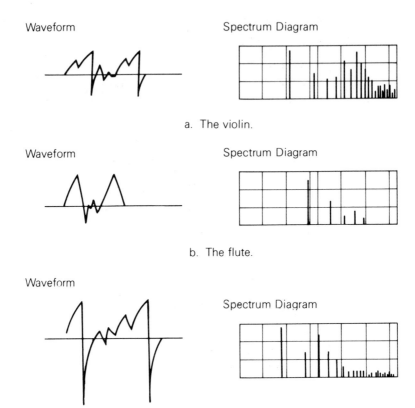

FIGURE 10-4. Wave forms and spectrum diagrams for common musical instruments.

more than 1 sine wave is present. The effect is known as heterodyning. With 2 sine waves of different frequency present in a distorting amplifier, it results in the following sine waves:

1. The 2 original sine waves at their respective frequencies.
2. A new sine wave whose frequency is the sum of the 2 original frequencies.
3. A new sine wave whose frequency is the difference between the 2 original frequencies.

In mathematical shorthand, the frequencies are:

> 1. f_1
> 2. f_2
> 3. $f_1 + f_2$
> 4. $f_1 - f_2$ (or $f_2 - f_1$, if f_2 is the higher frequency)

In audio these additional frequencies are undesirable and virtually impossible to get rid of, once they are produced, because they are nearly infinite in number. In other systems though, this same heterodyne action is most useful, and amplifiers are deliberately designed to be distorted to take advantage of it. These systems usually involve only 2 sine waves of different frequencies, as opposed to the infinite variety contained in music or speech. If necessary, the 3 unwanted frequencies can be easily filtered out using resonant tank circuits.

Study Problems

1. What are the frequency and bandwidth characteristics of:
   a. An audio amplifier?
   b. A radio-frequency amplifier?
   c. A video amplifier?
2. How are radio-frequency and intermediate-frequency amplifiers different?
3. Define distortion.
4. What is a harmonic?
5. What is the *basic* wave shape?
6. What is a complex wave shape (not a sine wave) composed of?
7. What is it that (mostly) makes a saxophone sound different from a clarinet?
8. What is intermodulation distortion?

NOISE

The Bare Facts

1. Electrical noise is produced by the random motion of electrons.

2. The noise generated in an electronic circuit increases as the temperature increases.

3. Noise is composed of an infinite number of sine waves at different frequencies.

4. Noise reproduced by a loudspeaker sounds much like a waterfall.
5. Too much noise masks the intelligence that is being processed. Try talking to someone standing near a good-sized waterfall.
6. Noise in television pictures appears as "snow" on the screen.

A waterfall generates true acoustical noise—an infinite number of sine waves at all frequencies. The noise is generated as a result of the random variations of water flow as it tumbles over rocks and through space. At a distance the sound of a waterfall is a soothing and restful background, but up close it makes it difficult to carry on a conversation.

Electrons moving through a wire or a resistor, transistor, or other device also have a degree of randomness to their motion. The random part of the electron movement is primarily the result of heat-induced motion. The higher the temperature of an electronic device, the more electrical noise is produced. Some noise is inevitable in any electronic system, but like the waterfall up close, there is a point where the noise level overrides the music, pictures, or other intelligence the system is trying to process.

In audio systems, electrical noise sounds very much like a waterfall at a distance. Sleep-inducing electronic gadgets are available that deliberately produce noise to help you relax and to help mask irritating sounds such as barking dogs and banging pots and pans. Electronic noise in television appears as "snow" on the TV screen. You have undoubtedly had experience with a picture that contained so much noise that it was difficult to make sense of the picture. In space-tracking and radio telescopes special amplifiers are used that are supercooled by liquid nitrogen to keep the noise level down. The signals from space are often so weak that a very little noise can make a signal unintelligible.

Musical instruments also produce noise, and the noise has much to do with the characteristic sound produced by the instrument. The sound of a clarinet, for example, is very difficult to synthesize electronically because of the complex way in which the noise level varies in relationship to the sound level and pitch. In music synthesizers, controlled amounts of noise are added to the tones generated, but the result is only a fair imitation of an acoustical instrument with complex noise behavior.

Study Problems

1. Define noise.
2. How is electronic noise related to temperature?
3. What causes electronic noise?

4. What does noise look like on a television screen?
5. Is noise important to the way a musical instrument sounds?

10-3 OSCILLATORS

The Bare Facts

1. An oscillator is an electronic generator.
2. It produces a continuous signal.
3. There are 2 basic kinds of oscillator: *relaxation* and *feedback*.
4. Relaxation oscillators are simple circuits, but are not overly stable and do not produce pure sine waves.
5. Feedback oscillators produce pure sine waves unless the amplifier is distorted.
6. Relaxation oscillators charge a capacitor through a resistor until a critical threshold voltage turns on a special snap-action electronic device.
7. Feedback oscillators use positive feedback.
8. Feedback oscillators consist of 2 parts: an amplifier and a feedback network.
9. The feedback network is responsible for selecting the frequency of oscillation.
10. If a feedback oscillator is to work, 2 criteria must be met: (a) The amplifier must have sufficient gain to make up for unavoidable losses in the feedback network and for all energy taken from the oscillator to perform its intended task. (b) There must be zero degrees of phase shift between the output and the input of the amplifier (positive feedback) at the desired frequency.
11. Feedback-oscillator and feedback networks fall into 2 basic classifications: (a) tuned (resonant) circuit networks, and (b) phase-shift feedback networks.
12. Every amplifier produces output noise. Somewhere in the noise the desired frequency exists. It is the feedback network's task to reject all but the desired frequency and pass only the desired frequency back to the amplifier input.

An oscillator is an electronic generator that produces a continuous sine (or complex) wave as long as the power is applied. In a radio or television transmitter, an oscillator produces a sine-wave signal that is amplified and delivered to a transmitting antenna. Oscillators also produce signals that control the tracing out of the received picture so that it exactly corresponds to the transmitter's original formation of the picture. These signals are transmitted along with the picture and sound information.

Oscillators produce electronic "tones" in electronic organs and synthesizers. The electronic vibrations move a loudspeaker cone that sets up vibrations in the air that our ear and brain interpret as sound. Oscillator applications make a very long list indeed. Perhaps you can suggest some applications.

Oscillators fall into 2 basic categories: relaxation types and feedback types. The relaxation oscillator is the simplest type, but relaxation oscillators produce a complex wave, usually a square wave, and they are less stable than the feedback types. Where stable sine-wave oscillations are required, feedback oscillators are the choice.

RELAXATION OSCILLATORS

The relaxation oscillator uses a special electronic device that has a critical voltage threshold at which it begins to conduct with a kind of snap action. There are several semiconductor devices that have this characteristic. Among them are 4-layer diodes, unijunction transistors, and diacs. The diac is a bidirectional 4-layer diode. Gas devices like the neon lamp also exhibit this snap-action characteristic.

Most relaxation oscillators charge a capacitor through a resistor until the capacitor reaches the critical turn-on voltage of the snap-action device. The snap-action device then suddenly turns on and rapidly discharges the capacitor. When the capacitor charge voltage drops below a second critical voltage, the device turns off and the capacitor begins charging for the next cycle.

Figure 10-5 is a hydraulic analogy of a typical relaxation oscillator.* A steady drip from the valve gradually fills the cup. When the weight of the water in the cup becomes sufficient to break the iron lever away from the magnet, the cup falls, dumping its contents into the tank. A pulse of water emerges from the output port. As soon as the cup is a bit more than half empty, the weight raises the cup to its original position, and the magnet again "grabs" the iron lever. The cycle begins anew. Water emerges from the output port as a single pulse of water each time the cup dumps. A graph of the output bursts would be an imperfect version of the square wave shown as the output in Figure 10-5b.

---

* The author built this mechanism as a classroom demonstration. It was crudely built, but worked fine.

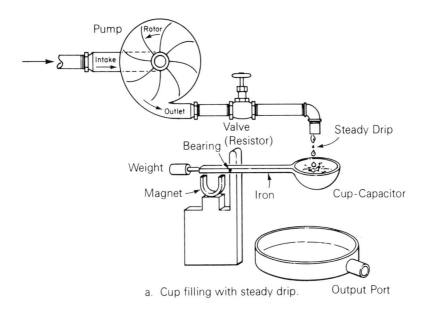

a. Cup filling with steady drip.

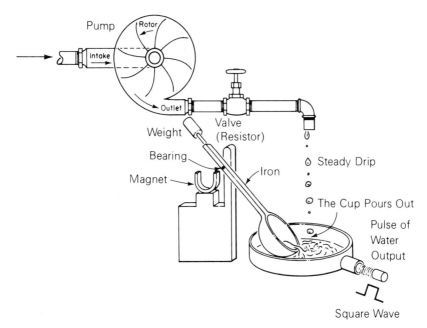

b. Weight of the water in the cup overcoming the pull of the magnet, causing the cup to dump its contents.

FIGURE 10-5. Hydraulic relaxation oscillator.

In the hydraulic analogy of Figure 10-5, the valve (resistor) and the cup (capacitor) function as a resistor-capacitor time-constant circuit. The rate of output pulses can be controlled by valve adjustment. The magnet is crucial because it provides a snap action. Otherwise the cup would gradually drop until it was draining at the same rate it was being filled and all action would cease. At best, it would dither up and down in a more or less unpredictable fashion.

The four-layer (Shockley) diode exhibits the essential snap-action characteristics for an electronic relaxation oscillator. The four-layer-diode relaxation oscillator is shown in Figure 10-6a. Figure 10-6b shows the same basic circuit using a neon lamp as the snap-action device.

Relaxation oscillators make particularly good timing devices for photographic timers, burglar-alarm delay timers, and the like. There is an integrated-circuit package, the 555 timer, that is a very sophisticated relaxation oscillator. It contains a built-in amplifier, and circuitry to improve stability and make the timing insensitive to power-supply variations. The 555 is rapidly becoming the most common relaxation-oscillator type.

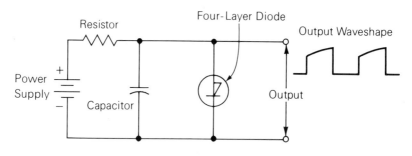

a. Four-layer-diode version.

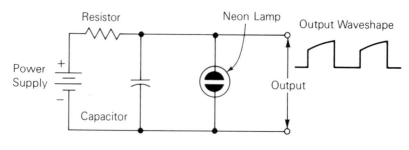

b. Neon-lamp version.

FIGURE 10-6. Electronic relaxation oscillators.

FEEDBACK OSCILLATORS

A feedback oscillator feeds a portion of the output signal back to the input. The feedback is positive. Both circuits shown in Figure 10-7 will oscillate, but the frequency of oscillation will be indeterminate.

Requirements for oscillation:

1. The signal voltage fed back from output to input must have zero degrees of phase shift with respect to the input signal by the time it arrives at the input.
2. The amplifier must have enough voltage gain to make up for the voltage loss in the feedback network.

In Figure 10-7 the amplifier will have zero-degree phase shift. If the value of the resistor (feedback network) is properly selected, the amplifier gain will be adequate to balance its losses.

Calling the resistors in Figure 10-7 *feedback networks* is a little like calling a bicycle wheel a *vehicle*. There is generally more to it. In Figure 10-7, Op-Amp A produces a phase shift of 180°. Op-Amp B also produces 180° of phase shift for a total of 360°, which brings us

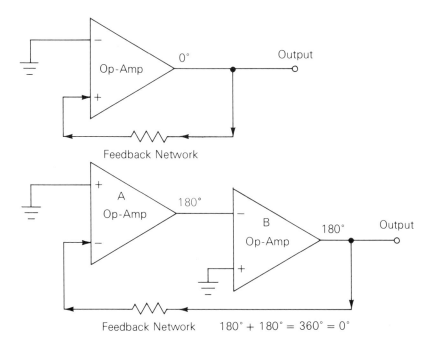

FIGURE 10-7. Simple oscillators.

back to 0°. 180° is negative feedback, while 360° or 0° is positive feedback. Positive feedback is required for oscillation.

In a practical oscillator, the output frequency must be predetermined. To accomplish this, the feedback network must do one of two things: It must either pass a single frequency and reject all others, or it must shift the phase to greater or less than 0° for all frequencies but one. In essence, it must ensure that the requirements for oscillation are met for one frequency, and one frequency only. Two methods are used for this purpose. The first method uses a resonant tank circuit, and the second uses a phase-shift circuit that produces 360° (or 0° for a noninverting amplifier) of phase shift only at one specific frequency.

Figure 10-8 illustrates how a tuned tank-circuit filter works as a feedback network. A feedback oscillator feeds a selected frequency around a loop from output to input and through the amplifier. The question at this point is, "Where did the signal that is circulating come from in the first place?" Every amplifier generates some noise, and signals of *all* frequencies are contained in that noise. We only need something in the feedback loop that will reject all but the desired frequency.

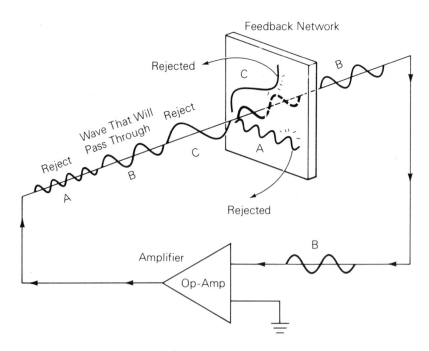

FIGURE 10-8. The feedback network.

In Figure 10-8, Waves A, B, and C represent 3 of the infinite number of frequencies contained in the amplifier-generated noise. Only Wave B can fit through the sine-wave-shaped aperture in the "feedback network." Waves A and C get rejected by the feedback network (along with all other unwanted frequencies). Only Wave (frequency) B is allowed to recirculate around the loop. The illustration in Figure 10-8 is somewhat oversimplified, but it makes the point.

In all cases where a transistor, field-effect device (FET), or tube is used as the amplifier, the amplifier will produce 180° phase shift. The feedback network must not only be able to select the proper frequency, but it also must provide an additional 180° of phase shift to achieve positive feedback. When an op-amp is used, we have a choice to use it as either an inverting (180° phase shift) or a noninverting (0° phase shift) amplifier. Depending on our choice, the feedback network will have to provide either 180° or 0° phase shift.

Figure 10-9 shows 2 oscillators that use a resonant tank circuit for frequency selection. Inductor L and Capacitor C form the resonant circuit. A second winding is wound on the same coil form as the inductor to act as a transformer secondary. The secondary winding can provide either 180° or 0° phase shift simply by switching the 2 secondary wires. The dots show which wires have the same phase. The op-amp has some advantages over the transistor or FET as an amplifier in an oscillator circuit. The most important one is its low output impedance. It can deliver more current to a load without loading the resonant circuit. Excessive loading of the tank lowers its Q and makes the oscillator less stable. A second advantage is that the noninverting input can be used for the oscillator feedback and some negative feedback can be added to the inverting input. The addition of negative feedback allows precise amplifier gain control and minimizes distortion of the output sine wave. Bipolar transistors, MOSFETs, and vacuum tubes can also be used as the amplifier for oscillator circuits.

There are perhaps a dozen common variations of oscillators with tuned transformer feedback networks, but they can all be reduced to the circuits shown in Figure 10-9.

CRYSTAL OSCILLATORS

A quartz crystal can be used as a highly stable, highly selective oscillator feedback network. The crystal is the electromechanical analog of a resonant tank circuit. Crystal-controlled oscillators are almost universally used where extremely precise frequencies must be generated. Figure 10-10 shows the schematic diagram of 2 basic crystal-controlled oscillators. There are also a number of variations of this circuit.

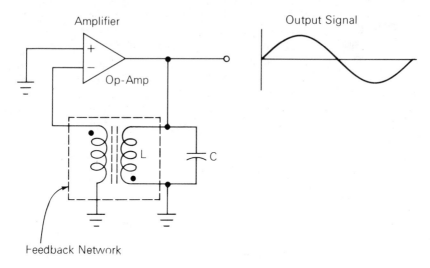

a. Op-amp version.

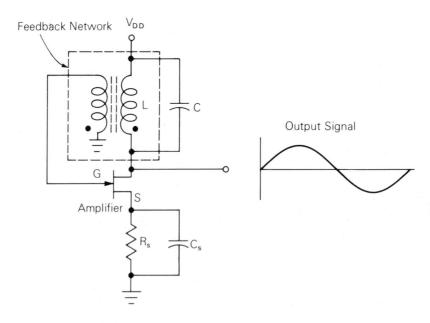

b. FET version.

FIGURE 10-9. Oscillators with tuned transformer feedback networks.

AMPLIFIER SYSTEMS 357

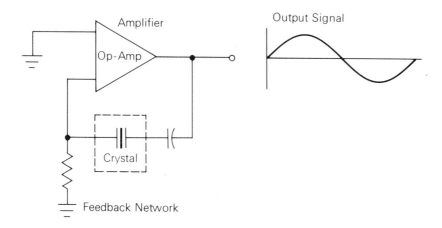

a. Op-amp version.

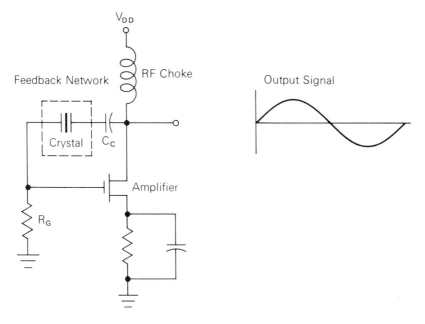

b. FET version.

FIGURE 10-10. Oscillators with quartz-crystal feedback networks.

PHASE-SHIFT OSCILLATORS

The coils required for resonant tank circuits at high frequencies are simple, compact, and inexpensive. The coils required for resonant tank circuits in the audio-frequency range become very bulky and very expensive. Crystals are not practical for this range of frequencies either. Low-frequency oscillators generally use resistor-capacitor feedback networks.

The phase-shift oscillator uses an inverting (180° phase-shift) amplifier. A resistor-capacitor network is designed to provide exactly 180° of phase shift at 1 and only 1 frequency. 180° + 180° = 360°, which brings us back to 0° of total phase shift, a necessary condition for oscillation. Figure 10-11 shows 3 versions of the phase-shift oscillators. Each of the 3 uses a different amplifier, but otherwise they are identical.

In order to understand how the feedback network operates, let's go back to some basic alternating-current theory (from Chapter 6). Notice that there are 3 sections to the feedback network in Figure 10-11a. Each section is designed to have 60° of phase shift at the desired frequency of operation. 60° + 60° + 60° = 180° total.

Figure 10-12 shows the vector diagram of a resistor-capacitor circuit with 60° of phase shift. The reactance of a capacitor is dependent on the frequency. The reactance vector will lengthen if the frequency is lower, and shorten if it is increased. Because the resistance is independent of frequency, the resistance-vector length will remain constant no matter what the frequency.

If you will examine Figure 10-12, you can see that if the reactance vector lengthens, the impedance vector will rotate clockwise to something greater than 60° of phase shift. If the capacitive reactance decreases, the vector will shorten, rotating the impedance vector to something less than 60°. With a carefully selected value of capacitance and resistance, only the desired frequency will produce the required 60° phase shift per feedback-network section. Only the desired frequency will meet the 180° (total for three sections) requirement for oscillation. A slight difference in frequency will result in only a small phase shift. But because there are 3 sections, each producing a small phase error, the phase error totals up to enough to prevent oscillation except at the desired frequency. The impedance (length of the impedance vector) will also vary as the frequency changes, but this has no real influence on the circuit's behavior. As in the case of the resonant tank-feedback oscillator, the correct frequency is selected out of all of the frequencies contained in the amplifier noise.

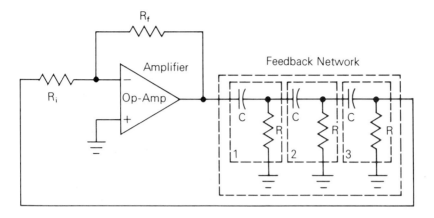

a. Op-amp version.

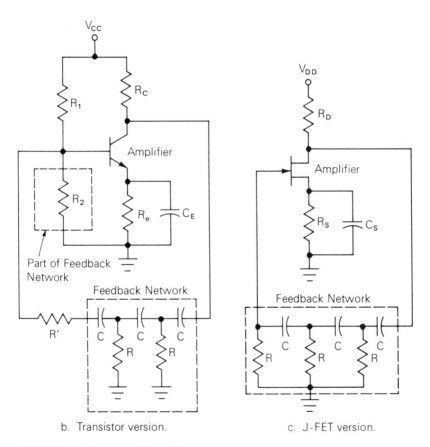

b. Transistor version.   c. J-FET version.

FIGURE 10-11. The phase-shift oscillator.

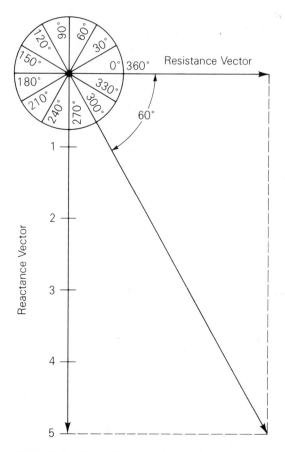

FIGURE 10-12. Phase-shift feedback-network vectors.

Study Problems

1. Place a piece of tracing paper over Figure 10-12 and find the phase angle when: (a) the reactance vector is lengthened by 1 unit; (b) the reactance vector is shortened by 1 unit.
2. List the requirements for oscillation in a feedback oscillator.
3. Why are phase-shift oscillators used when resonant tank-circuit oscillators work just as well?
4. Explain how relaxation oscillators work.
5. Explain in your own words why a snap-action element is required in relaxation-type oscillators.
6. When a stable frequency sine wave is required, what type of oscillator is used (relaxation or feedback)?
7. Where does the *original* signal come from in a feedback oscillator?

AMPLIFIER SYSTEMS 361

8. Do relaxation oscillators produce a sine-wave output signal?
9. Why is an extra winding often added to the coil in a resonant tank-circuit feedback network?

10-4 MODULATION

The Bare Facts

1. Modulation is the process of impressing music, speech, or other information on a radio wave.

2. The radio-frequency wave *carries* the information and is radiated into space by a transmitting antenna.

3. An antenna at the receiving end receives the carrier wave, along with its modulation, out of space.

4. Amplifiers increase the signal strength of the received wave.

5. A detector or demodulator extracts the original information from the carrier. The carrier is then "thrown away," having served its purpose.

6. The basic methods of modulation are amplitude modulation and frequency modulation.

7. All other methods, such as phase modulation, single-sideband (SSB) modulation, and so on, are variations of the two basic methods.

8. In amplitude modulation the strength of the carrier wave is varied in step with the music or other information. The carrier frequency does not change.

9. In frequency modulation, the *frequency* of the carrier wave varies in step with the music. The carrier strength does not vary.

Modulation is the process of impressing intelligence on some kind of carrier wave. For example, suppose that we desire to communicate over a beam of light. In a previous example we suggested that we could send code over a light beam by turning it on and off in a specified pattern. This idea can be carried further by varying the brightness of the light in the pattern of speech or music. The light beam is the carrier and the process of varying the brightness in step with the speech or music is called *modulation*.

When the strength (brightness) of the light beam is modulated, the process is called *amplitude* modulation. Frequency modulation is the other common form of modulation. In amplitude modulation the

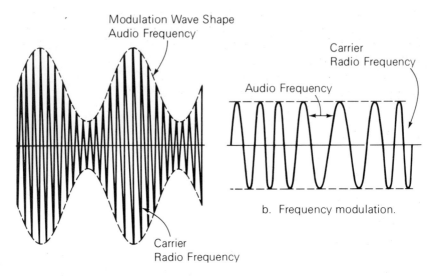

b. Frequency modulation.

a. Amplitude modulation.

FIGURE 10-13. Amplitude and frequency modulation.

frequency of the carrier remains constant and the strength is modulated with the information we desire to transmit. In frequency modulation the strength (amplitude) of the carrier remains constant and the frequency of the carrier is varied in step with information to be impressed on it. If we were to watch a frequency-modulated light beam we would see its brightness as being constant, but the color would vary in step with the information modulating it. Figure 10-13 illustrates amplitude and frequency modulation.

Study Problems

1. Define modulation.
2. In frequency modulation, what characteristic of the carrier wave is varied and what characteristic is constant?
3. In amplitude modulation, what characteristic of the carrier wave is varied and what characteristic is constant?
4. Assume you are watching 2 light beams projected on a motion-picture screen. You have been told that each is being modulated.
    a. Beam A is red. It does not change color, but the brightness varies. Is beam A amplitude- or frequency-modulated?
    b. Beam B is blue but its color shifts between a deep blue and a definite green. The beam's brightness seems fairly constant but you can't be sure. Is the modulation AM or FM?
5. Explain why you can't be sure that the brightness of Beam B is constant in Problem 4.

10-5 AMPLIFIER SYSTEMS

The circuit shown in Figure 10-14 is a modern audio amplifier. Op-amp 1 is called a *preamplifier*. It differs from other op-amps only in its lower-than-usual noise generation. Noise generated in later stages is less important because it gets amplified less. The noise generated earlier in the system (closer to the input) gets amplified by all succeeding stages.

The selector switch selects either phonograph or tuner inputs. The resistor R_d reduces the voltage from the tuner to prevent "blasting" and the necessity to readjust the volume when switching between tuner and phonograph. The tuner normally produces a higher output-signal voltage than the phonograph.

The components in the dashed box form a standard bass and treble tone-control circuit. Amplifier 2 is there primarily to make up for the losses in the tone-control circuit.

The negative-feedback loops for the op-amps have some capacitors in them to help reduce the noise and to narrow the bandwidth of the op-amps. Even inexpensive op-amps have a typical high-frequency response of 250,000 hertz or greater. A frequency response much greater than 25,000 hertz in an audio system invites problems and offers very few benefits. Even the purist finds it hard to advocate a frequency response greater than 50,000 hertz. Operating with excess bandwidth can easily lead to a tendency for the amplifier to oscillate, cause subtle forms of distortion, and produce additional noise. These hazards of excess bandwidth generally far outweigh any real or imagined improvement in quality that might result from better high-frequency response. Amplifier 3 is a power amplifier. It usually consists of a more-or-less conventional op-amp driving a special power-amplifier circuit. In some cases the power-amplifier circuit is not part of the integrated circuit (IC) and is made up of discrete components. The tendency up to 25 watts or so is to put the entire amplifier, op-amp driver, and power amplifier in a common package. At higher power levels it becomes more difficult to get rid of the heat produced in the power transistors, when they are on an IC chip. Figure 10-15 shows the circuit for the power-amplifier package (Amplifier 3 in Figure 10-14).

In the power section in Figure 10-15, Q1 and Q2 are complementary transistors, one NPN and one PNP. The circuit is called a *complementary symmetry amplifier*. Each transistor conducts for one-half of each cycle. Q2 conducts only on the positive half cycle and Q_1 conducts only on the negative half cycle. R_L, D_1, and R_d bias both transistors slightly on, to avoid zero-crossing distortion (see Figure 10-2c). The circuit is often called a *push-pull* circuit. The complementary symmetry circuit has become the industry standard for power amplifiers.

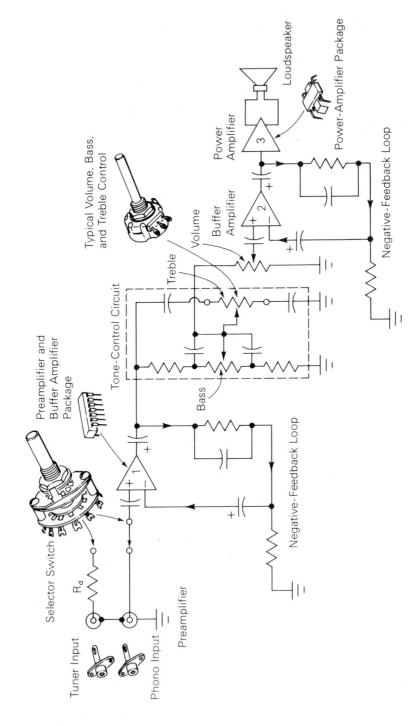

FIGURE 10-14. Modern audio-amplifier system.

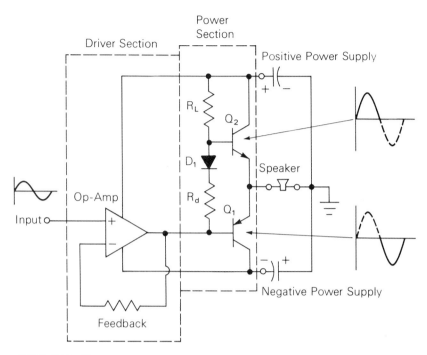

FIGURE 10-15. The power op-amp.

HEAT SINKS

Power transistors operate at fairly high current, and because they do have some internal resistance, they produce heat when operating. Internally generated heat, if allowed to build up, can alter the operating point and even destroy the transistor. It is essential to dissipate the heat rapidly enough to prevent too great an internal temperature rise. The heat transfer directly from the transistor case into the air is very poor because the case has a very low thermal resistance while the air has a very high thermal resistance (or thermal impedance). This large difference in thermal impedances (resistances) results in very poor heat transfer.

When maximum power transfer is required between two widely different electrical impedances, we install an impedance-matching transformer. When maximum heat transfer is required between two widely different thermal impedances, we install a thermal-impedance-matching transformer. It is called a *heat sink*. A heat sink is made of high-thermal-conductivity metal with a relatively large surface area. For moderate power levels, mounting the power transistors on the aluminum chassis that supports the circuit boards, controls, and so on, is adequate. At higher power levels we resort to commercial heat sinks that are designed to pack the greatest surface area into the smallest reasonable space. Figure 10-16 shows samples of commercially available heat sinks.

It is extremely important that the transistor case make contact over as large an area as the case permits. All burrs around mounting holes must be removed and the mating surfaces should be as flat as possible. Because all surfaces have some irregularities and pits, it is common practice to coat both surfaces with a high-thermal-conductivity silicone grease. Silicone grease is a pasty, usually white, messy substance, but it is very effective. When the transistor is bolted down, much of the silicone grease squeezes out, leaving only a very thin film between transistor and heat sink. This excess, squeezed-out grease may be cleaned off, but it is not a common practice in commercial manufacture.

Study Problems

1. What is a heat sink?
2. Why is a heat sink required for power amplifiers?
3. What is the electrical analog of a heat sink?
4. Why is silicone grease often used between power transistors and power op-amps and the heat sink?

FIGURE 10-16. Commercial heat sinks.

AMPLIFIERS IN RADIO RECEIVERS

The Bare Facts

1. The bandwidth of radio-frequency amplifier circuits is carefully limited by tuned resonant circuits.

2. The radio-frequency amplifier is a kind of preamplifier. It is always found in car radios, but not always in home-based units.

3. The oscillator-mixer is an oscillator circuit combined with a slightly distorted amplifier.

4. The mixer produces an output consisting of the radio-station frequency, the oscillator frequency, the sum of the 2 frequencies, and the difference between the 2 frequencies.

5. Resonant circuits eliminate all but the difference frequency, which is called the *intermediate* frequency.

6. The intermediate frequency contains all of the original modulated information.

7. The diode detector strips off half of the modulation envelope (each half contains all of the transmitted information).

8. A capacitor bypasses the IF (intermediate frequency) to ground, retaining the original audio information.

9. The audio information is fed to the input of an audio power amplifier and then to a loudspeaker.

10. An FM receiver requires a more complex detector and sometimes a stereo separation circuit, but is otherwise functionally the same as an AM receiver.

11. The special advantage of FM over AM is its noise-free sound reproduction.

Figure 10-17 is the schematic diagram of a modern AM radio receiver.

The RF Amplifier

The radio-frequency (RF) amplifier circuit consists of an integrated-circuit amplifier (Amplifier 1), the transformers T_1 and T_2, and 2 of the 3 gangs of the tuning capacitor. T_1 is called the *antenna coil* (or antenna transformer). It is wound on an iron ceramic (ferrite) rod that takes the place of an outside antenna. The tuning capacitor forms a circuit that resonates with the primary coil of the antenna coil. By adjusting the setting of the variable capacitor, the resonant frequency

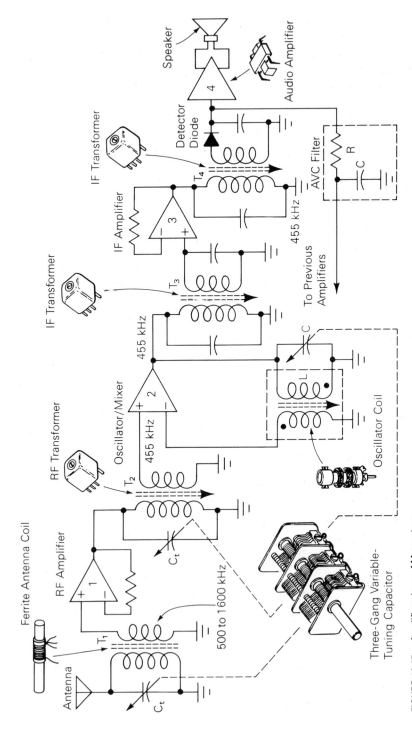

FIGURE 10-17. Amplifiers in an AM receiver.

can be set to match the transmitter frequency of any station in the standard AM broadcast band.

Amplifier 1 is basically an op-amp, but is designed to operate over a frequency range of from 535 kilohertz to 1605 kilohertz. Other than being designed to operate best at those frequencies, it differs very little from any other op-amp. Transformer T_2 also tunes the broadcast band and provides additional rejection of unwanted stations.

The Oscillator-Mixer

Amplifier 2 is the same type as Amplifier 1, but it serves two purposes. The inverting input, along with the oscillator coil and one gang of the tuning capacitor, form an oscillator like the one in Figure 10-9a. This is called the *local* oscillator. The noninverting input in this case is used as an input for the RF signal.

The oscillator frequency is varied along with the frequency of T_1 and T_2, but it is always 455 kilohertz higher in frequency than the incoming radio signal. The oscillator and incoming radio signals are mixed in the amplifier, which is deliberately allowed to produce some intermodulation distortion.

Four frequencies appear in the output of the amplifier:

1. The original radio signal frequency

2. The original oscillator frequency

3. The sum of the 2 frequencies

4. The difference between the 2 frequencies

Example:
  Assume that the incoming frequency is 1000 kilohertz.
  The frequencies coming out of the mixer are:

1. 1000 kilohertz (incoming signal)

2. 1455 kilohertz (oscillator frequency)

3. 2455 kilohertz (the sum of 1 and 2)

4. 455 kilohertz (the difference between 1 and 2)

Transformer T_3 is tuned to 455 kilohertz and has about a 10 kilohertz bandwidth. It effectively rejects Frequencies 1, 2, and 3, as well as the frequencies of all other stations in the broadcast band. In the radio-frequency stage there are other stations very close to the resonant frequency of the tank, and they tend to sneak through. At the new 455 kilohertz intermediate frequency, unwanted station frequencies are well away from the 455 kilohertz intermediate frequency and are totally rejected by the resonant circuits of T_3 and T_4. The new 455 kilohertz intermediate frequency still contains all of the original modulation. Amplifier 3 provides amplification for the 455 kilohertz IF signal.

The process of amplitude modulation produces 2 identical envelopes, one negative and one positive (see Figure 10-13a). In order to recover the modulated information, it is necessary to eliminate one of these so-called sidebands. The detector diode, functioning as a 1-way valve, accomplishes this task. The capacitor from the output of the diode to ground is selected to have a low reactance at 455 kilohertz and a high reactance in the audio-frequency range. It bleeds off the 455 kilohertz carrier frequencies but does not affect the audio frequencies. The audio frequencies are fed into the audio-amplifier integrated circuit (see Figure 10-15) and finally to the loudspeaker.

The output of the automatic-volume-control (AVC) filter in the dashed box is fed back to the previous amplifier stages. The AVC or AGC (automatic gain control) varies the bias on the amplifier stages according to the incoming signal strength. The gain is reduced for strong stations and increased for weak ones. Varying the bias on an op-amp varies the voltage gain. The AVC circuit keeps the sound level fairly constant for both strong and weak radio signals. The AVC circuit uses the detector diode to produce a DC voltage proportional to the AC audio voltage. It is essentially a half-wave power supply of the sort to be discussed in the next section.

This type of receiver is called a *superheterodyne*. *Heterodyne* is the name applied to the mixing of signals to produce the IF frequency. The *super* was perhaps an addition made by the inventor when he compared his receiver to the others available at the time.

FM Receivers

The FM receiver is essentially the same as the AM receiver up to the detector circuit, with the exception that the FM band covers a range of from 88 to 108 megahertz and the intermediate frequency is 10.7 megahertz. The process of extracting program information from the FM signal (detection or demodulation) is considerably more complex than that of detecting an AM signal. In an FM receiver the simple diode detector is generally replaced by one or two special integrated circuits. In a stereo receiver, two tasks must be performed at this stage. The program material must be extracted from the modulated carrier and the 2 stereo channels must be separated. The stereo receiver must then be provided with 2 audio amplifiers, 1 for each stereo channel.

One of the main advantages of FM is that noise can be almost completely eliminated. There are a great many sources of electrical noise in radio transmissions. Electrical noise is generated by amplifiers, neon signs, fluorescent lights, automobile ignition systems, electric motors, household lamp dimmers, microwave ovens, high-voltage power lines, and many other electrical devices. Because noise includes a very broad band of frequencies, some of it is certain to get through all of the tuned circuits in a receiver. Noise always modulates a signal as amplitude modulation, riding on top of the desired signal as illustrated in Figure 10-18.

With an AM signal, it is nearly impossible to remove the noise without removing at least part of the desired program information. A circuit to remove even a reasonable percentage of the noise would be

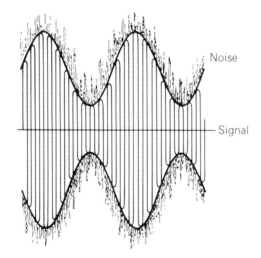

a. Amplitude-modulated signal and noise.

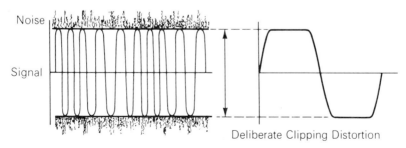

b. Frequency-modulated signal and noise.

FIGURE 10-18. The noise problem.

more complex than the receiver itself. In the case of FM, an amplifier (called a *limiter*) with deliberate clipping distortion (easy to do) can shave off most of the noise from all sources without damaging the original transmitted information (see Figure 10-18b).

There is another advantage to commercial FM broadcasting. The Federal Communications Commission has limited the audio-frequency range for AM broadcasts to a maximum of 5000 hertz, while allowing 15,000 hertz for FM broadcasting.

Study Problems

1. What is the purpose of the oscillator-mixer circuit?
2. What component in Figure 10-17 is used to select the desired station?
3. What frequencies are available at the output of the oscillator-mixer? Which of the 4 is used?
4. What is the purpose of Amplifier 3 in Figure 10-17?
5. What is the function of the detector diode in Figure 10-17?
6. What is the function of the AVC circuit in a radio receiver?
7. What are the differences between AM and FM receivers?
8. Why can FM receivers be made more noise-free than AM receivers?
9. List some noise sources that interfere with radio reception.
10. Explain why FM receivers can eliminate far more noise than AM receivers.

10-6 POWER SUPPLIES

The Bare Facts

1. Every electronic amplifier needs a source of power.
2. Circuits that convert the AC line voltage into DC at an appropriate voltage are called *power supplies*.
3. A transformer is used to step the AC line voltage up or down to the desired voltage.
4. Semiconductor (rectifier) diodes operate as 1-way valves to ensure that current flows in only one direction (DC).
5. Capacitors and resistors are used to smooth out the DC voltage variations that are present at the output of the diodes.
6. When more pure DC is required than can be economically provided by resistor-capacitor filters, an electronic regulator is added to the power-supply circuit.

7. When a constant output voltage is required, in spite of line-voltage variations and varying current demand from the supply, an electronic regulator circuit is used.

8. The simplest regulator available is the zener-diode regulator.

9. When larger currents are involved, or tighter restraints on output-voltage variations are required, a feedback regulator is used.

Every electronic device needs a source of power. The power usually required is direct current, either from a battery or from a circuit that converts household (or industrial) power-line alternating current into direct current at an appropriate voltage. This section is concerned with the conversion of AC power-line voltages into suitable DC voltages.

Figure 10-19 shows the 2 sections of a typical half-wave power supply. In Figure 10-19a the transformer steps the 120 volt line voltage up or down to the appropriate voltage. The diode, functioning as a 1-way electronic valve, allows current to flow in one direction only. This produces the direct-current pulse waveform called the *rectified* waveform.

At this stage in the power supply, current flows in only one direction, but the voltage is pulsating and not a steady direct-current flow. Such a pulsating voltage applied to an audio amplifier would produce an irritating *hum*, much like the sound of the electric alarm clock that destroys your sleep each morning. This sound is only slightly less annoying when you are fully awake.

The rectified output (Figure 10-19a) is fed into the filter network in Figure 10-19b (Terminal A to Terminal A, and Terminal B to Terminal B). Each pulse in the rectified output not only supplies power to the amplifier system connected to the supply, but also charges Capacitors C_1 and C_2. During the half of the AC waveform that has been cut off by the diode, power must still be delivered to the amplifier(s). During the half cycle in which the diode cuts off the current, the capacitors discharge their stored energy and supply the need. The result is only a small ripple in the direct current, as shown in the filter output waveform. Two capacitors and a resistor perform better than a single, much larger capacitor because of the time delay between the two capacitors. The result is less ripple with less cost. Capacitors are fairly expensive. Some circuits are more tolerant of a little ripple than others. Op-amps and complementary power amplifiers tend to be fairly tolerant.

When hum is still a problem, or when the amplifier's current demand varies widely, a regulator stage is added. Some years ago large

374 CHAPTER 10

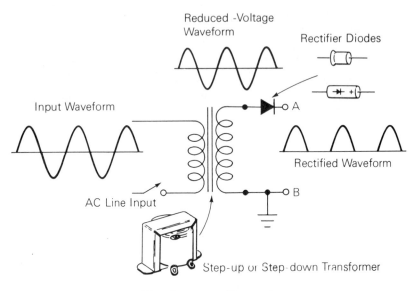

a. Transformer-rectifier section.

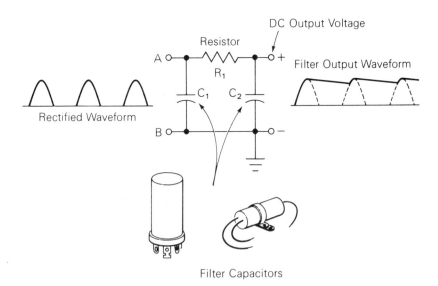

b. Filter section.

FIGURE 10-19. Basic half-wave power supply.

inductors and large capacitors were used to answer these problems because electronic regulator circuits were far too expensive. Since the invention of transistors and integrated circuits, electronic regulators have become far cheaper than capacitors. However, in many cases a full-wave rectifier can provide adequate filtering without regulators. Let's look at how that works.

Figure 10-20 shows 2 versions of a full-wave rectifier. The circuit in Figure 10-20a requires either 2 separate secondary windings on a

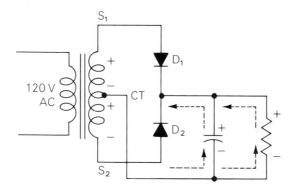

a. Full-wave circuit with center-tapped transformer.

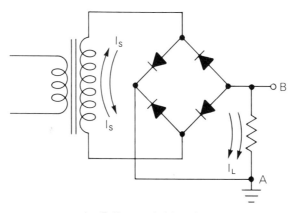

b. Full-wave bridge circuit.

c. Full wave and half wave compared.

FIGURE 10-20. The full-wave rectifier.

transformer or a center tap (CT) in a single winding. The results are the same, but a winding with a center tap is cheaper to manufacture.

When the anode of D_1 is positive, current flows out of the center tap, up (see arrows) through the load and back to S_1. When the cycle reverses (second half-cycle), S_2 becomes positive with respect to the tap. Current flows out of the tap, up through the load, and back to S_2. One diode conducts on each half-cycle and the other is in the off condition (open circuit). The full-wave shape is shown in Figure 10-20c, Waveform 2. The bridge circuit in Figure 10-20b can provide a full-wave output without using a center-tapped transformer. This is a frequently preferred circuit because diodes are so cheap, and transformers with a center tap and the desired secondary voltage are not always as available as untapped transformers. The cost of the 2 circuits is about the same. The 2 extra diodes in Figure 10-20b add something to the cost, and so does the tapped transformer in Figure 10-20a. In Figure 10-20c, the improved filtering with full-wave rectification is illustrated by the area between pulses. The size of the capacitors for a given amount of ripple is proportional to the area between pulses. Because the area in Waveform 2 is so much smaller, less capacitance is required for a given filter effectiveness. Capacitors are much more expensive than a tap on a transformer or two extra diodes, when better filtering is needed.

REGULATORS

The simplest common regulator is the zener-diode regulator. The zener diode is operated in a reverse-biased avalanche mode. A regulator maintains a constant-voltage output in spite of varying input voltages and current demands.

The important thing about the avalanche mechanism is that the voltage across the device is essentially the same, regardless of the current through it. In that regard it is much like a battery with its constant voltage, except that it does not require frequent replacing or recharging. It is unlike a battery in that it cannot supply power on its own to a circuit. Figure 10-21b shows a simple but effective zener-diode voltage regulator. The zener regulator is effective only for low-power power supplies.

The Feedback Regulator

When higher current or better filtering is required, a negative-feedback voltage-regulator circuit is called for. The circuit is shown in Figure 10-22. Transistor Q1 is the control, or pass, transistor. It serves as an automatic variable resistor that keeps the voltage across the output terminals constant at some desired value. The op-amp

AMPLIFIER SYSTEMS   377

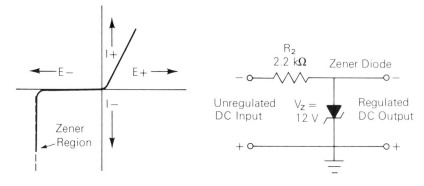

a. Zener-diode conduction curve.

b. Zener-diode regulator circuit.

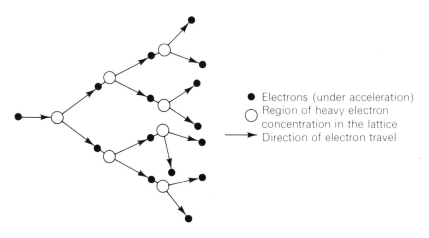

c. The avalanche mechanism.

FIGURE 10-21.   Zener-diode voltage regulator.

samples the output voltage. The noninverting input of the op-amp is set at a fixed reference voltage determined by the zener diode. A predetermined fraction of the regulated output voltage is supplied to the inverting input of the op-amp. The difference between the reference voltage and the regulator output sample appears at the op-amp output and controls the base current of Q1.

Any increase in the regulator output voltage is inverted by the op-amp and turns the control transistor down, bringing the regulator output voltage back to where it should be. Any decrease in regulator output voltage is inverted by the op-amp, which turns the control transistor up, again bringing the regulator output voltage back *up* to

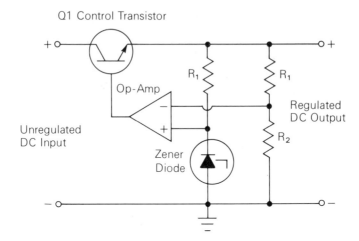

FIGURE 10-22. Feedback regulator.

where it should be. The regulator action is so fast that it also levels out ripple variations more effectively than adding large, expensive capacitors. Power regulators require adequate heat sinks.

DUAL POWER SUPPLIES

Op-amps normally require $+10$ to $+15$ volts and -10 to -15 volts for proper operation. Figure 10-23 shows one of several schemes to provide dual voltages for op-amps (and other circuits) using only one transformer. The transformer is by far the most expensive part of a power supply. An additional regulator can be provided for each output voltage if needed.

Study Problems

1. Why are power supplies needed in electronics?
2. Define *ripple*.
3. What does the transformer do in a power-supply system?
4. What does the diode rectifier do in a power supply?
5. What is the function of the filter network in a power supply?
6. What are the functions of a regulator in a power supply?

10-7 SWITCHING-AMPLIFIER CIRCUITS

One of the most common thyristor switching circuits is the one used to dim lights or control the speed of drill motors, blenders, and similar household appliances. If you will remember, thyristor devices can be turned off only by interrupting the anode-cathode current. When a thyristor is used with alternating current, it turns off

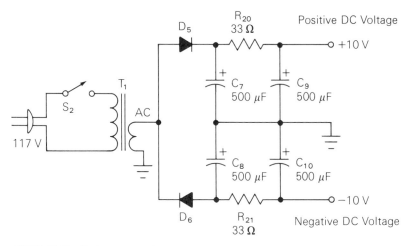

FIGURE 10-23. Dual power supply.

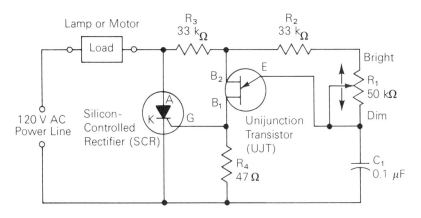

Note: A = anode; K = cathode; G = gate; E = emitter.

FIGURE 10-24. SCR lamp dimmer and motor speed-control circuit.

each time the AC wave form passes through zero. The technical term for turning a thyristor off is *commutation*. When the AC line takes care of turning the thyristor off, the thyristor is said to be *self-commutating*. The circuit in Figure 10-24 is that of an incandescent-lamp dimmer or motor speed control. The unijunction transistor (UJT) is connected as a relaxation oscillator. R_2 prevents the potentiometer (R_1) from damaging the circuit when R_1 is in the zero ohms position. R_2 also ensures the proper range for R_1 to keep the control action smooth.

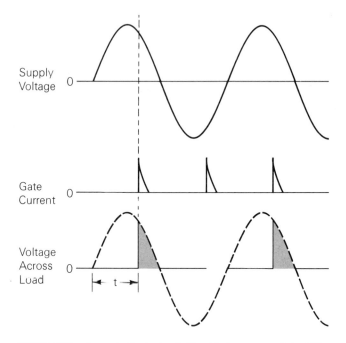

FIGURE 10-25. Power to the load varied by delaying turn-on of the SCR.

How the circuit works:

1. Capacitor C_1 charges through Resistors R_1 and R_2.
2. When the capacitor is charged to a high enough voltage the UJT fires.
3. When the UJT fires it turns the SCR on.
4. The lamp (load) lights and remains lit until the AC sine-wave voltage goes through zero. The silicon-controlled rectifier (SCR) then shuts off and awaits a new gating pulse from the UJT oscillator.
5. The brightness of the lamp is proportional to the percentage of each (half) cycle during which the SCR is turned on.

The actual control of the *on*-time percentage is accomplished by delaying the SCR turn-on until later in the cycle. Figure 10-25 explains. The greater the time, t, in Figure 10-25, the dimmer the light will be. For the brightest light the UJT oscillator must turn on at the start of the AC cycle.

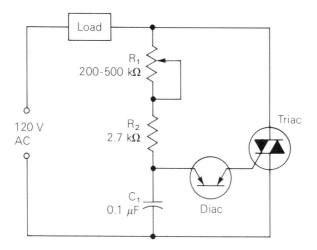

FIGURE 10-26. The diac-triac switching-amplifier control system.

The maximum brightness of a lamp (the load) in the circuit of Figure 10-24 is about half normal because it stays off for the entire negative half-cycle. The SCR is good for only one direction. Many of the commercial circuits of this type (Figure 10-24) have a switch connected to the control shaft that switches the SCR out and gives the lamp (or motor) the full line voltage. This jump from half to full power does not seem to bother most consumers. However, with the invention of the triac, circuits like the one in Figure 10-26 have become almost universal in home lamp dimmers and motor speed controls. Industrial applications often involve some special problems that make the SCR circuit in Figure 10-24 a preferred way to go.

The triac circuit in Figure 10-26 works on exactly the same principle as the SCR circuit in Figure 10-24, except that the diac trigger diode turns the triac on for *both* half-cycles. Since both diac and triac are bidirectional, the circuit in Figure 10-26 can provide a range of control from almost zero to full brightness (or speed).

Chapter 11

A Toy Robot. He follows a light beam and responds to four verbal commands. (Courtesy PW and associates)

ENERGY CONVERSION DEVICES FOR ELECTRONIC SYSTEMS

ENERGY CONVERSION DEVICES FOR ELECTRONIC SYSTEMS 383

11-1 INTRODUCTION

In this chapter we will look at devices and systems that convert energy from one form into another. These converters are often called *transducers*. Microphones convert acoustical energy into electrical energy. After electronic amplification, a loudspeaker converts the electrical back into acoustical energy. Electrical energy can be converted into light and heat, or light and heat can be converted into electrical energy. We will also examine one of the greatest advances in communications since radio, the laser.

11-2 LASERS AND LIGHT SOURCES

The Bare Facts

1. All light sources operate by either spontaneous or stimulated emission.

2. When energy is absorbed by an electron, it is raised to a higher energy state. When it falls back to its normal (ground) state, it gives up its stored energy in the form of electromagnetic radiation (emission).

3. In all devices except lasers, electrons drop back to the ground state in a random fashion due to thermal agitation.

4. The light output from all light producers except lasers consists of many frequencies (wavelengths) and phases. This kind of mixed light is called *incoherent* light.

5. Lasers are optical-feedback oscillators. Electrons are stimulated to fall to the ground state by photons of light fed back by reflective surfaces in the laser tube.

6. The light output from a laser consists of single-frequency waves, all in phase. This kind of light is called *coherent*.

7. Electrons can move up to higher energy steps only in discrete jumps (called *quantum* jumps). When electrons drop to ground state, the emission can be at only one frequency for any given quantum level.

8. Electrons can be *pumped* up to higher levels by another (incoherent) light source or by using an electrical current.

9. Incoherent-light sources can consist of incandescent filaments, gas discharges, or semiconductor light-emitting diodes.

10. Lasers use the same mechanisms, with the exception of incandescent filaments. Pumping by using another light source is exclusively a laser technique.

The word *laser* stands for *l*ight *a*mplification by *s*timulated *e*mission of *r*adiation. The words *amplification* and *stimulated* are the keys to the differences between the laser and all other light producers. Other light sources work by *spontaneous* emission, which generally rules out amplification.

In any electronic device a certain number of electrons absorb heat, light, or some other kind of energy. They are energized to a high-energy state for a time. They soon give up that energy and drop back to the more normal low-energy or *ground* state. Nothing much observable happens. However, if we pass enough current through a wire to raise it to a white-hot state, a great many electrons are absorbing heat energy. When each falls back to the ground state it gives off its absorbed energy in the form of light energy.

What we commonly refer to as the *optical* or *light* spectrum covers a lot of wavelengths (and frequencies) that the human eye cannot see. Figure 11-1 shows the optical spectrum. The shaded bandpass curve is the part of the spectrum our eyes respond to. The left side of the range extends down to the microwave and radio broadcast frequencies. The upper (right) end extends to high-energy radiations such as gamma radiation and cosmic rays.

We can see only about an octave out of the 20 or more octaves in the radiant energy spectrum. All the visible-light producers except the laser generate a broad band of light frequencies. These devices produce what is essentially a band of optical noise. Only the laser produces a single-frequency light output. The laser produces coherent waves, while other light sources produce incoherent waves.

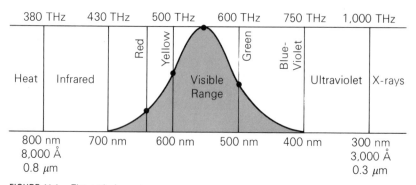

FIGURE 11-1. The optical spectrum.

Figure 11-2 illustrates the difference between the two. The coherent waves in Figure 11-2a are all of the same frequency and in phase. This is the kind of wave produced by an electronic oscillator. The laser is an electronic feedback oscillator tuned to an optical wavelength.

Incoherent waves are made up of many frequencies and phases. Incandescent lamps, flourescent lamps, and so on produce this limited-bandwidth, incoherent, optical-noise kind of output. The limited bandwidth is the main difference between ordinary random electron noise and optical noise. A second difference is the fact that certain optical frequencies will be missing within the band. We will examine the reasons for the missing frequencies shortly.

The output bandwidth is also dependent on the kind of "bottle" in which we place the light source. The household incandescent lamp, for example, produces a fair amount of ultraviolet radiation, but it simply does not pass through the glass bulb. The glass absorbs it and converts it into heat.

SPONTANEOUS AND STIMULATED EMISSION

All light sources except lasers operate on the principle of spontaneous emission. When energy strikes an electron, it can absorb that energy by increasing the radius of its orbit around the nucleus. It can move only in discrete steps, however. It must absorb exactly enough energy to move up to an allowed energy state. If an electron in the first allowable orbit is struck by enough radiant energy to move it beyond that orbit but not all the way to the second allowable orbit, it will absorb no energy at all.

An electron in this higher energy state is only partially stable. It takes only a very little energy loss to cause it to fall to its original (ground) state. When the electron falls, it gives up its stored energy as a specific frequency (wavelength) of radiant energy. Each kind of

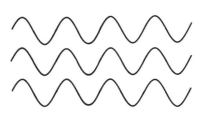
a. Coherent radiation.

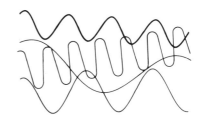
b. Incoherent radiation.

FIGURE 11-2. Coherent and incoherent radiation.

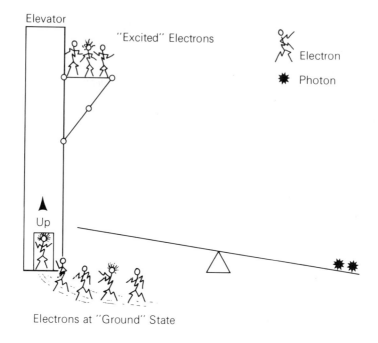

FIGURE 11-3. Electrons at ground and excited states.

atom has its special energy content for each level. In visible-light terms, each level represents a specific color. The process of raising electrons to an elevated or high-energy state is called *pumping*. Figure 11-3 illustrates the pumping action. A photon is a bundle of energy corresponding to a specific frequency.

In all light-emitting devices except the laser, emission is spontaneous. Electrons are "shaken down," more or less at random, by heat energy. The output light (photons) is incoherent, consisting of a number of different frequencies. Spontaneous emission is illustrated in Figure 11-4. Electrons are knocked off the platform in a random fashion by thermal activity, resulting in an output of incoherent light.

LASERS

The laser is a feedback oscillator. The feedback path is traveled by photons instead of electrons, but the result is the same. The length of the path that the photon must travel is critical. In a laser, like an organ pipe, it is the physical dimensions of the system that determine the resonant frequency. Figure 11-5 illustrates the laser action.

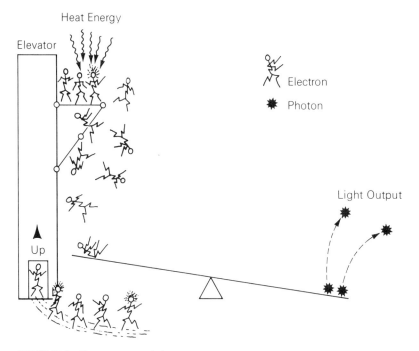

FIGURE 11-4. Spontaneous emission.

In Figure 11-5a, electrons are raised to a higher energy level. An electron falls as a result of thermal agitation. This is analogous to the noise energy that starts an electronic oscillator going. An optical-feedback path sends photons back to stimulate the fall of excited (elevated) electrons. They are dumped when stimulated by the feedback photons, and release their energy in the form of additional photons. Some of the newly released photons are fed back to stimulate the fall of still more electrons (see Figure 11-5b). The light output is coherent. The platform in Figure 11-5 is assumed to snap back instantly to its horizontal position after it has been dropped. Figure 11-5 is an illustration, not a true analogy. Before we explore lasers in more depth, suppose we take a closer look at some common garden-variety light sources.

INCANDESCENT LAMPS

The ordinary light bulb, the one Edison invented, uses a high-resistance wire of tungsten or a nickel chromium alloy. The resistance wire (filament) is enclosed in a glass bulb. The air has been pumped out and exchanged for an inert gas, nitrogen. The air has

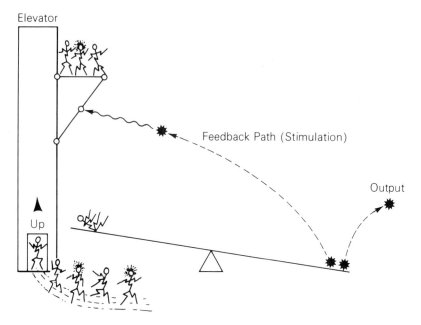

a. "Noise" starting the action.

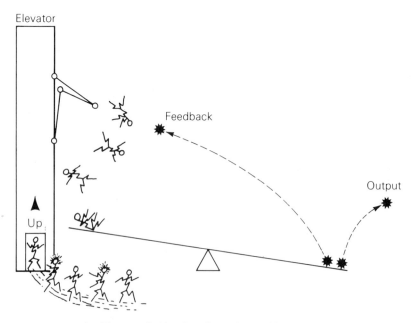

b. Photons fed back to keep the oscillation going.

FIGURE 11-5. Laser action.

been evacuated because the filament would rapidly oxidize (rust) at a white-hot temperature. The air has been replaced by nitrogen at atmospheric pressure to permit the use of a thin glass shell. Thin glass passes more light. However, without the internal nitrogen pressure the bulb would have to be made of thick glass so atmospheric pressure could not crush it.

At the high temperature generated by passing current through a high resistance, electrons are pumped to higher energy levels. As electrons are shaken loose by thermal agitation, they fall to ground level, emitting their stored energy as a broad spectrum of light. An incandescent lamp produces mostly heat and infrared radiation, with only about 18 percent of the total emission within the visible spectrum. You can place your hand near an incandescent lamp and note the large amount of heat.

FLUORESCENT LAMPS

Fluorescent lamps are a particularly good example of energy transformation. The fluorescent lamp is a gas-discharge device. The inert gases such as neon, and mercury vapor, which behaves as an inert gas, have some very special characteristics. An inert gas is normally a nonconductor, but at some critical potential (voltage), electrons are literally torn free from the atom. The freed electrons race toward the positive electrode, and the much heavier gas atoms, stripped of electrons, become positive and move at a more leisurely pace toward the negative electrode. Figure 11-6 illustrates the situation and Figure 11-7 is a graph of inert-gas conduction behavior.

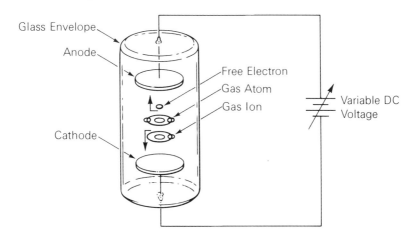

FIGURE 11-6. Glow discharge tube.

Figure 11-7 indicates that there is only a very small current (the dark current) flowing through a gas-discharge tube (like the one in Figure 11-6) until a critical threshold voltage is reached. The dark current is the result of light and other background radiation in the environment, and amounts to only a few millionths of an amp. At the critical threshold voltage the device suddenly enters the glow discharge phase. At this point the current through the device increases to about a thousand times that of the dark current and a visible glow appears around the cathode (negative electrode of the device). As the voltage across the device is increased, the glow changes color from red-orange to blue-violet. Along the glow discharge part of the curve the current may increase tenfold or more. Suddenly (and often not very predictably) the color slips up into the violet, with much ultraviolet that we cannot see. The current through the device rises rapidly to a value of many, often destructive, amps.

The avalanche mechanism is like the avalanche mechanism in semiconductors. Avalanche in gas devices is far less manageable

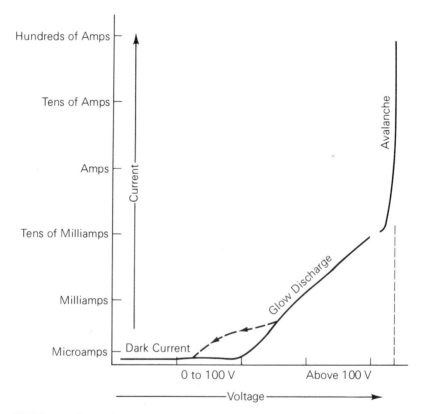

FIGURE 11-7. Gas conduction curve.

than in semiconductors, and the avalanche conduction area is generally avoided in gas devices. The dashed lines in Figure 11-7 indicate that once the device has begun to glow (discharge), the voltage must be reduced to a much lower value before the glow extinguishes. The relaxation oscillator in Figure 10-6b depends on this difference in turn-on and turn-off voltage. Figure 11-8 shows two typical neon gas lamps.

The fluorescent lamp diagrammed in Figure 11-9 is a 2-step energy-conversion device. When the gas is conducting in the glow

FIGURE 11-8. Neon lamps.

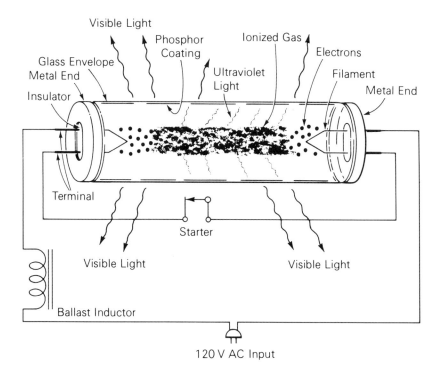

FIGURE 11-9. The fluorescent lamp.

discharge region, the bulk of its radiation is in the invisible, ultraviolet range. The inside of the tube is coated with a phosphor. The phosphor absorbs photons in the ultraviolet range. When the phosphor electrons fall back to ground state, they give up energy in the form of visible light.

The filaments emit electrons to enable the tube to start conducting at 120 volts. Without these extra electrons the tube would not go into glow discharge reliably at 120 volts. The starter turns the filaments off after the tube begins conducting. The glow will be sustained at a much lower voltage than the tube's normal firing voltage, so the tube will continue to glow after the filaments are turned off. Filaments have a relatively short life, so turning them off as soon as the tube is started drastically increases the life of the lamp. The ballast coil prevents powerline voltage surges from driving the glow discharge up into the destructive avalanche region, and is an essential part of the system.

THE RUBY LASER

The ruby laser is shown schematically in Figure 11-10. The ruby rod is an aluminum oxide compound doped with chromium atoms. The rod is a manufactured ruby, carefully grown, cut, and polished. The chromium impurity atoms are pumped up by the gas-discharge flash tube wound around the rod. As the chromium atoms fall back to ground state, they produce a characteristic red light.

One end of the rod is mirrored to be totally reflective. The other is partly reflective and allows a percentage of the light to escape from the rod. Most of the light is reflected back and forth within the rod to provide the necessary feedback stimulation to produce coherent

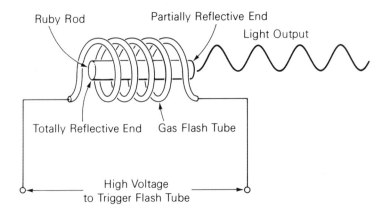

FIGURE 11-10. The ruby laser.

light. The curvature of the round rod allows only a very little light to escape through the sides. The chromium atoms are pumped up at regular intervals by delivering a pulse of high voltage to the flash tube.

There are two basic classifications of lasers: pulsed and continuous. Continuous lasers produce a steady beam and work without rest periods. High-power continuous lasers often require liquid cooling. As the internal temperature is raised, spontaneous emission, the optical analog of electronic noise, increases. Early lasers had to be cooled by very cold liquid nitrogen, and some still do require low-temperature liquified gas coolants. Most low-power lasers can now operate at normal environmental temperatures without elaborate cooling systems.

GAS LASERS

Gas lasers have become common and some have the advantage that the light color can be changed by pumping out the gas mixture and pumping in a new one. The pulsed gas laser operates much like the ruby laser. The ruby is replaced by a gas-filled glass tube. Continuous gas lasers can be pumped simply by passing an electric current through the gas and operating it in its glow discharge region.

SEMICONDUCTOR LASERS AND LIGHT-EMITTING DIODES (LEDS)

Gallium arsenide and gallium-arsenide-phosphide compounds have a compound crystal structure like that of silicon and behave as true semiconductors. These compound semiconductors do not make very good transistors, but they do make good light-emitting junction diodes (LEDs). The light-emitting diode is used for indicator lights, calculator numerical displays, and computer display systems. Figure 11-11 shows the construction of an LED.

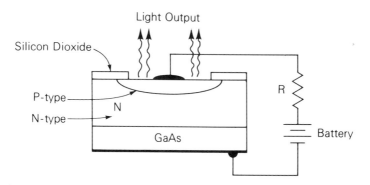

FIGURE 11-11. Light-emitting diode.

The LED is basically a PN junction diode, operating in a forward-biased mode. Recombinations of electrons and holes release the stored energy in the excited electrons. The P-type gallium arsenide must be made very thin for efficient light output. Electrons are pumped up by an externally supplied direct current, a battery in Figure 11-11. The resistor, R, is necessary to limit the current to a safe value.

The semiconductor laser is essentially the same device as the LED except that surfaces are carefully ground and polished to reflect (feedback) light back into the device to provide the necessary stimulation. Semiconductor lasers are easily amplitude-modulated by simply varying the current through them. Semiconductor lasers are currently being used in experimental communication systems in which glass fibers replace wires in telephone cables. A single glass fiber (no larger than a telephone wire) can replace thousands of conventional wires.

Study Problems

1. Explain the difference between spontaneous and stimulated emission.
2. What does the word *laser* stand for?
3. List three basic types of lasers.
4. What does the gas flash tube do in a ruby laser?
5. What are the similarities and differences between the light-emitting diode and the semiconductor laser?
6. Define coherent and incoherent light.
7. What kind of light (coherent or incoherent) is produced by lasers?
8. Explain how the fluorescent tube operates.
9. What is the function of the ballast coil in a fluorescent lamp?
10. What is the purpose of the phosphor coating on the inside of a fluorescent tube?
11. List the three states of conduction in a gas-discharge device.

11-3 CONVERTING LIGHT INTO ELECTRICAL CURRENT

The Bare Facts

1. The silicon solar cell is the most common device used to generate electrical energy directly from light energy.

2. Most other light-sensitive devices do not generate electrical current. They only vary the current from a battery or other power source as the light varies.

3. The silicon solar cell is a silicon junction diode with a very large surface area, and with the N-type material thin enough to be almost transparent.

4. Silicon solar cells have practical efficiencies of about 14 to 18 percent.

5. The silicon photodiode operates on the same principle as the solar cell, but it is much smaller, is reverse-biased, and is intended for light detection instead of power generation.

6. Vacuum and gas-filled phototubes are gradually being replaced (in most applications) by photodiodes and phototransistors.

7. Bulk photoconductive cells are made from selenium, cadmium sulphide, cadmium selenide, and lead sulphide.

8. Bulk photoconductive cells are cheap, but applications are limited because they respond comparatively slowly to changes in light intensity.

9. Bulk devices are also called *light-dependent resistors* (LDRs).

10. Selenium bulk cells can be self-generating, like silicon solar cells, but their output power is low. They are used as self-generating cells in light meters used in photography.

11. Bulk cells are used in counting devices, photoelectric intruder alarms, and so on.

12. Photodiodes and phototransistors are used in computer paper-tape readers, receivers for laser transmitters, sound motion-picture playback equipment, and other applications that require fast response to light changes.

The current interest in alternative power sources has brought the silicon solar cell to public awareness. The solar cell is one of the very few reasonably efficient devices for converting light directly into electricity. There are a number of other photosensitive devices that are not useful for power generation, but that find many applications in electronics, photoelectric relays, motion-picture sound, receivers for fiberoptic-laser communications, and so forth.

THE SILICON SOLAR CELL

The energy reaching earth from the sun is more than 500 times greater than the world's current energy consumption. The solar energy striking the average tennis court in 10 minutes is equivalent to that contained in a gallon of gasoline: some 36 kilowatt hours.

In reality we never get those 36 kw from a gallon of gasoline, nor can solar cells deliver more than a fraction of the available energy. If we use that gallon of gasoline to run an internal-combustion engine to drive a generator, the engine will develop power at only about 12 percent efficiency. The engine will produce a power output of only about 4.3 kw from the gallon of gasoline. Subtract the generator's loss of about 5 percent and the total system's power output is about 4 kw.

For silicon photocells the maximum theoretical efficiency is about 25 percent and the maximum practical efficiency is on the order of 18 percent. Few modern homes have enough land for a tennis court, let alone space for the 5 or so courts that would be required to power a home when low light level periods are taken into account. Large solar-cell farms in otherwise unused desert areas and smaller, local solar-cell farms are promising prospects once the manufacturing cost is reduced to a level competitive with other energy sources.

Figure 11-12 is a diagram of the silicon photocell. The solar cell in this figure is a silicon PN junction diode with a *very* thin layer of N-type crystal. The P-type part of the diode is much thicker and provides nearly all of the mechanical strength of the cell.

When a photon of light with the proper amount of energy strikes the face of the cell it penetrates the thin N-layer, splitting off an electron and driving it up into the N-layer. The hole that is formed when the electron is freed migrates down into the P-layer. The dislodged electron upsets the charge balance in the N-layer and an electron moves freely through the external circuit in an effort to reunite with a hole in the P-layer. It can perform useful work on the way.

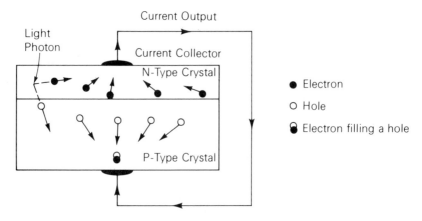

FIGURE 11-12. The silicon solar cell.

Photons with greater than the required energy also form electron-hole pairs at the junction. The excess energy is dissipated as heat. Photons with less than enough energy to free electrons simply pass through the cell.

Solar cells are classed as photovoltaic or self-generating cells. There are 3 other important classifications: photoemissive, photoresistive, and photoconducting diodes and transistors.

THE PHOTOEMISSIVE CELL

Figure 11-13a details the construction of a photoemissive tube. Figure 11-13b illustrates how the device works. The glass envelope is either evacuated or filled with an inert gas at low pressure. A photon of light strikes the sensitive cathode, dislodging an electron. Once freed, the electron is attracted to the positive anode. The only electrons available for current flow are those released by light striking the sensitive cathode. The more light, the larger the current in the external circuit. For many years this photoelectric tube has been the mainstay in industrial applications and in the reproduction of sound on motion-picture film. It is gradually being displaced by smaller and more rugged phototransistors.

PHOTODIODES AND TRANSISTORS

The photodiode operates on the same principle as the silicon solar cell. There are some small differences in physical construction between photodiodes and solar cells. The photodiode is much smaller and is intended to operate at much lower light intensities than the solar cell. The photodiode is also reverse-biased to enhance its low-light-level response characteristics.

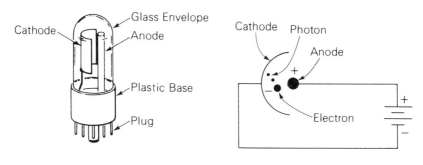

FIGURE 11-13. Vacuum or gas phototube.

A phototransistor is a properly biased transistor with a significant amount of junction area exposed to incident light. The base-collector diode operates as an ordinary photodiode and provides the base current, which is amplified in the transistor. No external base input is required and the base current is dependent on the amount of light striking the phototransistor. The sensitivity is greatly enhanced by the transistor's current gain (beta).

PHOTORESISTIVE CELLS

Selenium, cadmium sulphide, cadmium selenide, and lead sulphide are light-sensitive substances. The resistance of a film of these materials drops when light strikes the film. These so-called *bulk* photo devices require no junction or elaborate impurity control. They are sensitive and cheap, but respond too slowly to changes in light intensity for many applications. They find applications in photoelectric parts counters and anti-intrusion alarm systems. Lead sulphide responds well to infrared radiation and is used in a number of heat-seeking military systems.

Study Problems

1. Make a sketch and explain how the silicon solar cell works.
2. Do other light-sensitive devices generate an electric current? Which ones?
3. Explain the operating principle of bulk photocells.
4. What are the differences between silicon photodiodes and silicon solar cells?
5. List several materials used in bulk photocells.
6. In what way is a phototransistor superior to a photodiode?
7. Why are bulk cells not suitable for sound on film reproduction?
8. List some applications for each of the following photocell types (try to think of some not mentioned in the text):
   a. Bulk photoconductors (light-dependent resistors).
   b. Photodiodes or phototransistors.
   c. Silicon solar cells.

11-4 SOUND CONVERSION DEVICES

The Bare Facts

1. A microphone converts sound (acoustical) energy into an alternating electrical current.
2. A loudspeaker converts audio-frequency alternating currents into sound (acoustical) energy.

ENERGY CONVERSION DEVICES FOR ELECTRONIC SYSTEMS 399

3. The following types of microphones are in common use:

Type	Application
Carbon granule	Telephone
Ceramic	Home entertainment and public address
Dynamic	Home entertainment, public address, and mobile radio communications
Ribbon (velocity)	Studio recording
Capacitance	Studio and public address

4. The following are microphone principles of operation:

Type	Principle of Operation
Carbon	Diaphragm compresses loose carbon granules. Variable resistance. Requires power source.
Ceramic	Diaphragm compresses and relaxes piezoelectric crystal. Generates an alternating current.
Dynamic	Diaphragm moves a coil in a magnetic field (same principle as rotary power generators). Generates an alternating current.
Ribbon	Same principle as dynamic microphone. Uses thin ribbon instead of wound coil. Requires built-in impedance step-up transformer.
Capacitance	Diaphragm moves one plate of a capacitor, emptying and filling the capacitor in step with sound waves. Requires power source.

5. Loudspeakers are reciprocating electric motors. Their construction is very similar to that of the dynamic microphone. They operate on the same coil-in-magnetic-field principle as rotating electric motors.

6. Phonograph pickups are generally either dynamic or crystal types.

THE CARBON MICROPHONE

A microphone converts acoustical vibrations into electrical-current variations. The earliest practical microphone was the carbon-granule microphone. This microphone is still used in most modern telephones. Figure 11-14 shows how the carbon microphone is constructed. Sound waves push and pull the diaphragm, the analog of the human eardrum, back and forth. The metal cup attached to the diaphragm compresses or relaxes the loosely packed carbon granules. Compressing the granules lowers the resistance between the diaphragm and the carbon-granule container. Relaxing the granules increases the resistance. The resistance varies in step with the moving diaphragm.

A battery (or other voltage source) drives current through the microphone. The "electrical signal output" voltage varies in concert with the sound waves moving the diaphragm. This varying voltage is suitable for input to an electronic amplifier.

The carbon microphone is simple, sensitive, rugged, and produces large output-voltage variations for any reasonable sound-wave strength. However, it tends to be somewhat (electrically) noisy and has a limited high-frequency response. The limited high-frequency

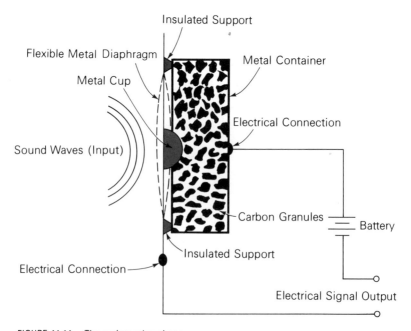

FIGURE 11-14. The carbon microphone.

response is adequate for speech, but quite unacceptable for music. The carbon microphone is a sound-varied, variable resistor.

THE CRYSTAL MICROPHONE

The quartz crystal, used in oscillators, exhibits an effect known as the piezoelectric effect. An electrical potential can deform the crystal, and conversely, a deformation of the crystal can produce an electrical potential (voltage). There are other materials that exhibit the piezoelectric effect, including certain salt crystals and special ceramic materials.

If a diaphragm is mechanically coupled to a piezoelectric crystal, we have the makings of a microphone. Theoretically, quartz could be used, but as a practical matter quartz is just too stiff. Special ceramic slabs have become the preferred material for microphones. Figure 11-15 shows the construction of a ceramic microphone.

The ceramic microphone generates an output voltage across the 2 conductive plates. No battery is required. Ceramic microphones are the most common type used with home-entertainment tape recorders and the like. Their frequency response is adequate for most music requirements in the home-entertainment field. They are not normally considered to be of broadcast or recording-studio quality, however.

When an audio signal voltage is applied across the 2 conductive plates, the ceramic microphone becomes an earphone. For earphone service the diaphragm is modified and a different case style is used.

THE DYNAMIC MICROPHONE

The dynamic microphone is an electromechanical generator that operates on the same principle as rotating power (AC) generators.

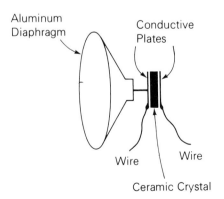

FIGURE 11-15. The ceramic microphone.

Sound creates a changing air pressure that moves a lightweight aluminum diaphragm back and forth. A coil (voice coil) attached to the diaphragm moves with it in a permanent-magnet field. The motion of the coil (generator armature) generates a small alternating voltage that follows the sound variations striking the diaphragm.

THE LOUDSPEAKER

The loudspeaker is a reciprocating electric motor, constructed like the dynamic microphone except on a much larger scale. The aluminum diaphragm is replaced by a much larger, soft paper cone. An alternating-current sound signal is applied to the voice coil (motor armature) that drives the coil back and forth in response to the sound-signal voltage from the audio amplifier. The voice coil is attached to the paper cone, which moves air, setting up the appropriate sound waves.

The speaker and microphone sometimes are used interchangeably. For example, most intercom systems use a small loudspeaker as both speaker and microphone. Microphones are rarely used as speakers because they tend to be too delicate and the diaphragm is too small to move much air. Dynamic earphones, however, are essentially modified dynamic microphones. Figure 11-16 illustrates the construction of a typical loudspeaker. The construction of the dynamic microphone is very much like that of the speaker.

THE VELOCITY OR RIBBON MICROPHONE

A studio version of the dynamic microphone uses a thin aluminum ribbon stretched between the poles of a magnet instead of a wound voice coil. The principle of operation is the same as that of a dynamic microphone. The ribbon is far less massive than the diaphragm-coil arrangement, which improves the high-frequency response. The ribbon microphone is more delicate, and because of the low resistance of the ribbon a built-in step-up transformer is required.

THE CAPACITANCE MICROPHONE

At one time the capacitance microphone was called a *condenser* microphone, and the term is still somewhat current. Capacitors used to be called *condensers*—an analog of *steam* condensers that proved to be an incorrect analog. The capacitor microphone is basically a variable capacitor, whose capacitance is varied by the use of a mechanical coupling between the diaphragm and one of the capacitor plates. The capacitor microphone is shown in Figure 11-17.

ENERGY CONVERSION DEVICES FOR ELECTRONIC SYSTEMS 403

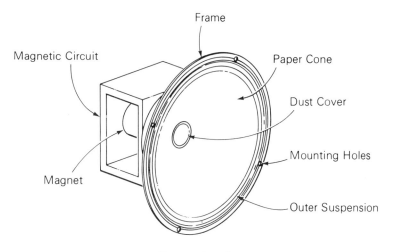

a. Perspective view.

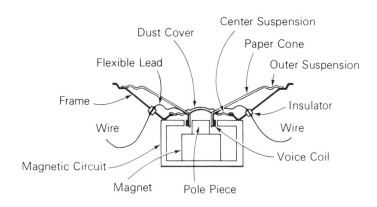

b. Cutaway view.

FIGURE 11-16. The loudspeaker.

A capacitor can store a given number of electrons when a specific voltage is applied. If its capacitance is increased, it will draw more electrons from the power source. If its capacitance is decreased it must return some electrons back to the power source. The analogy in Figure 11-18 illustrates the situation.

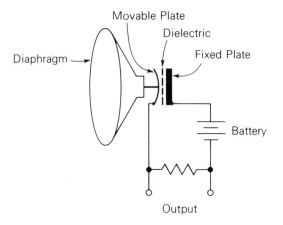

FIGURE 11-17. The capacitor (condenser) microphone.

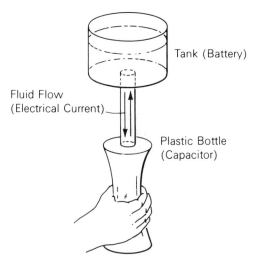

FIGURE 11-18. How the capacitor microphone works.

When the diaphragm is driven back and forth by the pressure of the sound waves, it changes the capacitance, forcing current back and forth through the resistor in Figure 11-17. This alternating-current voltage can be fed into the input of an audio amplifier. The capacitor microphone requires a power source of some sort and must be built with some precision. It is not an inexpensive microphone but its performance is definitely commercial studio quality.

PHONOGRAPH PICKUPS

The typical phonograph pickup cartridge is little more than a microphone with a groove-tracking needle instead of a diaphragm. Ceramic and dynamic types have dominated the industry. Carbon-granule versions are not used because of their inability to reproduce the high frequencies required for good music reproduction. Capacitance versions have been tried but have proved too complex and delicate for the service.

Study Problems

1. Why is a carbon microphone not suitable for music?
2. Explain how the dynamic microphone works.
3. Explain how the loudspeaker works.
4. Match the items in the following two columns:
    1. Dynamic microphone
    2. Carbon microphone
    3. Speaker
    4. Capacitor microphone
    5. Ribbon microphone

    a. Piezoelectric
    b. Telephone
    c. Variable resistance
    d. Electromagnetic generator
    e. Electromagnetic motor
    f. Requires power source
    g. Velocity microphone
    h. Requires matching transformer

Note: There may be more than one lettered item (a, b, c . . .) for each number.

11-5 THERMOELECTRIC DEVICES

Thermoelectric devices convert heat into electric current.

THERMISTORS

The thermistor is a mixture of oxides such as manganese, nickel, iron, cobalt, titanium, and magnesium. The thermistor is not doped nor is it a junction device. The themistor is a heat-dependent resistor. The resistance generally, but not always, decreases as the temperature increases. With an external power source connected, the current will increase in the circuit as the temperature increases. Principal applications involve temperature measurement and control.

THERMOCOUPLES

The thermocouple is formed as a welded junction of 2 wires. Generally one wire is iron and the other is an alloy called *constantan*. When the junction is heated, a small direct current is generated. A large number of junctions connected together in series and (or) parallel make a thermopile that can hold open a valve in a gas heater. The pilot flame heats the pile, generating enough electricity to hold the valve open. If the pilot flame goes out, the valve shuts off the gas supply.

Thermocouples will cool when an electric current is passed through the junction. The components are still fairly expensive for most applications, but thermoelectric cooling has been used to advantage in military and space vehicles.

Chapter 12

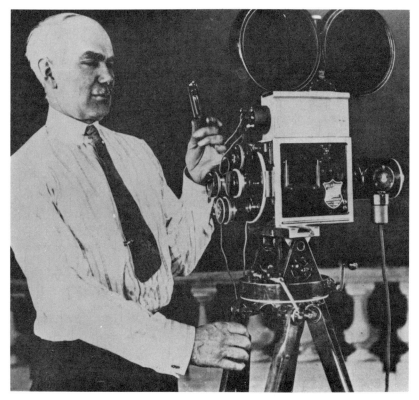

Reproduced with permission of A.T.&T. Co.

THE RECORDING AND REPRODUCTION OF SOUND

12-1 INTRODUCTION

Electronics has brought the recording and reproduction of sound from its scratchy beginnings on an Edison cylinder to concert-hall realism in your living room. Electronics is also used to amplify acoustical instruments and to alter their sound. Electronics has made organs available to individuals who could not otherwise afford one, and has provided us with musical instruments that never before existed. Portable electronic pianos, music synthesizers, and even computer-controlled musical instruments are becoming commonplace in the music world.

In this chapter we will examine both the acoustical and electronic aspects of recording and reproducing sound. The two aspects cannot be separated. The most electronically perfect music system available will produce poor results if the loudspeaker is not placed in a proper box. That box is actually an acoustical transformer and is probably the most complex of all transformers.

Loudspeaker-enclosure design is a controversial subject among both music lovers and acoustical engineers. It is still part science, part art, and mostly compromise. Some 25 years ago an acoustical engineer of the author's aquaintance built what he and many other experts felt was the nearly perfect loudspeaker enclosure. The enclosure consisted of a steel-reinforced concrete horn. The 2 meter (about 6 feet) square mouth entered through the wall of his living room. The rear of the horn tapered back to the loudspeaker, extending over 10 meters (30 feet) into his backyard. The horn weighed several tons. This was a no-compromise approach and used the best of modern acoustical know-how. It was still controversial. Some called it the ultimate in speaker enclosures; others were less impressed.

Room acoustics play an important part in the recording and reproduction of sound. Electronic devices exist that can make up for most of the inevitable compromises in acoustic design. We will look at some practical techniques for getting the best performance out of your personal sound equipment. These techniques differ very little from commercial broadcast and studio techniques, and are actually derived from commercial experience.

12-2 THE DECIBEL

The Bare Facts

1. The decibel (dB) is the standard unit for sound intensity measurement.
2. Because the bel is a very large unit, the decibel is more often used.
3. The decibel is one-tenth of a bel.

4. The decibel is always a ratio.

5. The human ear responds to a very large range of sound intensities.

6. In order to cover this wide range, the ear exhibits a nonlinear response to sound intensity.

7. The average ear can just discern a 3 decibel increase in sound-power intensity. A trained ear can detect a 1 dB increase.

8. Three decibels represent a doubling in sound power.

9. The decibel scale reflects the human ear's nonlinear response.

The unit of measurement for the intensity of sound is the bel, named after Alexander Graham Bell, the inventor of the telephone. The bel is too large a unit to be very practical, so one-tenth of a bel, the *decibel*, is universally used. The abbreviation for decibel is dB.

The sound intensities that we are exposed to range from the level our ears can barely detect to the threshold of pain, an intensity 1,000,000,000,000 times louder than a barely perceptible sound. In order to accommodate this enormous range of sound levels, the human nervous system has a built-in automatic volume-control system. This automatic volume-control system reduces the intensity, as our brain perceives it, as the volume level increases. The result is similar to the behavior of the pupil in the eye, which contracts for high light levels and allows less light to reach the visual receptors. Human hearing does much the same thing for sound intensities, although the actual mechanism is not well understood. The decibel (dB) scale reflects the human ear's response to this extreme range of sound levels.

Table 12-1 summarizes the range of sound intensities. Most sound measurements using decibel measure define zero decibels as 10^{-12} watts per square meter, which is approximately the human threshold of hearing. Because of the tremendous range of intensities involved, the decibel scale has been constructed using powers of 10. Decibels are always a ratio.

Table 12-2 shows the decibel values for various power, voltage, and current ratios.

For the mathematically interested:

> $dB = 10 \log_{10} P_1/P_2$
> where dB is the number of decibels, $\log_{10}$ is the base 10 logarithm, and P_1/P_2 is the ratio of two power levels.

TABLE 12-1 THE DECIBEL SCALE

Example	Decibels	Relative Loudness
Threshold of Pain	120	1,000,000,000,000
Jet Engine	110	100,000,000,000
Riveting Machine	100	10,000,000,000
Train	90	1,000,000,000
Heavy Street Traffic	80	100,000,000
Diesel Truck	70	10,000,000
Busy Street	60	1,000,000
Vacuum Cleaner	50	100,000
Typical Office	40	10,000
Quiet Office	30	1,000
Typical Home	20	100
Whisper	10	10
Threshold of Hearing	0	0

TABLE 12-2 THE RELATION BETWEEN DECIBELS AND POWER AND CURRENT OR VOLTAGE RATIOS

Power Ratio	Decibels	Current or Voltage Ratio	Decibels
1	0	1	0
2	3.0	2	6.0
3	4.8	3	9.5
4	6.0	4	12.0
5	7.0	5	14.0
6	7.8	6	15.6
7	8.5	7	16.9
8	9.0	8	18.1
9	9.5	9	19.1
10	10	10	20
100	20	100	40
1,000	30	1,000	60
10,000	40	10,000	80
100,000	50	100,000	100
1,000,000	60	1,000,000	120

> For voltage and current ratios:
> dB = 20 $\log_{10}$ V_1/V_2 (voltage)
> dB = 20 $\log_{10}$ I_1/I_2 (current)

Key parts of the decibel scale are easy to remember:

1. Each decibel increase or decrease means about a 25 percent increase or decrease in power level.

2. A power-level increase of 3 decibels corresponds to a doubling of power. Power ratio = 2.

3. A power level decrease of 3 decibels indicates that the power level has been reduced to half of its value.

$$\text{Power ratio} = \frac{1}{2} \text{ (or 0.5).}$$

How to Use the Decibel Chart (Table 12-2):

Example 1:

What is the decibel power gain of an amplifier with an input power of 25 watts and an output power of 100 watts?

1. Find the power ratio: Power ratio = 100/25 = 4.
2. Look up the power ratio (4) on Table 12-2.
   Answer: The amplifier has 6 decibels of power gain.

Example 2:

What is the loss in decibels in an audio cable when the input power is 10 milliwatts and the output power is 5 milliwatts?

1. Find the power ratio: Power ratio = 10/5 = 2.
2. Look up the power ratio on Table 12-2.
   Answer: The audio cable has a loss of 3 decibels.

Study Problems

1. An amplifier has an input power of 1 watt and an output power of 8 watts. What is the amplifier power gain in decibels?
2. A voltage amplifier has a voltage gain of 100. What is the voltage gain in decibels? (*Hint:* Voltage gain is the voltage ratio.)

12-3 THE GREAT AUDIO POWER DEBATE

The Bare (Controversial) Facts

1. Excessive sound levels over a period of time can result in permanent hearing loss.

2. The excessive-volume-level habit in music listening has come about in an attempt to compensate for poor loudspeaker systems.
3. Modern speakers are quite good within certain practical contraints.
4. The best speaker available is a poor performer unless it is placed in a proper enclosure.
5. Most available speaker boxes are poor acoustical devices and cannot get the most out of even an expensive loudspeaker.
6. Good speaker enclosures are available, but with a few exceptions they are expensive and large.
7. Good speaker enclosures have been produced at a reasonable price, but have not done well in the marketplace because the public is generally unaware of how important the right enclosure is.
8. There are many misconceptions about speaker enclosures. Most of them will be discussed in this chapter.

The automatic-volume-control system in the human ear has another feature that protects the hearing system from persistent environmental noise. The volume-control system permanently reduces hearing capability for certain frequencies where sound levels exceed safe values for prolonged periods. The phenomenon became obvious during World War II, among both military personnel and factory workers. Riflemen developed a dead spot at certain frequencies in their hearing. Men who fired larger guns lost hearing in a different frequency range, and aircraft riveters developed a still different loss. The hearing loss in many cases proved to be irreversible.

An example of the importance of acoustical considerations is the high volume used in the playing and performance of "rock" music. This started because early high-fidelity music fans (particularly classical music fans) believed that a very high volume was necessary to achieve concert-hall realism. The demand for more powerful amplifiers led to a power race among manufacturers, so that when rock became popular, outrageously powerful amplifiers were already available. And still the power war escalated, producing a generation of more or less hard-of-hearing young people.

The situation turned out to have a tragicomic aspect. The early classical fans turned up the volume to begin with because of the poor speaker enclosures available. Bass instruments were washed out by most affordable speaker enclosures. Normally, bass instruments produce higher sound pressures, but less (subjective) loudness than instruments in the middle and high-frequency ranges. In order for a speaker enclosure to have a good smooth bass response, it must usually be quite large. Large enclosures are bound to be much more expensive (and take up more space) than smaller ones. To make up for the low acoustical power in the low-frequency range, fans simply cranked up the power for all ranges. The bass instruments sounded better and the human ear compensated for the extra loudness of the middle and high frequencies.

Then came rock, where the beat was of primary importance, demanding even greater emphasis on the bass guitar and consequently demanding more amplifier power. To compound the problem, amplifiers got larger, and since amplifiers and speakers had to be portable, speaker-box size often got smaller so the equipment could be more easily transported. The smaller speaker enclosures reduced the bass response and demanded still higher power levels.

The problem came about initially because of inadequate speaker enclosures and got out of hand because people did not understand decibels and the response of the human ear. Increasing power in a brute-force attempt to make up for an acoustical deficiency was self-defeating both economically and in terms of health. If the extra money spent to purchase increasingly powerful amplifiers had been spent in the first place on better enclosures, the whole problem might have been avoided.

A few years ago a small company produced a relatively large horn type of enclosure, cleverly folded into a small package. It was made of plastic foam with an outside covering of wood veneer or plastic laminate. The idea of folding a large horn into a small space is not new. Nearly all of the very best enclosures are folded horns, but they involve a great deal of intricate woodwork and are quite heavy. The plastic-foam version eliminated the costly woodwork by simply molding the intricate folds in plastic foam, and the whole assembly was very light. The plastic-foam enclosures were remarkable performers, but they did not sell. The buying public simply did not understand the problem and continued to buy overpowered amplifiers and put speakers into laughable enclosures.

The average human ear can barely detect a difference of 3 decibels in acoustical power. This means that the real acoustical power must be doubled to be barely discernible to the ear. Replacing your 50 watt amplifier with a 100 watt amplifier provides a barely noticeable difference!

If you bought a new 100 watt guitar amplifier, with the amplifier and speaker in the same box, the amplifier would be larger, take up more space in the box, and probably reduce the efficiency of the enclosure. The actual acoustical power increase may be less than discernible to the ear. Psychologically, after spending a couple hundred dollars on the new unit, it will be impossible to convince you that you have purchased almost zero increase in power, as the human ear perceives it. But the fact is that you have purchased very little, if any, additional hearable power in stepping up from 50 to 100 watts.

THE MORAL:

Brute power is a very poor substitute for proper acoustical equipment, and it can be hazardous to your health.

12-4 ACOUSTICAL TRANSFORMERS

The acoustical transformer is found in every area of sound reproduction. The vibrations set up in the air by the reeds in woodwind instruments, loudspeaker diaphragms, and so on, are almost invariably a poor impedance match to the free-air environment where the sound is heard. Even auditoriums and outdoor band shells function as acoustical transformers.

The horn is the nearly universal acoustical transformer shape. The small end is always near the sound generator, and the horn flares out from there into the larger space where listeners congregate. Every theater stage forms a small focal point and flares out until the row of seats in the back is much wider than the stage. You can come up with a number of reasons for this shape, including just plain tradition as well as the desire to accommodate larger paying audiences who can all see the stage. But none of these answer. Many performers over the centuries have wanted a larger stage on which to perform. Stage managers have almost always ignored such requests. Perhaps they could give no reason except costs and tradition. The cost argument probably would not stand up because a rectangular building would almost always be cheaper than the traditional flared shape. The reason is that the flared structure forms an acoustical horn (transformer), and people can hear the actors or musicians on the stage from anywhere in the theater. The tradition of a flared-shape theater dates back to long before acoustics was a science or even a developed art. Figure 12-1 illustrates a number of horn-shaped acoustical transformers.

In the case of the theater, people sit within the horn, and the theater shape can be considered to be a continuation of the small horns on the various instruments. The acoustical equivalent of two transformers in tanden is illustrated in Figure 12-2.

Stringed instruments require a different acoustical transformer design. Unlike a diaphragm or reed, a plucked string moves very little air. The enclosure (sounding box) for guitars, violins, and the like is a

THE RECORDING AND REPRODUCTION OF SOUND 415

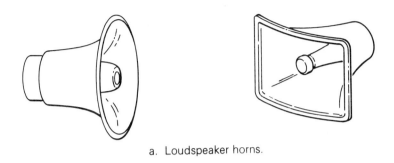

a. Loudspeaker horns.

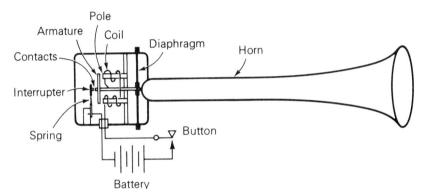

b. Automotive horn.

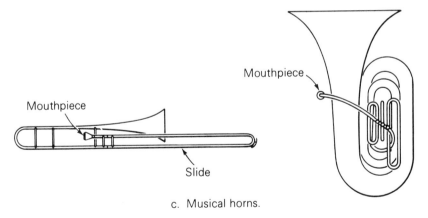

c. Musical horns.

FIGURE 12-1. Horns.

complex system of acoustical resonant circuits. The design problems in such a system are far more complex than those of designing a simple horn. The design of these instruments has been an evolutionary one. The violin design required thousands of years to reach the perfection of a Stradivarius.

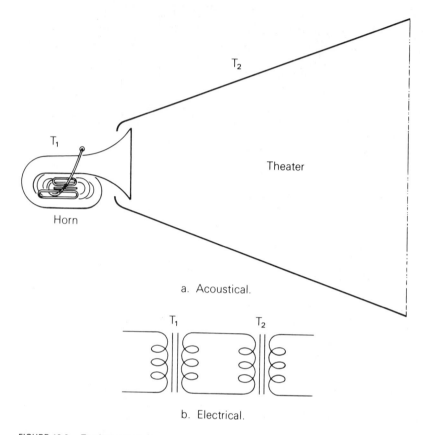

FIGURE 12-2. Tandem transformers.

The violin is the most acoustically perfect system known to man. Computers and other modern technology have been used in an effort to develop a better violin, but without success. Improvements are possible in other members of the string family such as the bass viol, the cello, and the guitar, but the physical changes would tend to make them nearly unplayable.

The best loudspeaker enclosures are horns or folded horns. Designs based on the resonant systems used in stringed instruments are also used in speaker enclosures. These designs take up less space (and are much less expensive) but rarely approach the performance of a well-designed horn.

LOUDSPEAKER ENCLOSURES

Loudspeaker enclosures come in 3 basic types: the folded horn, the closed box, and the ported-box reflex enclosure. There is not much

THE RECORDING AND REPRODUCTION OF SOUND 417

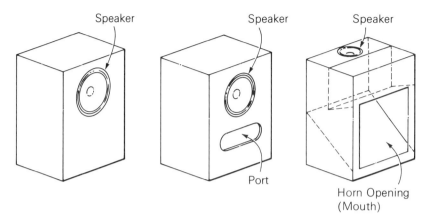

a. The closed box.   b. The base-reflex (ported) enclosure.   c. The folded horn.

FIGURE 12-3. Types of speaker enclosures.

question that the folded horn, properly designed, is the best of the three, but its cost is usually prohibitive and it, too, is a compromise system. Figure 12-3 illustrates the 3 types of enclosure.

Some Speaker-Enclosure Facts:

Fact: The box should contribute no vibrations of its own to the system. Concrete would be better than wood but is generally impractical. Stiffness of the panels is the prime concern.

Fact: Speakers, ports, and braces should all be placed slightly off center to reduce standing waves and resonant vibrations in the box.

Fact: Braces should be placed parallel to the long dimension of the panel.

Fact: The front of the box should flare away like a horn or have a spherical shape in front. As a compromise, front corners should be rounded or beveled, and any projections should be eliminated. Sharp corners can cause diffraction of the sound and produce a rough quality.

Fact: The addition of sound-absorbent materials actually increases the effective size of the box and improves the low-frequency response. It does tend to kill some brightness that is a result of internal reflections and that should not be there in the first place. This synthetic brightness is a form of distortion and causes listener fatigue.

Fact: If a bass-reflex enclosure booms, it is more often because the box is the wrong size or the speaker is poor. Too small a speaker magnet often produces this effect. A closed box tends to be more tolerant of cheap speakers than a bass-reflex enclosure.

Fact: It is not possible to have both good bass response and high efficiency in a small box. One must always be traded off for improvement in the other. Speakers will almost always lose some efficiency in small enclosures (there goes some more expensive amplifier power). With a proper design, a larger box is almost certain to be superior in both bass response and efficiency.

ACOUSTICS IN AUDITORIUMS, THEATERS, AND RECORDING STUDIOS

Acoustical problems in auditoriums, theaters, and studios are many and variable, and good acoustics for all situations are difficult to obtain. The shape of the hall, sound dampening to reduce reverberation, and speaker and microphone placement are all involved. Variables such as audience size and distribution also complicate the problem.

The control of sound reflections (reverberation) is one of the most difficult of the problems. Early radio studios and recording studios were designed to have almost no reverberation because it was easier to eliminate it altogether than to control it. A total lack of reverberation takes much of the life out of a musical presentation, so special reverberation chambers and spring reverberation units were used to mix controlled amounts of reverberation into the music before it was recorded or transmitted. The disadvantage of this scheme was that musicians and vocalists in a dead room could not hear each other very well. This made it difficult for vocalists and musicians to stay on pitch.

A modern approach to the control of reverberation, without eliminating it, involves curved surfaces that diffuse the reflected sound in random directions. The principle is much the same as that of the frosted glass used to diffuse light in light fixtures. A combination of polycylinders and diffusospheres can be placed experimentally for maximum results. Figure 12-4 illustrates the use of these techniques.

ACOUSTICAL FEEDBACK

Those microphone squeals that cause so much trouble are a case of positive feedback. They are oscillations with an acoustical feedback network. Directional microphones, directional column-type speakers, and special chambers built into microphones are useful in coping with the problem.

THE RECORDING AND REPRODUCTION OF SOUND 419

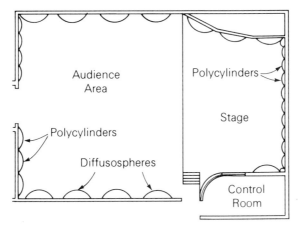

a. Top view.

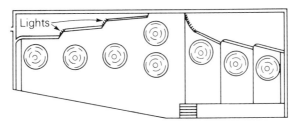

b. Side view.

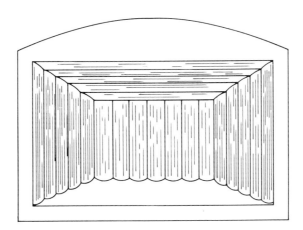

c. Perspective view of polycylinders.

FIGURE 12-4. Polycylinders and diffusospheres.

The most powerful tool for solving many acoustical problems, including acoustical feedback, is an electronic device called a *graphic equalizer*. Such devices have been used in broadcast and recording studios for many years, but only in the last few years have they been affordable in other applications. Modern integrated circuits have made them available within the price range of even the smallest theaters and small bands.

The equalizers contain from 5 to 10 amplifiers and filter networks. The audio-frequency response curve is divided into 10 segments with a control, an amplifier, and a special filter network for each segment. Each small segment over the audio range can be adjusted by the operator without affecting any other frequencies. Each small band of frequencies can be made much louder than the rest or it can be nearly eliminated, as the situation requires.

In the case of severe feedback, with a little experimentation the operator can quickly find the frequency of oscillation. He can reduce the overall gain at that frequency and stop the feedback. The entire frequency response curve can be tailored to the peculiarities of the room, the speakers, and so on. Minor adjustments can even be made for varying audience size and distribution.

12-5 RECORDING

DISC RECORDING

Disc recording is the oldest and still one of the most common recording methods. Among the various methods disc recording has the advantage of a well-developed, cheap mass-production technology.

Suppose we look at the methods of production of modern disc recordings. The process starts in the recording studio. Vocalists and musicians often record separately to allow the audio engineers to achieve a near-perfect balance between the music and vocal. The recording is done on magnetic tape. After the recording session is finished, the sound engineers blend music and vocal, and edit the tape to produce the finished product. The tape is then copied onto a plastic-coated aluminum disc, using a recording lathe. A lead screw drives a cutting head (Figure 12-5) across the disc, cutting a spiral groove from the outside to the inside of the disc. The recording head carries a cutting tool (the stylus) that vibrates as the electrical signal varies. The disc produced by the recording head is called the *original*. It is metalized and then electroplated. The thin plating, a perfect negative replica of the original (Figure 12-6b), is removed from the original and backed by a solid metal disc. This is the master.

The process is repeated to get a positive replica. Again the thin plated sheet is stripped off and backed by a solid metal plate. The

THE RECORDING AND REPRODUCTION OF SOUND   421

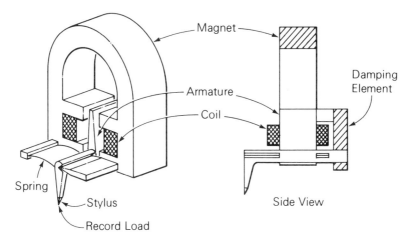

FIGURE 12-5. The disc recording head.

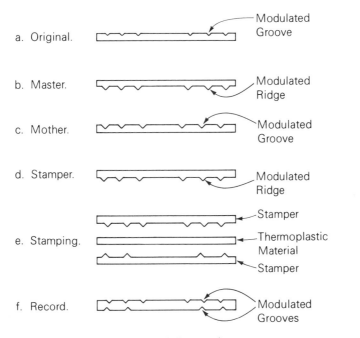

FIGURE 12-6. Stamping commercial disc records.

assembly is called the *stamper*. Two stampers, each carrying the musical selections for one side of the record, are used to press the grooves into a hot thermoplastic disc. See Figure 12-6 for the detail of the steps.

For stereo recordings, the recording stylus is driven by 2 recording coils, one for each channel. The 2 recording coils are placed at an angle to provide separation between the 2 channels. The pickup used to play back the recording also contains 2 ceramic cartridges coupled to the playback needle at the same angle used in the recording head.

MAGNETIC-TAPE RECORDING

Magnetic recording uses a plastic tape coated with iron oxide or some other easily magnetized material. The tape is driven by a capstan shaft and rubber pinch wheel (puck). The tape is not driven by the reel because the tape speed would vary with the amount of tape on the reel and the tightness with which it was wound. Figure 12-7 shows the components of a typical reel-to-reel tape deck.

Stereo machines use 4 tracks, 2 on each half of the tape. Each pair of tracks form 2 stereo channels. The information is impressed on the tape by a dual transformer-like magnetic head. There is a tiny air gap in the transformer core. The tape is pressed against the gap by a pressure pad and completes the magnetic circuit. This results in the application of a strong magnetic flux in small segments on the part of the tape that is in the gap at each instant.

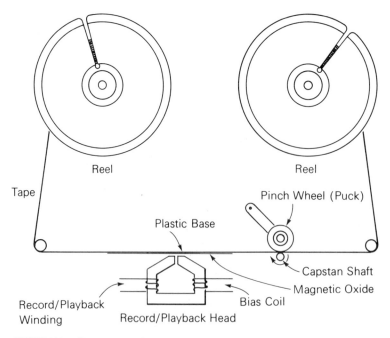

FIGURE 12-7. The tape recorder.

The audio signal applied to the head varies the strength of the field across the air gap, which aligns the magnetic domains in the tape's magnetic coating. When the tape leaves the head it is permanently magnetized with a magnetic strength that is proportional to the electrical signal (head) current that was present when the tape was passing over the head.

The record playback head has an extra winding on its core. A bias oscillator signal (about 40 to 80 kilohertz) is applied to this winding to dislodge the magnetic domains. Shaking the domains loose before the recording head sets them in new positions greatly improves the recording fidelity and reduces distortion. The same oscillator signal may also be applied to a separate erase head at a greater field strength to ensure complete erasure of old program material.

The recording head is often used for playback. The bias oscillator is disconnected during playback and the head is switched from the amplifier output to the input. The head now becomes a small AC generator as the magnetism induced from the tape into the core varies. Figure 12-8 is a block diagram of the tape-recorder electronics.

OPTICAL RECORDING

Commercial motion-picture film uses a modulated light source to expose the film sound track. There are two methods of sound-track recording: variable density and variable width. Figure 12-9 illustrates the two methods.

Electromechanical systems using tiny mirrors have been used to modulate light since the beginning of sound-track recording. It is not practical to modulate an incandescent lamp by varying the voltage to it because the filament cannot heat and cool fast enough to respond to the higher audio frequencies. Light-emitting diodes, however, can be easily modulated by varying the voltage, and they can change light levels rapidly enough to respond to well beyond the highest audio frequencies. Light-emitting diodes are now replacing mechanical systems in commercial equipment. Optical sound tracks have also been brought into the price range of the home movie enthusiast as a result of the invention of the light-emitting diode.

12-6 INTERCONNECTING AUDIO EQUIPMENT

One of the most troublesome problems in audio electronics equipment is hum and noise pickup due to the improper use of interconnecting cables and grounding methods. There are no easy solutions to some of the more stubborn problems, but a number of techniques have been developed that avoid nearly all of the common hum and noise problems.

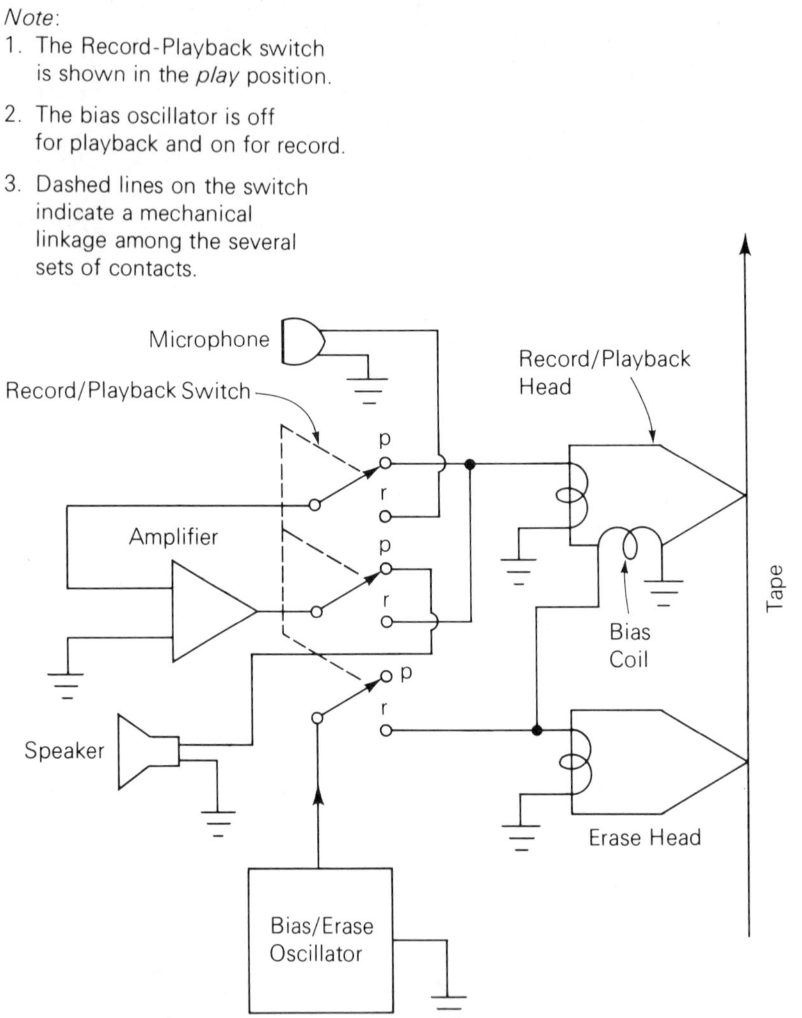
Note:
1. The Record-Playback switch is shown in the play position.
2. The bias oscillator is off for playback and on for record.
3. Dashed lines on the switch indicate a mechanical linkage among the several sets of contacts.
Microphone
Record/Playback Switch
p
r
p
r
p
r
Amplifier
Speaker
Bias/Erase Oscillator
Record/Playback Head
Bias Coil
Erase Head
Tape

FIGURE 12-8. Tape-recorder electronics.

SHIELDED CABLE

All cables leading into an amplifier input must be special shielded cable. Shielded cable consists of an insulated center conductor surrounded by a braided or twisted-wire tubing. The outside braided shell is the shield and is always connected to the chassis ground or common. Shielded cable is difficult to work with and can break, short, or cause other problems if it is not used correctly.

Speaker cable does not require shielding and can be run over long distances with little or no problem. When speaker wires are connected to screw-type terminals, spade lugs should be used. The wire

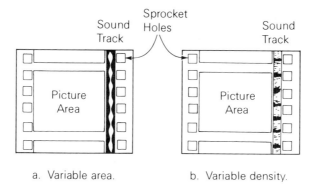

FIGURE 12-9. Optical sound tracks.

must be multistrand wire to avoid work-hardening breakage, but this leads to the possibility of a loose strand bridging the 2 terminals. A short across the terminals can destroy expensive power transistors in the amplifier if it does not have built-in short-circuit protection. At best, a single-strand short will cause low (or no) power to be delivered to the speaker. Properly connected spade lugs eliminate this danger.

All joints in audio cables and wires must be soldered. The solder must be designed for electronics soldering. Acid-core solders corrode equipment and cause electrical shorts. The solder should be a tin-lead alloy with at least 60 percent tin.

12-7 ELECTRONIC REVERBERATION

Synthetic reverberation always involves a time delay in the sound. This can be accomplished in several ways. In radio a few years ago, bare basement rooms were equipped with a loudspeaker and microphone, and the sound was allowed to bounce around the smooth walls. A slightly later technique involved lengths of garden hose or surgical tubing with a small speaker at one end and a microphone at the other. The sound traveled through 20 to 100 meters of tubing before it was added to the original sound. A number of these systems, with varying lengths of hose or tubing, provided considerable control over reverberation time delays. The coiled tubing also took up much less space than the earlier reverberation chambers.

Coiled springs driven by an electromagnetic driver (similar to a disc recording head) and picked up by a modified phonograph pickup at the receiving end are a common reverberation unit in home-entertainment equipment. Coiled springs provide the delay and are quite compact, but they do introduce some spring vibrations into the original sound.

426 CHAPTER 12

The latest reverberation system is completely electronic. The heart of the unit is a charge-coupled integrated-circuit device. At the time of this writing, the latest systems are somewhat expensive for home-entertainment use, but if they start selling in adequate quantities the price will come down drastically.

Figure 12-10 illustrates these reverberation systems.

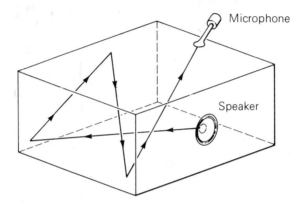

a. Reverberation chamber.

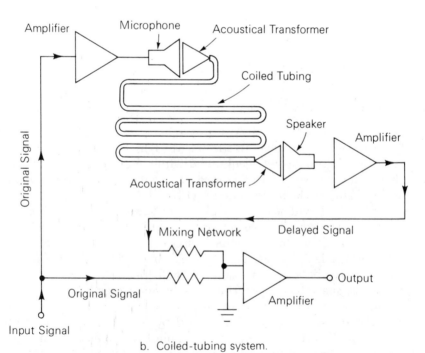

b. Coiled-tubing system.

FIGURE 12-10. Reverberation systems.

THE RECORDING AND REPRODUCTION OF SOUND

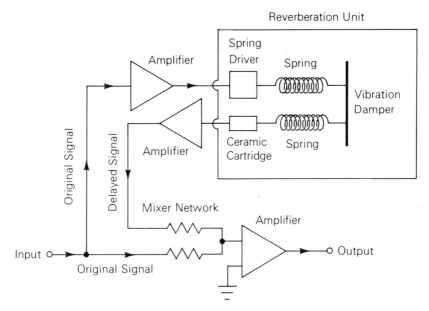

c. Spring reverberation unit.

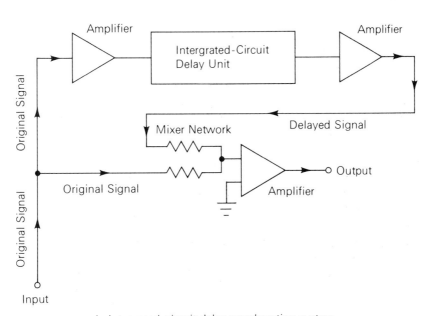

d. Integrated-circuit delay reverberation system.

FIGURE 12-10. (continued)

Study Problems

1. Why are theaters constructed like a horn with the audience sitting in the mouth?
2. What makes the folded-horn type of loudspeaker enclosure so expensive?
3. What increase in acoustical power is required before the human ear can detect that increase? Give the value in decibels, and as a power ratio (1, 2, 4, . . . times the increase).
4. What is the electrical analog of a horn?
5. What is the purpose of the shield in shielded microphone cable?
6. What is acoustical feedback?
7. What techniques are used to minimize acoustical feedback?
8. Why are a shaft and a pinch roller instead of a reel drive used to move the tape in a tape recorder?
9. List some methods commonly used to produce synthetic reverberation.
10. List the three basic types of loudspeaker enclosures.

Chapter 13

TELEVISION

13-1 INTRODUCTION

Television is very much a part of contemporary life. It is a very complex system, particularly when color is involved. Simpler color systems are possible, but the technology is fairly well fixed by Federal Communications Commission decisions that were made before color TV was developed. The FCC required that color broadcasts be compatible with black-and-white sets. The term *compatible* means that a color broadcast must produce an acceptable picture on a black and white set. The FCC also established specifications for TV signals and the maximum bandwidth that a TV signal is allowed. A great deal of ingenuity was required to squeeze the extra signals necessary for color into the already crowded bandwidth.

Artists have long known that any color (or a good approximation) could be produced by mixing 3 primary colors. Color television uses this principle. Because of the limited bandwidth available, the 3-color information had to be encoded at the transmitter and decoded at the receiving end. The result was very complex receiver circuitry. In this chapter we will cover the important concepts but necessarily leave out some of the more complex details.

13-2 HOW TELEVISION PICTURES ARE MADE

The Bare Facts

1. Each television picture consists of 525 lines containing up to 150,000 light and dark spots (picture elements) for each picture.

2. Sixty "still" pictures per second are displayed to provide the illusion of motion.

3. Each picture is scanned in the same order as words on the printed page. Each of the up to 150,000 elements of varying brightness is presented in sequence, in the same way a child's eye might scan a printed page a letter at a time.

4. At a normal viewing distance the human visual system "sees" a complete picture and not individual dots.

Television, like motion pictures, takes advantage of two peculiarities of human vision. The human visual system assembles discrete pictures or picture elements into a continuous image if individual pieces are presented rapidly enough. The human visual system also forms a continuous image when a large but finite number of individual picture elements are presented. The photo that starts this chapter illustrates the phenomenon. The right-hand side of the picture is recognizable, but even from a distance the individual picture elements are easily visible. On the left-hand side, however, you would need a magnifying lens to see the individual elements because they are smaller and there are more of them.

The right-hand side of the photograph seems coarse because it lacks enough elements to resolve the fine detail. The individual elements tend to blend as the viewing distance is increased. If you look closely at a television picture, you can see some individual picture elements and individual lines across the screen. At the proper viewing distances the picture blends into a continuous whole.

THE PICTURE TUBE

The Bare Facts

1. In the picture tube a stream of electrons is fired at a phosphor-coated screen. The phosphor emits light when struck by fast-moving electrons.

2. The beam is produced by a filament-cathode assembly similar to that in a vacuum tube.

3. The electron beam is traced up, down, and across the screen by a coil assembly called the *deflection yoke*.

4. The yoke slips over the outside of the neck of the tube.

5. The picture tube is evacuated to prevent collisions between electrons and air molecules.

6. The glass envelope must be very thick to keep the atmospheric pressure from crushing it.

7. The high-voltage anode requires 10,000 to 30,000 volts.

8. The color picture tube contains 3 electron guns, one for each of the 3 colors: red, blue, and green.

9. The screen has spots of phosphor arranged in a triangular group of 3, with one dot for each of the colors.

10. There are approximately 342,000 dots for each color, making a total of about 1,026,000 phosphor dots.

11. A thin, perforated metal plate, called the *aperture* or *shadow mask*, ensures that each beam hits the right color spot anywhere on the screen.

12. In some color tubes the phosphors are placed on the screen in stripes arranged in groups of 3 (one for each color).

13. When the phosphors are laid down in stripes, the shadow mask contains slits instead of holes.

The picture tube takes advantage of the fact that certain phosphors glow when struck by high-speed electrons. The wide part of the bell-shaped glass "bottle" in Figure 13-1 has a more or less flat glass cover. This "screen" is coated on the inside with a phosphorescent material. Extending from the narrow end of the bell is a small-diameter glass tube (the neck), which contains a filament-cathode structure and a control grid much like those found in ordinary vacuum tubes. Within the neck there are also electrodes that concentrate the electrons into a pencil-point-like beam directed at the center of the screen.

An external set of electromagnets deflect the beam up or down and to the left or right. This external electro-magnetic assembly is called a *deflection yoke* and is slipped over the neck of the tube. The inside of the "bell" is sprayed with a conductive coating that serves as the vacuum-tube plate (anode). A button connector on the outside of the bell connects to the internal anode coating. The outside of the bell is also coated with aquadag conductive coating. The outside aquadag is connected to the chassis ground. The inside coating, the picture-tube glass shell, and the outside coating form a filter capacitor for the 30,000 volt anode power supply. A wire from a 30,000 volt power supply connects to the button on the bell of the tube (see Figure 13-2). This built-in capacitor can hold much of the 30,000 volt charge for several days and can be a serious shock hazard for the unwary.

The face-plate of the picture tube is about a centimeter (one-half inch) thick, and the rest of the tube is also made of very thick glass. The air is pumped out of the tube, so that almost the entire weight of the atmosphere exerts a crushing force of nearly 2 metric tons on the face of a 21-inch picture tube.*

THE COLOR PICTURE TUBE

The RCA-type color picture tube has 3 electron guns, one each for red, blue, and green primary colors. One gun is aimed at the red, one at the green, and one at the blue dots. The electron guns are directed through a perforated metal mask to ensure that each beam will hit the proper color phosphor dot on the screen. Figure 13-3 shows how

---

*Changing a picture tube is a very dangerous job for the home mechanic, and is NOT recommended.

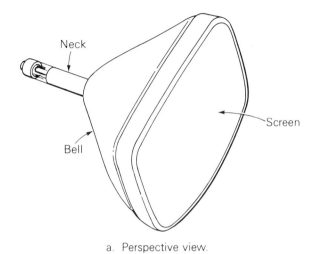

a. Perspective view.

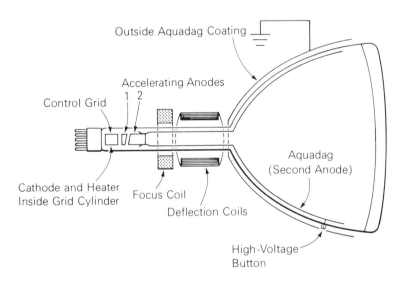

b. Construction.

FIGURE 13-1. The picture tube.

the mask works. In some color picture tubes the color phosphors are laid down in narrow stripes instead of triads of dots, and the mask has slits instead of holes to direct the beam.

As the electron beams leave the guns, they are swept from left to right across and down the screen, in the same fashion as we read the printed page, by the deflection yoke. This deflection coil assembly is

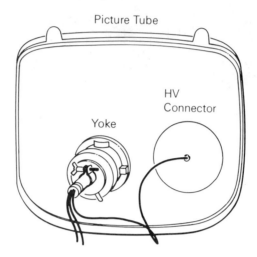

FIGURE 13-2. The high-voltage anode connection.

driven by changing electric currents from electronic sweep circuits that cause the electrons to trace out the picture on the screen.

The phosphor dots glow red, green, or blue, according to their kind when struck by the electron stream. Each group of three primary colors is resolved by the eye into a dot of one color. Sixty complete pictures or frames are traced out each second. These individual still pictures are displayed rapidly enough to present the illusion of continuous motion in the same way motion-picture film gives us the illusion of motion.

THE TELEVISION CAMERA

The Bare Facts

1. The *vidicon* is one of the most popular camera tubes.

2. The vidicon contains a target consisting of a thin layer of photoconductive material and a transparent conductive layer.

3. The resistance of the photoconductive layer diminishes with an increase in light intensity. An electron beam scans the target, producing a current that is proportional to the light intensity at the spot on the target being struck by the beam.

4. A lens system focuses an image on the target.

5. A color camera uses 3 camera tubes, one for each color: red, blue, and green.

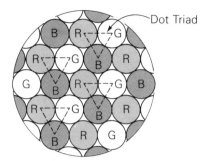

Phosphor pattern with 342,000
Sets of Red, Green, and Blue Dots
There is a total of 1,026,000 dots
0.25 millimeters (0.01 inch) in diameter.

a. Phosphor dot patterns on the screen.

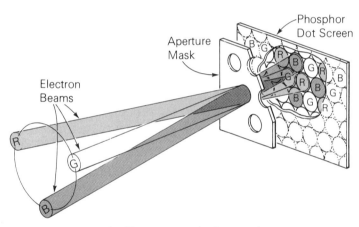

b. The aperture shadow mask.

FIGURE 13-3. The color picture tube.

There are several television-camera tubes, but the vidicon has become established as the television workhorse.

Figure 13-4 shows the construction of the vidicon. An image of the scene is projected on the approximately 1 square centimeter target by a lens system. The target consists of 2 layers: a transparent conductive layer, and a film of bulk photoconductive material. An electron beam scans the target. The resistance varies from about 2 megohms for brightly lit elements to about 20 megohms for dark elements (see Figure 13-5b). When the electron beam is directed at a specific picture element, the beam current is determined by the resistance of the

436 CHAPTER 13

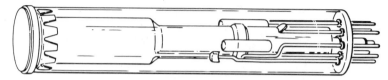

a. What the vidicon looks like.

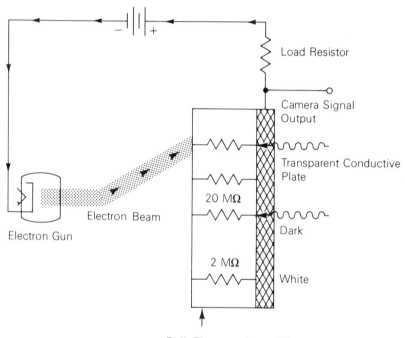

b. How the vidicon works.

FIGURE 13-4. The vidicon.

element being sampled by the beam. The resistance of that element depends on the amount of illumination at that spot.

The color camera uses 3 camera tubes, corresponding to the red, green, and blue picture-tube guns. The colors from the scene are separated by a set of 3 prisms as shown in Figure 13-5. A filter in front of each vidicon ensures color purity.

The 3 colors are not transmitted separately because black-and-white receivers must be able to show a black-and-white picture even though it is transmitted in color. As a result of this compatibility

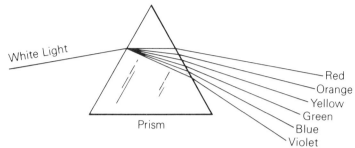

White light, when passed through a glass prism, breaks down into the colors of the spectrum.

a. The prism.

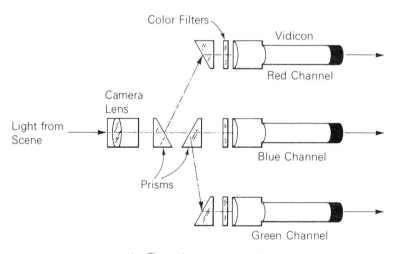

b. The color-camera system.

FIGURE 13-5. The color camera.

requirement, the color signals are mixed at the transmitter, and extra circuitry must be included in the color receiver to separate the 3 colors out of the composite signal.

SCANNING THE PICTURE

The Bare Facts

1. At the TV transmitter, 2 special oscillators sweep the camera-tube beam over the target area.

2. At the receiver, 2 special oscillators sweep the picture-tube beam over the screen.

3. The transmitter produces synchronizing pulses that keep the receiver scanning oscillators in exact step with those at the transmitter.
4. As the beam scans the picture tube, other signals from the transmitter control the picture-tube electron guns.
5. The electron guns can be varied so that the light on the screen ranges from dark to bright.
6. The electron beam traces out 262½ lines across the screen. The beam then returns to the top of the screen and traces out 262½ additional lines. The second group of lines falls in between each pair of lines previously traced. The process is called *interlaced scanning*.
7. Interlacing reduces flicker.
8. Each group of 262½ lines is called a *field*.
9. Two fields are required to make a frame (one complete still picture).
10. Thirty complete pictures are transmitted each second to produce the illusion of motion.

Two oscillator circuits deliver varying currents to the 2 sets of coils in the deflection yoke. The beam moves from left to right at a steady rate. At the end of the line the beam moves very rapidly back to the left and down one line. The beam then traces out the next line. While the beam moves across the screen, the control grid in the picture tube varies the intensity of the beam according to the picture information. The result is a line consisting of light and dark elements. At viewing distance the human visual system assembles these scanning lines with their varying brightness into a complete picture. The scanning operation is illustrated in Figure 13-6. Each picture (frame) is traced out in 2 successive operations. All of the odd-numbered lines are traced in one operation and the even-numbered lines in the second. The technique is called *interlacing*, and is used to reduce flicker. Figure 13-7 illustrates interlaced scanning. Both the odd and even interlaced fields trace out 262½ lines. The complete picture (one frame) contains a total of 525 lines.

Thirty complete frames (60 interlaced fields) are displayed each second. In motion-picture films 48 views of the scene are shown each second. Actually there are only 24 frames, but each frame is displayed twice to reduce flicker. A complete television picture is composed of up to 150,000 individual elements.

TELEVISION 439

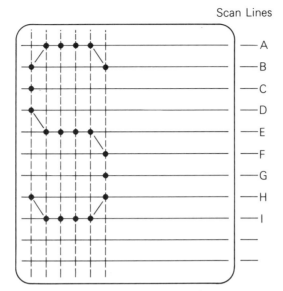

FIGURE 13-6. Simplified scanning illustration.

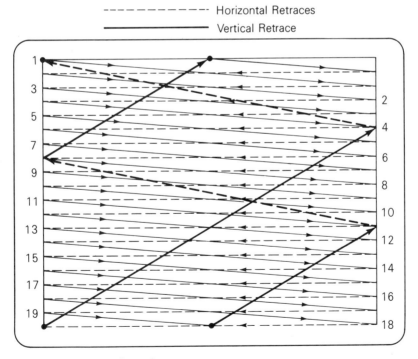

FIGURE 13-7. Interlaced scanning.

SYNCHRONIZATION

The two oscillators in the TV receiver that drive the beam across and down the screen to trace out the picture must follow oscillators that drive the beam in the camera tube. If the system is to work, the picture tube must display each picture element as the camera tube views that particular element in the scene. At the television transmitter, special circuits generate synchronizing pulses that control the scanning of the camera tube. These same pulses are transmitted as part of the television signal to keep the picture-tube scanning exactly in step with the camera-tube scanning.

The horizontal-scanning oscillator runs free at approximately 15,750 hertz and the vertical-sweep oscillator runs free at a frequency near 60 hertz. The synchronizing pulses are used to pull the oscillators into exact synchronization with camera scanning oscillators at the transmitter.

When the picture rolls vertically, the receiver vertical-sweep (scanning) oscillator has fallen out of sync with the transmitter. Generally, this means that the oscillator's free-running frequency has drifted beyond the range where the synchronizing pulses can lock it in. The vertical hold control can be used to bring the oscillator frequency within the locking range.

Diagonal slices of the picture indicate a loss of horizontal synchronization. The horizontal hold control is used to bring the horizontal-scan oscillator frequency into locking range.

Here are some definitions of important color TV terms:

1. White: A mixture of red, blue, and green in the following proportions:

    Red 30 percent
    Green 59 percent
    Blue 11 percent

2. Hue: the color.

3. Saturation: the degree to which a color is *not diluted* by white light. A vivid red is a highly saturated color, where a light pink is far less saturated.

4. Chrominance: color information.

5. Luminance: brightness information (all that is necessary for black-and-white reproduction).

Study Problems

1. How many TV picture frames are transmitted per second?
2. A frame is composed of how many fields?
3. What is the purpose of interlaced scanning?
4. How many scanning lines are there per frame?
5. What is the horizontal scanning frequency?
6. What is the vertical scanning frequency?
7. What part of the TV signal keeps the receiver scanning oscillators in step with those of the transmitter?
8. What type of modulation is used for the picture signal?
9. What type of modulation is used for the sound?
10. What part of the TV receiver converts the picture signal into light?
11. What is the composition of the 2 layers of the vidicon target?
12. Why are 3 camera tubes used in a color TV camera?
13. What are the 3 phosphor colors used in color TV picture tubes?
14. What is the purpose of the shadow (aperture) mask in a color TV picture tube?
15. Approximately how many red phosphor dots are deposited on the screen of a color tube?
16. What is the purpose of the aquadag coating on the inside of a picture tube?
17. Why is the outside of the picture-tube bell also coated with aquadag?
18. What kind of safety hazard is posed by the 2 aquadag coatings?
19. Define hue.
20. Define saturation.

13-3 THE TELEVISION RECEIVER

Figure 13-8 shows the picture signal and synchronizing pulses. Figure 13-9 is a block diagram of a color-TV receiver. The shaded blocks identify the sections common to both black-and-white and color sets. Let's examine the function of each block. In the following explanation the numbers in parentheses correspond to the numbers of the blocks in the figure.

The Tuner (2)

The antenna (1) receives all channels and feeds them to the tuner (2). The tuner uses tuned resonant circuits to reject all channel frequencies but the one selected. Within the tuner are the channel-selection tuned circuits, a local oscillator, an RF amplifier, and a mixer to produce the 45.74 megahertz intermediate frequency. The tuner cir-

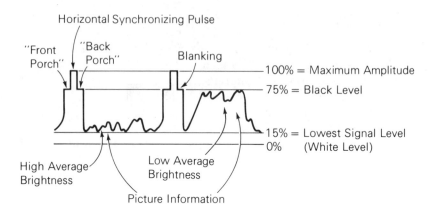

a. The complete signal.

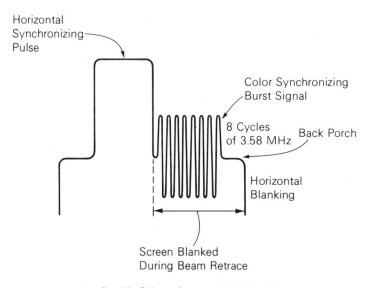

b. Detail of the color synchronizing burst.

FIGURE 13-8. Composite TV signal.

cuit is nearly identical to the tuning circuit in the AM and FM receivers in Chapter 10. The 45.75 megahertz intermediate frequency contains 2 parts: an amplitude-modulated (AM) picture carrier and a frequency-modulated (FM) sound carrier. The picture carrier contains black-and-white video information, synchronizing pulses for the scanning circuits, color information, and a color-circuit synchronizing burst.

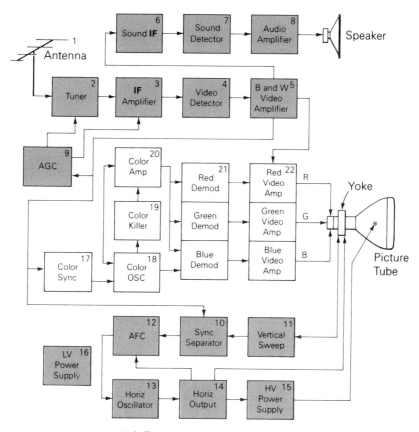

FIGURE 13-9. Television block diagram.

The Intermediate-Frequency Amplifier (3)

The IF amplifier amplifies all of these signals. This requires a bandwidth of about 4 megahertz.

The Video Detector (4)

The video detector removes the IF carrier from the picture, sound, and synchronizing signals.

The Video Amplifier (5)

The video amplifier amplifies the video signal, and distributes it to sound, synchronizing, and picture-tube circuits. Special tuned circuits ensure that sound information is not delivered to the picture tube or synchronizing circuits. Other tuned circuits keep picture information from getting into the sound IF amplifier. The synchronizing pulses are arranged so that they can be used to blank the picture-tube screen each time the scan retraces to the left side of the screen (see Figure 13-8).

The Sound IF Amplifier (6)

The sound IF amplifier amplifies a 4.5 megahertz subcarrier that carries the FM sound information. The sound detector (7) is a standard FM detector that extracts the sound signal from the subcarrier.

The Audio Amplifier (8)

The audio amplifier amplifies the sound signal and delivers audio power to the loudspeaker.

The Automatic-Gain-Control Circuit (9)

The AGC circuit rectifies and filters the composite picture and synchronizing signal. It develops a DC control voltage that automatically adjusts for the varying signal strengths of the various channels. The tuner and IF amplifier are both controlled by the AGC voltage.

The Sync Separator (10)

The sync separator rejects the picture information, keeping only the synchronizing pulses. It then separates the horizontal and vertical synchronizing pulses and sends them to the appropriate sweep (scanning) oscillators.

The Vertical-Sweep (Scan) Oscillator (11)

The vertical-sweep oscillator is synchronized by the pulses from the sync separator. Synchronization of the vertical oscillator is relatively easy because small errors in the sweep frequency are tolerable. The output currents from the vertical-sweep oscillator drive the yoke coil, on the picture-tube neck, to move the beam up and down.

The Automatic Frequency Control (AFC) (12)

The AFC circuit compares the actual horizontal-oscillator frequency with the frequency of the horizontal synchronization pulses and produces a correction voltage if the horizontal oscillator begins to drift off frequency. This elaborate synchronizing circuitry is necessary because horizontal synchronization is very critical and the time available to correct any scanning frequency error is very short.

The Horizontal Oscillator (13)

The horizontal oscillator generates the horizontal scanning signal. Its frequency is synchronized to the sync pulses by the automatic frequency control (AFC) circuit (12).

The Horizontal Output Circuit (14)

The horizontal output steps up the power level in order to drive horizontal coils in the deflection yoke. The deflection yoke is coupled to the horizontal output through a special transformer called a *flyback* transformer. The flyback transformer also has an extra step-up winding that provides the high voltage (30,000 or so volts) required by the picture-tube anode.

The High-Voltage Power Supply (15)

The high-voltage power supply rectifies and filters the high-voltage pulses from the flyback transformer and delivers a high-voltage direct current to the picture-tube anode.

The Low-Voltage Power Supply (16)

The low-voltage power supply is a conventional household power-line-operated power supply that provides direct-current power to all of the electronic circuits.

The Color Sync Circuit (17)

The color sync circuit extracts the color synchronization burst from the composite video signal.

The Color Oscillator (18)

The color oscillator is a precise crystal oscillator. The color sync circuit keeps the crystal oscillator in phase with the transmitter. In order to separate the color signals, a precise phase reference is required. The color oscillator provides this. Because the transmitted signal is so crowded with other information, it is not practical to transmit a continuous signal for this purpose. Instead, about 8 cycles are transmitted in a short burst during the retrace period. This color burst keeps the oscillator built into the receiver in phase with the transmitter reference oscillator.

The Color Killer (19)

The color killer turns off the color amplifier (20) when the picture is transmitted in black and white. The color sync burst is not transmitted with a black-and-white picture. The color killer keeps the color amplifier turned on as long as the color sync burst is present, and turns the color amplifier off in its absence. The color killer eliminates a noise problem that tends to produce a confetti-like colored snow during a black-and-white program.

The Color Amplifier (20)

The color amplifier amplifies the color (chroma) signal and delivers the amplified signal to the 3 color demodulators.

The Color Demodulators (21)

The outputs from the 3 cameras at the color TV transmitter are encoded as complex phase-shifted signals and impressed on the carrier as the color information. This elaborate coding is necessary to squeeze the color information into the very crowded TV signal. The demodulators decode the carefully mixed-up signals and produce independent signals to turn on each of the 3 (red, blue, and green) picture-tube guns at the proper time.

The Color Video Amplifiers (22)

The color video amplifiers amplify the demodulator signal voltages to provide a large enough signal voltage to drive the control grids in the 3 picture-tube electron guns.

Study Problems (refer to Figure 13-9)

1. What is the function of the tuner?
2. Where in the circuit is the sound signal separated from the picture and sync signal?
3. What is the function of the automatic-gain-control circuit?
4. What is the function of the sync separator?
5. What is the function of the vertical-sweep circuit?
6. What is the source of the 30,000 volts DC required for the aquadag anode of the picture tube?
7. What is the purpose of the color killer?
8. Why is the color oscillator necessary?
9. What is the function of the red, blue, and green color demodulators?

Chapter 14

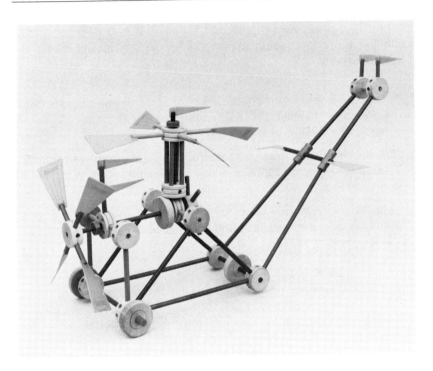

Digital ICs, like Tinkertoys, work together in an almost infinite number of combinations.

DIGITAL ELECTRONICS

14-1 INTRODUCTION

Digital electronics is the fastest-growing technology on the electronics scene. Modern integrated circuits have made computers available at extremely low cost. Full-fledged microcomputers can be purchased for less than $1,000. Only a few years ago this kind of computing power would have cost up to $250,000 to purchase, or $1,000 (or more) per month to rent. Look who is climbing aboard the digital bandwagon:

1. Computer hobbyists (called *hackers*)* have opened up an unexpected new market for microcomputers. Computer clubs are springing up all over the country.
2. The automotive industry is turning to microcomputers to control engine ignition, fuel flow, braking, and automotive-system trouble shooting.
3. Domestic appliances such as microwave ovens and TV game units are being microcomputer-controlled.
4. Electronic tasks that were once performed by analog circuits are being taken over by digital circuits.
5. Pocket calculators provide a remarkable amount of computing power for a few dollars. Many of these calculators contain more computing power than the early multi-million-dollar giant computers.

All of this is but the tip of the iceberg, and new digital wonders enter our lives almost daily.

Digital circuits often use a large number of integrated-circuit packages. The integrated-circuit packages are mounted on a printed circuit board with flat copper traces replacing traditional wires. The circuit board, in digital terminology, is called a *printed circuit* (PC) *card*. The card has "fingers" etched along an edge that form the male half of the plug-socket arrangement. The card is plugged into a special socket called an *edge connector*. Edge-connector sockets are either wired, often with a flat "ribbon" of plastic-encased wires, or plugged into a card-interconnecting circuit board (known as a *mother* board) to interconnect individual cards.

Figure 14-1 shows a typical digital-circuit PC card.

---
* Slang.

DIGITAL ELECTRONICS 449

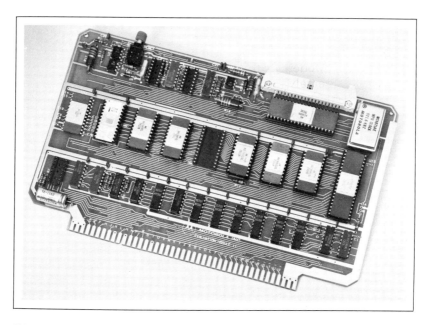

FIGURE 14-1. A typical computer card. (Courtesy Motorola, Inc.)

14-2 WHAT IS DIGITAL LOGIC?

The Bare Facts

1. Digital logic is a systematic arrangement of ultra-high-speed electronic switches called *gates*.

2. There are three basic kinds of gates:
   a. AND gates
   b. OR gates
   c. Inverters (or NOT gates)

3. The most popular kind of gate in modern digital circuits is a composite gate called the NAND gate.

4. The NAND gate is an AND gate with a built-in NOT gate. NOT-AND equals NAND.

5. The combination of an OR gate and a NOT gate (NOR) is also commercially available.

6. The NOR gate is not as common as the NAND gate.

7. There are usually several gates in an integrated-circuit package.

8. A combination of gates in a particular arrangement can perform nearly any digital task.

9. Logic circuits can be divided into 3 classes:
   a. Control
   b. Memory
   c. Computation

10. Logic circuits in integrated-circuit packages are classified according to the number of gates in a package.

11. The number of gates in a package can range from 1 or 2—to many thousands.

Digital logic is a systematic arrangement of super-high-speed electronic on-off switches called *gates*. Nearly all digital systems, no matter how complex, use only 3 kinds of gates; the AND gate, the OR gate, and the inverter. Nearly any computer task can be performed using various combinations of these 3 basic gates. Digital logic is used to control vending machines, traffic lights, and many other machines.

Digital logic is a true-false, go-no-go system. Complex decisions can be made by combining a large enough number of true-false circuits. A true-false test can evaluate a student's knowledge of sophisticated concepts if enough true-false questions are asked. It may take hundreds of true-false questions to test the student's understanding of a concept that could be tested by a single essay question. For the same reason it takes a lot of digital true-false decisions to get a relatively simple result.

Logic operations fall into 3 categories: control, memory, and computation. A computer does computations but it also has electronic switches (logic) to route data from memory to the arithmetic unit or from memory to the outside world, and to perform a number of similar data-routing tasks.

Logic systems designed for control purposes may also have some computing logic and memory. For example, a traffic-light controller might count the cars in several directions and make calculations to determine how fast and when to change the lights. Memory could also be included to remember when traffic is usually light, on Sunday mornings for example, and to remember that traffic is particularly heavy on certain holidays.

DIGITAL INTEGRATED CIRCUITS

Nearly every circuit in digital computers, electronic scales, traffic controllers, and other digital devices is composed of a very few basic kinds of gate interconnected to perform specialized functions such as memory, arithmetic, and so on. Any digital system is mostly gates, and there are normally a great many of them.

An integrated circuit contains anywhere from 6 to 10,000 gates on a wafer (chip) of silicon about ½ centimeter square. The individual gates are interconnected by microscopic printed wiring to make circuits that do specific jobs, such as counting, or decoding digital information. Photographic masks and high-temperature automated chemical techniques permit almost molecule-by-molecule formation of the circuitry.

INTEGRATED-CIRCUIT CLASSIFICATIONS

Integrated circuits are classified according to the number of gates on the chip.

Small-Scale Integration (SSI):

This is an integrated-circuit package with 12 or fewer logic gates. It doesn't matter how they are interconnected on the chip, or what kind of circuit they form.

Medium-Scale Integration (MSI):

This is an integrated-circuit package with between 12 and 100 logic gates on the chip.

Large-Scale Integration (LSI):

This is an integrated-circuit package with over 100 gates on the chip. 10,000 gates on a chip are now fairly common, with up to one-half million on a chip in the near-production stage. LSI chips are larger than SSI and MSI chips. Transistors and digital integrated circuits are made of either bipolar (junction) or MOS transistors.

Bipolar Transistor

The bipolar transistor switches at a much higher speed than MOS transistors, but takes much more space on a chip and consumes a great deal more power. The bipolar transistor is used only in SSI and MSI packages (up to 100 gates) because of its size and power requirements.

Metal-Oxide Semiconductor (MOS)

The MOS-FET transistor is simpler and less expensive to fabricate, very much smaller, and consumes very little power compared to a conventional bipolar transistor. MOS transistors are used where 100 to 10,000 (or more) gates are required on a chip. MOS transistors, however, switch at much lower rates than bipolar transistors.

LOGIC FAMILIES

A logic family is a group of compatible functional integrated-circuit packages. Like Tinkertoys or Leggo, they can be assembled into a variety of arrangements. They just "snap" together. Each family is based on a specific transistor type. The gates in a given family can be assembled in a nearly infinite variety of circuit configurations.

Logic-Family Interfacing

When subsystems based on 2 different logic families must work together in a system, some extra hardware is often required. The special components required to join different families are called *interface* components.

The most popular logic families are fairly compatible with one another, but there are some minor differences. Like the situation shown in Figure 14-2, the solutions to these problems are simple but essential.

Commercial Logic Families

The Transistor-Transistor-Logic (TTL) Family: The TTL is the most popular small-scale integration type. It is based on bipolar transistors and is fast but power-hungry.

P-MOS and N-MOS Families: There are 2 types of metal-oxide semiconductor (MOS) transistor—the P type and the N type. The P-MOS transistor is the basis of the P-MOS gate family. The N-MOS family is based on the N-MOS transistor. Both types are used when 100 or more gates on a chip are required. The MOS is slower than the TTL, but is easy on power.

(C-MOS) Complementary-Metal-Oxide Semiconductors: The C-MOS is a complementary symmetry MOS transistor circuit, using one P-MOS and one N-MOS transmitter.

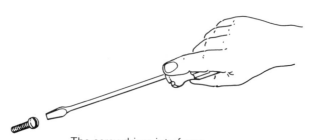

The screwdriver interfaces the hand to the screw. Hand and screw are not completely compatible without the interfacing screwdriver.

FIGURE 14-2. Interfacing.

Integrated Injection Logic (I^2L): The most exciting new logic family. I^2L is nearly as fast as TTL and as miserly with power as MOS. Its high speed, low power, small size, and simple construction make it suitable for large-scale integration. I^2L is based on a radical upside-down bipolar transistor design.

Emitter-Coupled Logic (*ECL*): ECL is a very fast but complex, relatively expensive gate circuit. Its use is restricted to SSI and MSI packages. It is primarily used in large, expensive computers and military systems.

THE HIERARCHY OF DIGITAL SYSTEMS

Figure 14-3 shows a division into 2 basic categories: memories and combinational logic circuits that control system operation and perform arithmetic operations. The bottom half of Figure 14-3 shows a division into microprocessor and mini and large machines. So far, LSI has been used less in larger computers because of its slower speed. Integrated injection logic and faster MOS devices promise to make LSI minicomputers possible. It is also a current trend to use several minis or micros to make up very large computing systems.

Any logic system by itself, be it a computer or automotive system controller, is like a brain in a bottle—useless. The brain must be able to communicate with the outside world and often interact with it. Logic systems operate at power levels too low to directly drive displays, typewriters, motors, and so on. The logic system may also put out its data or commands in a code that is not compatible with the device it must operate in the outside world.

Special devices that enable the logic system to operate things outside the computer are called *interface* devices. In the past, each interface problem was a special (and expensive) engineering problem. Standard off-the-shelf (and inexpensive) integrated-circuit devices have become available to solve many common interface problems.

Study Problems

1. How many gates are to be found in each of the following integrated-circuit packages?
   a. SSI
   b. MSI
   c. LSI
2. What is the most popular (composite) gate type?
3. Draw the logic diagram of the AND-NOT equivalent of a NAND gate.
4. What 2 gates form the NOR (composite) gate?
5. List the 3 broad categories of logic circuits.

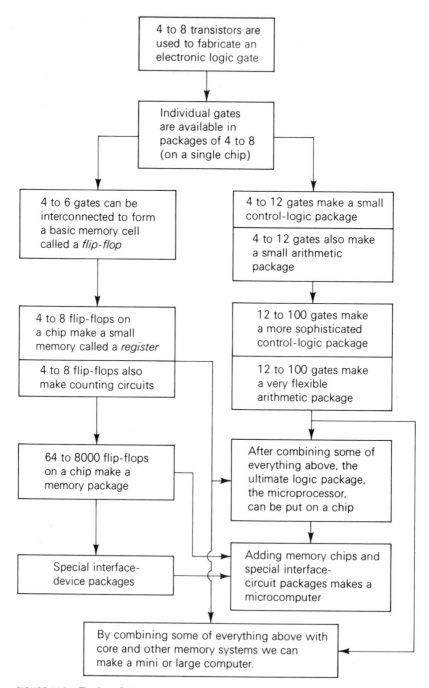

FIGURE 14-3. The bare facts.

14-3 ELECTRONIC LOGIC GATES

The Bare Facts

1. Truth tables are used to describe the true-false functions of logic gates and circuits.

2. The AND gate produces a true (logic 1) output only when *all* inputs are true (logic 1).

3. The OR gate produces a true (logic 1) output when *any* or *all* inputs are true.

4. The inverter converts a false (logic 0) input into a true output, and a true input to a false output.

5. In most modern logic circuits a true condition is represented by +5 volts and a false condition by zero volts.

6. The most popular gate type is a combination of an AND gate and an inverter. It is called the NAND gate.

7. A combination of an OR gate and an inverter is called a NOR gate.

8. NOR gates are much less common than NAND gates.

Electronic gates are high-speed versions of ordinary mechanical switches. Modern gates can switch from off to on or on to off in the time it takes a beam of light to travel about 3 meters. The speed of light is about 300,000,000 meters (186,000 miles) per second, which means that logic gates can switch in a period of approximately 10 nanoseconds (1 nanosecond = 0.000,000,001 or 10^{-9} seconds).

TRUTH TABLES

Tables that show the results of all possible combinations of true and false conditions in logic gates are called *truth tables*. Because all digital conditions are either true or false, they are called *binary* (two state), and digits 1 and 0 are used to represent the true and false conditions.

THE AND GATE

The AND operation is equivalent to switches in series. All possible combinations for the AND gate with 2 inputs are listed in Figure 14-4. The truth tables show that when both A and B are false, f is false; when A is true and B is false, f is false; when A is false and B is true, f is false; and when both A and B are true, f is true. All possible

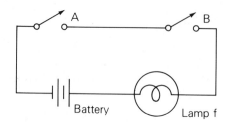

a. The AND switching circuit.   b. AND-gate symbol.

Switch A	Switch B	f (Lamp)
Open	Open	Not lit
Open	Closed	Not lit
Closed	Open	Not lit
Closed	Closed	Lit

A	B	f
0	0	0
0	1	0
1	0	0
1	1	1

A	B	f
F	F	F
F	T	F
T	F	F
T	T	T

c. Truth tables.

FIGURE 14-4. The AND gate.

combinations for the 2 input AND gate are listed in the truth tables in Figure 14-4. The letter f is the gate output symbol and means *function*. In the modern TTL family, 0's and 1's are represented by the voltages:

$$+5 \text{ volts} = \text{Logical 1} = \text{True} = \text{H}$$
$$0 \text{ volts} = \text{Logical 0} = \text{False} = \text{L}$$

A 1 (+5 volts) is often called *high* and 0 volts *low*. Truth tables are often written in the form shown in the left-hand truth table in Figure 14-4. The right-hand table is the more common general purpose truth table form.

THE OR GATE

All possible combinations for the OR gate with 2 inputs are listed in the truth tables in Figure 14-5. The tables show that when both A and B are false (L), f is false (L); when A is true (H) and B is false (L), f is true (H); when A is false (L) and B is true (H), f is true (H); and when both A and B are true (H), f is true (H).

THE INVERTER

The inverter is an inverting amplifier that performs the NOT function. If a 1 goes in, a 0 comes out; if a 0 goes in, a 1 comes out. An A

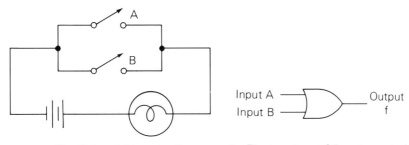

a. The OR switching circuit.

b. The two-input OR-gate symbol.

A	B	f
0	0	0
0	1	1
1	0	1
1	1	1

A	B	f
L	L	L
L	H	H
H	L	H
H	H	H

A	B	f
F	F	F
F	T	T
T	F	T
T	T	T

c. Truth tables.

FIGURE 14-5. The OR gate.

in produces an $\overline{A}$ (NOT A) out. The NOT operation is often called the *complement* operation. Figure 14-6 illustrates inverter operation.

Summary of inverter functions:

1. If $A = 0, \overline{A} = 1$
2. If $A = 1, \overline{A} = 0$
3. If $A = \text{High}, \overline{A} = \text{Low}$
4. If $A = \text{Low}, \overline{A} = \text{High}$

THE NAND GATE

Figure 14-7 shows the NAND-gate symbol, its AND-inverter equivalent circuit, and its truth table.

TTL LOGIC GATES

Transistor-transistor logic (TTL) is the most popular small-scale (SSI) and medium-scale (MSI) logic form. It is available in several package styles, but the dual in-line package (DIP) shown in Figure 14-8 is the most common.

A simplified version of the TTL NAND gate is shown in Figure 14-9a. A more complete version is shown in Figure 14-9b. Diodes

458 CHAPTER 14

Note: The circle at the end of the triangle signifies an inversion.

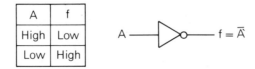

a. Using one inverter.

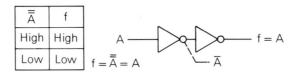

b. Using two inverters.

c. Using three inverters.

FIGURE 14-6. The inverter.

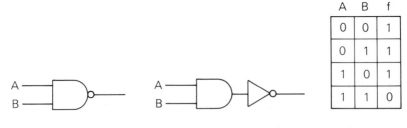

a. NAND gate symbol. b. AND-inverter equivalent of the NAND gate. c. The NAND truth table.

FIGURE 14-7. The NAND gate.

have been added to the inputs and to protect against negative voltages and a complementary symmetry power amplifier has been added to the output to provide enough power to allow it to drive up to 10 other gates. The odd-looking transistor connected to the gate inputs

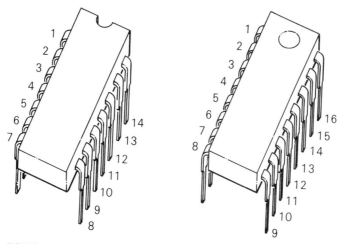
FIGURE 14-8. The dual in-line package (dip).

is called a *multi-emitter* transistor. Actually it is an integrated-circuit technique for fabricating diodes.

All members of the TTL family require a power-supply voltage of +5 volts and ground. Logic levels are standard +5 volts = 1, and 0 volts = logical 0. There is some tolerance built in so that logic levels need not be perfect. Figure 14-10 shows the logic levels for standard TTL logic. Figure 14-11 shows the photomicrograph of an integrated-circuit logic chip. The wires in the photo are less than half the diameter of a human hair. The inverter has only one input and one output. Other gates have only one output but can have (theoretically) any number of inputs.

Study Problems

1. Draw the symbol for each of the following gates:
   a. OR   b. AND   c. Inverter   d. NAND
2. Make a truth table for each of the gates listed in Study Problem 1.
3. Match the items in the following two columns:
   1. +5 volts   a. True
   2. 0 volts   b. False
                       c. Logic 1
                       d. Logic 0
4. All gates have only one output. True or false?
5. Can an inverter have more than one input?
6. Can an AND, OR, or NAND gate have more than one input?

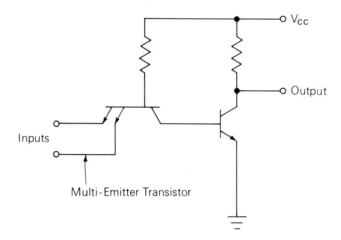

a. Simplified basic TTL circuit.

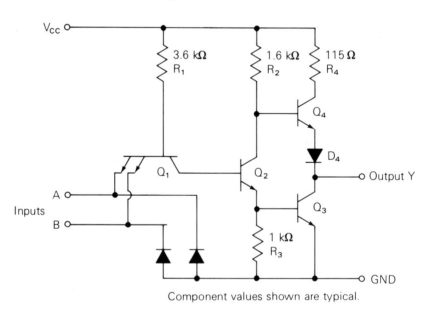

Component values shown are typical.

b. Standard logic circuit.

FIGURE 14-9. The TTL logic-gate schematic diagram.

DIGITAL ELECTRONICS   461

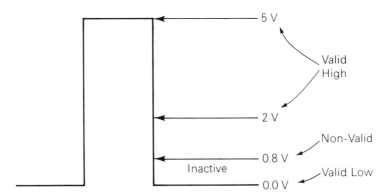

FIGURE 14-10. TTL logic levels.

FIGURE 14-11. An integrated-circuit chip and case.

14-4 BOOLEAN ALGEBRA

Boolean algebra is a special digital algebra. Boolean is far simpler than ordinary algebra because it has no fractions, roots or powers, and involves no difficult proofs. The truth table is the only proof required in Boolean algebra. Let's look at some symbols and definitions and then try an example. You will find the symbols and definitions in Table 14-1.

Boolean equations are very closely related to real digital circuits. Let's look at an example that demonstrates how they work together.

The problem: We want a digital circuit that will sound a buzzer in a car when the following conditions exist:

1. The seatbelt is NOT fastened
  AND
2. The ignition switch is on
  AND
3. The gearshift is in Drive
  OR
  Reverse

TABLE 14-1 BOOLEAN SYMBOLS AND EQUATIONS

Operation	Symbol	Read as
OR *example:*	$+$ $A + B$	OR A OR B
AND *example:*	$A \cdot B$[1] or (AB)[2]	AND A AND B
Complement	$\overline{A}$	not A or A not
Complement	$\overline{1}$	not 1 or 1 not
Complement	$\overline{A + B}$	not (A OR B) or (A OR B) not
Equation	$\overline{1} = 0$	not 1 equals 0
Equation	$\overline{0} = 1$	not 0 equals 1
Equation	$f = A + \overline{B}$	f equals A OR not B
Equation	$f = \overline{A} + \overline{B}$	f equals not A OR not B
Equation	$f = \overline{A}B$	f equals not A AND B
Equation	$f = (AB) + (CD)$[2]	f equals A AND B OR C AND D

[1] The dot is often left out.
[2] Parentheses are optional.

You will find the solution using both ordinary switches and logic gates, along with the proper Boolean equation, in Figure 14-12 (on the following page).

14-5 NUMBERS FOR DIGITAL SYSTEMS

The Bare Facts

1. The human number system is based on 10 fingers (digits).

2. Digital systems use gates, flip-flops, and memory devices that have only 2 possible states.

3. Because there are only 2 states (the equivalent of 2 fingers) available, digital systems use a base of 2.

4. The number system that uses the base 2 is called the *binary* system.

5. Digital systems also use number bases that can be encoded in binary.

THE BINARY NUMBER SYSTEM

We humans do our arithmetic in the base 10 system, probably because we have 10 fingers (digits) to count on. Because computers are true-false-type gates, the computer must do arithmetic in some system that contains only 2 digits. Table 14-2 compares the binary (base 2) number system to our base 10 system.

When we write numbers in our familiar decimal system, we don't have to think about what it means. It has become automatic. When we must deal with an unfamiliar number system, it is necessary to give the operation some thought. The following example explains what we actually mean when we write the number 425 in our decimal system.

Example:

$10^2 \quad 10^1 \quad 10^0$
$4 \quad\quad 2 \quad\quad 5 = 4 \times 10^2 + 2 \times 10^1 + 5 \times 10^0 = 400 + 20 + 5$

We do not need to write all of this out because we have learned to think in base 10.

The binary system works the same way except that the column headings are multiples of 2 instead of 10, and only 2 digits, 0 and 1, are available. Binary numbers are written with a subscript 2 to iden-

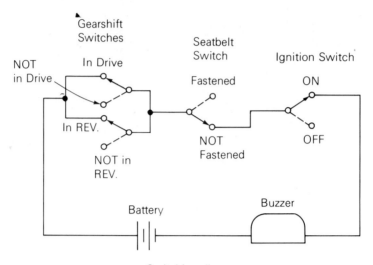

a. Switching diagram.

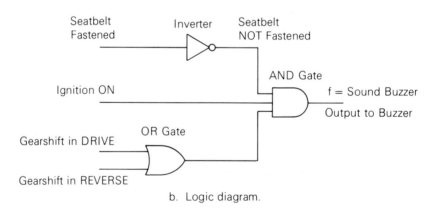

b. Logic diagram.

Sound buzzer (f) = ignition ON (I) **AND** seatbelt not fastened ($\bar{S}$)
AND [gearshift in drive (D) **OR** in reverse (R)]

c. The Boolean equation in words.

$$f = I \cdot \bar{S} \cdot (D + R)$$

d. In Boolean shorthand using the symbols in parentheses.

FIGURE 14-12. A seatbelt buzzer.

DIGITAL ELECTRONICS 465

TABLE 14-2 BINARY AND DECIMAL NUMBER SYSTEMS COMPARED

a. Base 10 (decimal).

Name of position	Thousands	Hundreds	Tens	Units
Position value in decimal form	1000	100	10	1
Use of radix to form value of each position	10 × 10 × 10	10 × 10	10	$\frac{10}{10}$
Position value in exponential form	10^3	10^2	10^1	10^0

b. Base 2 (binary).

Name of position	Sixteens	Eights	Fours	Two	Units
Position value in decimal form	16	8	4	2	1
Use of radix to form value of each position	2 × 2 × 2 × 2	2 × 2 × 2	2 × 2	2	$\frac{2}{2}$
Position value in exponential form	2^4	2^3	2^2	2^1	2^0

tify them as base 2 numbers. The following example shows how the number 41 in decimal is written and interpreted in binary:

$$\begin{array}{cccccc} 2^5 & 2^4 & 2^3 & 2^2 & 2^1 & 2^0 \\ 32 & 16 & 8 & 4 & 2 & 1 \\ \hline 1 & 0 & 1 & 0 & 0 & 1 \end{array}$$ Exponent value
Decimal value
Binary number

Reading out, we have $32 + 8 + 1 = 41_{10}$.
Thus $101001_2 = 41$ in our base of 10.

HEXADECIMAL

The hexadecimal number system (base 16) is frequently used in digital computers because it is a medium-sized base that is easy for people to deal with and is also directly related to binary. Four binary digits (bits) can stand for 16 hexadecimal digits. The first 10 (0 through 9) are the same as our base 10 digits, but an extra 6 are required for hexadecimal (hex). The letters A through F have been selected to represent the extra hex digits because they are symbols already available on any typewriter or computer printer. The hexadecimal system is further explained in Table 14-3, parts a, b, c.

TABLE 14-3 NUMBER SYSTEMS AND CODES

a. The hexadecimal system and its binary codes.

Structure of the Hexadecimal (Radix-16) System

3	2	1	0	Position
16^3	16^2	16^1	16^0	Exponential form
4096	256	16	1	Decimal value

Example:

0	B	A	3	Number in radix 16
0	+ 2816	+ 160	+ 3	= 2979_{10}

b. Additional hexadecimal symbols.

Decimal Equivalent	Symbol
10	A
11	B
12	C
13	D
14	E
15	F

c. Binary-coded hexadecimal.

Hex	Binary			
0	0	0	0	0
1	0	0	0	1
2	0	0	1	0
3	0	0	1	1
4	0	1	0	0
5	0	1	0	1
6	0	1	1	0
7	0	1	1	1
8	1	0	0	0
9	1	0	0	1
A	1	0	1	0
B	1	0	1	1
C	1	1	0	0
D	1	1	0	1
E	1	1	1	0
F	1	1	1	1

TABLE 14-3 (continued)

d. The binary-coded octal system.

Structure of the BCO System

Position	3	2	1	0
Position value	512	64	8	1
BCO value in exponential form	8^3 $2^2 2^1 2^0$	8^2 $2^2 2^1 2^0$	8^1 $2^2 2^1 2^0$	8^0 $2^2 2^1 2^0$

Example:

000	001	011	101	(BCO number)
0	1	3	5	(octal equivalent)

e. Octal digits.

Octal Digit	BCO Equivalent
0	000
1	001
2	010
3	011
4	100
5	101
6	110
7	111

f. The binary-coded decimal system.

Structure of the BCD System

Position	3	2	1	0
Position value	Thousands	Hundreds	Tens	Units
BCD value in exponential form	$2^3 2^2 2^1 2^0$	$2^3 2^2 2^1 2^0$	$2^3 2^2 2^1 2^0$	$2^3 2^2 2^1 2^0$

g. Four decimal numbers and their BCD equivalents.

Decimal Numbers	BCD Equivalent
001	0000 0000 0001
123	0001 0010 0011
546	0101 0100 0110
879	1000 0111 1001

OCTAL

The octal system (base 8) is also convenient for human use, and is also related to binary. Octal expressed in binary digits is called *binary-coded octal* and is abbreviated BCO. Octal is less satisfactory in most cases because it can express only 8 of our 10 decimal digits in one column. The octal system and BCO equivalents are shown in part e of Table 14-3.

BINARY-CODED DECIMAL (BCD)

The BCD code shown in Table 14-3, parts f and g, is identical to binary-coded hex except that the hex digits A through F are forbidden in the BCD code. The BCD code is used in pocket calculators and other small devices.

14-6 MAKING LOGIC GATES DO SOMETHING (AN EXAMPLE)

The Bare Facts

1. A truth table is a specifications table for digital circuits.
2. The truth table describes exactly how the logic gates are to be interconnected.
3. Once the task to be performed has been defined on a truth table, the logic diagram is easy to draw.

A truth table describes exactly what the logic circuit that goes with it must do. Once the truth table exists, it serves as a guide for writing a simple equation. This simple equation in turn tells you exactly how to draw the logic diagram. Let's try an example. Suppose we build a circuit to add 2 binary digits. Binary digits (BInary digiTs) are called *bits*. First we must state the rules for binary addition and then translate those rules into a truth table. The binary addition table and its equivalent truth table are shown in Table 14-4. In this case we are going to use a logic circuit to perform arithmetic.

Compare the addition table in Table 14-4 with the 2 truth tables. In order to write an equation, the terms are found by reading out the rows in the truth table that have a 1 in the column. A 0 under a letter heading is read as a NOT. A 1 is read as the letter itself.

Example:

A	B	f
1	0	1

$f = A \cdot \overline{B}$ (A AND not B)

The letters are connected by a dot (AND) symbol. Table 14-4b shows the terms to the right of the table. The terms are: $(\overline{A} \cdot B)$, $(A \cdot \overline{B})$. To complete the equation, place an OR(+) sign between the terms, add an equals sign and an f, and we get:

$$f = (\overline{A} \cdot B) + (A \cdot \overline{B})$$

Read the formula as: function = *not A and B or A and not B*.

TABLE 14-4 BINARY ADDITION TABLE AND TRUTH TABLES

a. Addition table.

	Sum	Carry
0 + 0 =	0	0
0 + 1 =	1	0
1 + 0 =	1	0
1 + 1 =	0	1

b. Sum truth table.

A	B	f	
0	0	0	
0	1	1	$\bar{A} \cdot B$
1	0	1	$A \cdot \bar{B}$
1	1	0	

c. Carry truth table.

A	B	f	
0	0	0	
0	1	0	
1	0	0	
1	1	1	$(A \cdot B)$

To draw a logic diagram, each term becomes an AND gate and the plus symbol becomes an OR gate. Figure 14-13a shows how this works.

The modern standard (TTL) logic gate is the NAND gate. The NAND gate is an AND gate with a built-in inverter. The built-in inverter comes about because a built-in common emitter amplifier is provided in each gate circuit. An AND gate and inverter can do anything that can be done using AND, OR, and inverter gates. AND-OR-NOT circuits can always be converted into NAND circuits simply by changing all gates (both AND and OR) into NAND (see Figure 14-13b). The truth table for the carry operation proves to be the truth table for an AND gate. An AND gate is all that is required for the carry operation.

THE EXCLUSIVE-OR GATE

The adder circuit that we just put together is one that solves so many digital problems that manufacturers provide the circuit in Figure 14-13 in a package called an *exclusive*-OR gate. Figure 14-14 shows the special symbol for the exclusive-OR.

Figure 14-15 shows the exclusive-OR logic diagram with inputs interconnected and inverters added to carry out the NOT operations demanded by the bars over the inputs shown in Figure 14-13. Both

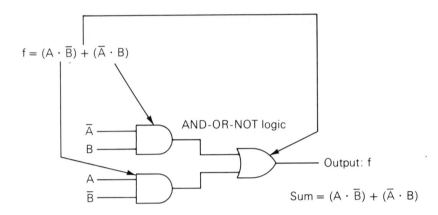

a. The adder using AND-OR-NOT logic.

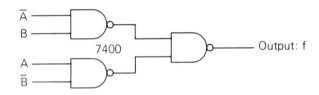

b. The adder circuit in NAND logic.

FIGURE 14-13. The circuit and equation.

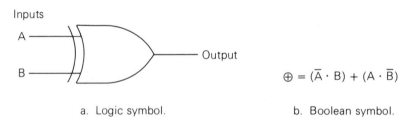

a. Logic symbol.   b. Boolean symbol.

FIGURE 14-14. The exclusive-OR symbol.

circuits are identical, but logic diagrams are often drawn as shown in Figure 14-15. The exclusive-OR gate always has 2 and only 2 inputs.

Study Problems

1. What is the purpose of a truth table?
2. Draw the symbol for the exclusive-OR gate.

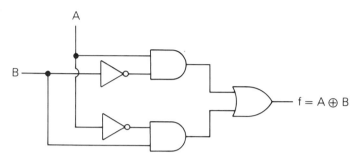

FIGURE 14-15. The exclusive-OR logic diagram as it would normally be drawn.

3. Compare the truth tables of the ordinary OR and exclusive-OR gates.
4. Why is the binary number system used in digital systems instead of our familiar base 10 system?
5. What is the base number for the hexadecimal system?
6. Why do we need the symbols A, B, C, D, E, and F in the hexadecimal system?
7. List one application for the exclusive-OR gate.

14-7 FLIP-FLOPS

The Bare Facts

1. The flip-flop is the basic memory element.
2. The basic flip-flop consists of 2 NAND (or NOR) gates arranged so that one, but only one, gate always has a high output.
3. The flip-flop acts like a toggle switch. Once placed in one of its two positions it stays there until it is actuated again.
4. The Type D flip-flop and the Type J-K are the two most popular types.
5. There is also a Type D flip-flop latch that has a limited purpose.
6. Both Type D and Type J-K flip-flops can be used to count, store data, or shift data from one flip-flop to another.

One of the simplest flip-flops consists of 2 NAND gates with the output of one fed back to the input of the other. Figure 14-16 shows the circuit.

The flip-flop is an electronic version of a toggle switch. Like the switch that turns on the lights in your house, once it is turned on, it

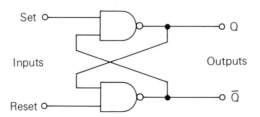

FIGURE 14-16. The basic flip-flop.

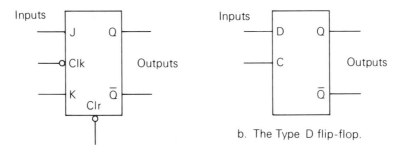

a. The J-K flip-flop.

b. The Type D flip-flop.

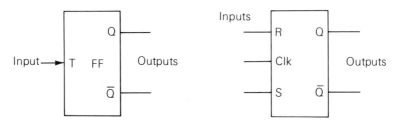

c. The Type T toggle flip-flop.

d. The clocked reset/set flip-flop.

FIGURE 14-17. Flip-flop symbols.

stays on until it is turned off. There is no need to stand there and hold it in the *on* position to keep the light on. It "remembers" that the lever was moved to the *on* position. Ordinary gates are analogous to a push-button that responds only while it is being pushed.

In the flip-flop in Figure 14-16, one gate is always on and the other one is off. A momentary signal applied to one of the inputs can cause the gates to exchange conditions. The gate that was previously off turns on, and the feedback switches the gate that was previously on to the off state.

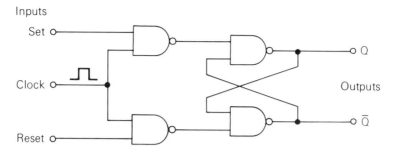

FIGURE 14-18. Flip-flop with clock input.

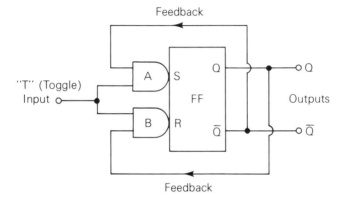

FIGURE 14-19. Type T flip-flop with steering gates.

The complete logic circuits for the various kinds of flip-flops are available in manufacturers' data manuals. Logic diagrams use the symbols in Figure 14-17 whenever a flip-flop is required.

CLOCKED FLIP-FLOPS

Nearly every flip-flop application requires that it change state only when the clock tells it to. A clock input is provided by the addition of 2 extra NAND gates to the basic flip-flop as shown in Figure 14-18.

THE TYPE T FLIP-FLOP

The Type T flip-flop shown in Figure 14-19 toggles with each input pulse. The first pulse sets Q to high. The second pulse sets Q back to low. It takes 2 input pulses to get one output pulse. The circuit divides by 2 (or counts by 2). The 2 outside feedback lines and Gates A and B steer the input signals to the flip-flop to produce the desired toggle action. The T type is the counting flip-flop.

THE J-K AND TYPE D FLIP-FLOPS

Simple flip-flops are used in memory arrays, but for most other systems the more complex master-slave J-K or Type D is required to avoid timing errors. The J-K and Type D can be used as a set-reset, or counting (toggle) flip-flop. Figure 14-20a shows the simple J-K flip-flop and 14-20b shows the master-slave version.

For high-speed operation, the master-slave arrangement used in J-K and Type D flip-flops is necessary to provide time for each flip-flop in a circuit to change states and settle down to a steady condition. The settling time must be allowed, or false counts and unreliable operation occur at high counting rates.

FLIP-FLOP APPLICATIONS

Flip-flops are used as temporary memory, either as a single unit to remember one binary digit (bit), or arranged on a memory chip for

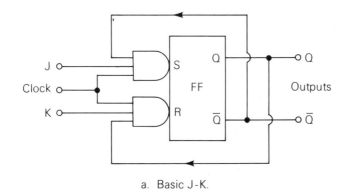

a. Basic J-K.

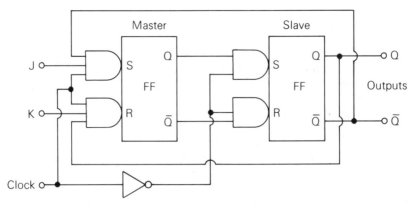

b. The master-slave J-K.

FIGURE 14-20. J-K flip-flops.

larger-scale storage. TTL type flip-flops are used in memories up to 100 bits or so. Flip-flops using enhancement MOS-FET transistors are used for arrays of up to 16,000 eight-bit groups. We will examine memory applications shortly.

COUNTERS

Flip-flops can be cascaded to count by 2's (binary), and by adding additional control logic they can be made to count by 10's, 12's etc. There are a number of variations on the basic binary counter circuit, but they are all variations of the circuit shown in Figure 14-21.

SHIFT REGISTERS

Flip-flops are also used for a kind of memory that can transfer stored bits from flip-flop to adjacent flip-flop. The shift register does not count. It transfers all data one flip-flop to the right for each clock pulse. The shift register diagram is shown in Figure 14-22.

Commercial shift-register integrated circuits are available with over 1000 flip-flops in an IC package. They are also available with extra control-logic gates that allow both right and left shift and the capacity to load all flip-flops simultaneously.

DECODERS

It is often necessary to convert the binary output of a counter into a standard decimal (0, 1, 2, 3, 4, 5, 6, 7, 8, 9) count or to some special code such as the code required to light the proper segments in 7-segment displays (see Figure 14-23b). Decoders are also used in memory systems. A binary address is decoded to select a particular cell out of the many in an array of memory cells.

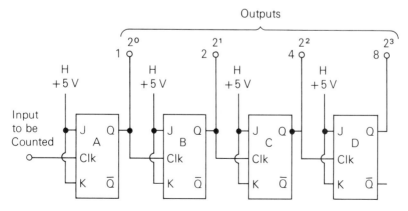

FIGURE 14-21. The binary counter.

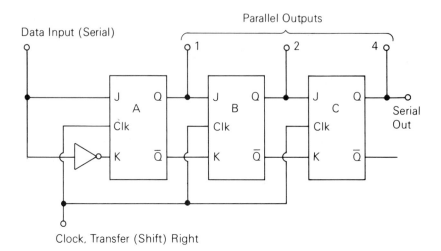

FIGURE 14-22. The shift register.

Decoders such as the binary-to-decimal decoder in Figure 14-23 are available as standard MSI integrated-circuit packages. A number of other commonly needed decoders are also available in standard DIP packages. The decoder is the electronic equivalent of a multiposition rotary switch that distributes combinations of several inputs to individual one-at-a-time outputs.

MULTIPLEXERS

The multiplexer is an electronic rotary switch that selects any one of several one-at-a-time inputs, and outputs the data in some binary code. Keyboard encoding is an example. A single decimal key can be pressed and the multiplexer translates that key closure into the desired binary output—number or coded number. Programmable decoders and multiplexers are available that allow the device to work in any code required for any nonstandard jobs that might come up.

Study Problems

1. Draw a logic diagram of a basic NAND-gate flip-flop.
2. List the 2 most common flip-flop types.
3. In what kind of circuit is a toggle flip-flop used?
4. What is a shift register?
5. Why is a decoder often needed in conjunction with a counter—that is, what is the purpose of the decoder?
6. Draw the symbols for the following flip-flops:
    a. The J-K.   b. The Type D.   c. The toggle.

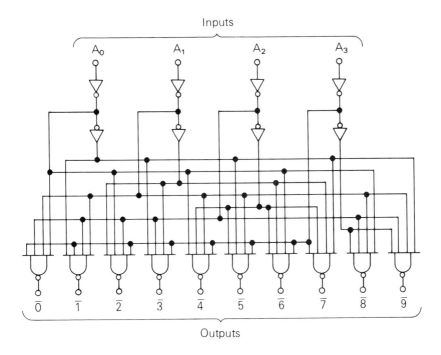

a. Binary-to-decimal decoder.

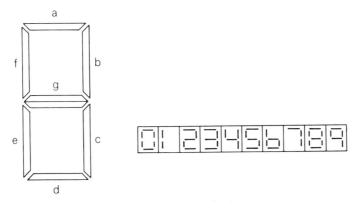

b. Seven-segment display.

FIGURE 14-23. Decoders.

14-8 MEMORY SYSTEMS

The Bare Facts

1. There are several kinds of memory mediums:
   a. Semiconductor memories consisting of flip-flops or special capacitor-MOS-FET storage cells
   b. Moving magnetic recording surfaces such as tape, discs, and drums
   c. Miniature stationary magnetizable cores
   d. Solid-state devices with movable magnetic domains called *bubbles*
   e. Punched paper tape and punched cards.
2. Random-access semiconductor memories (RAMs) can be written into or read out of.
3. Data can be placed in a RAM, left for as long as necessary, and erased to make room for new data.
4. Any cell in a RAM can be read or written into, in a definite amount of time.
5. Read-only memories (ROMs) contain "prerecorded" data such as often-used tables and other information.
6. ROMs cannot be written into. They can only be read.
7. Most semiconductor RAMs "forget" when the power is shut off. ROMs do not forget.
8. Shift-register memories have the flip-flops or MOS-FET cells arranged in series like cars on a train track.
9. Data must be pushed through a shift-register memory until the desired data arrive at the output before the data can be read.
10. The serial nature of the shift-register memory makes access time to a given block of data variable and longer than the access time of RAMs and ROMs.
11. Bubble memories and charge-coupled memories are very-high-density shift-register memories.

In digital systems, devices that can store information in the form of the electronic equivalent of 0's and 1's are memory devices. A cassette tape recorder can store a string of 0's and 1's in the form of 2 different tones, one tone for 0 and another for 1. Semiconductor devices can store 0's and 1's in the form of conducting or nonconducting transistors, or other special devices that can be held either on or off.

Tape recorders and magnetic-disc and drum recorders can be used as mass-storage devices. Punched paper tape and punched cards are other mass-storage methods.

MAGNETIC-CORE MEMORY

Thousands of tiny donut-shaped cores strung on wires are used in many computer main (RAM) read-write memories. In Figure 14-24a, current flowing through the wire magnetizes the core. In Figure 14-24b, after the current has stopped flowing the core remains permanently magnetized. To read the core a reverse current is sent through the wire. If the core was storing a 1 (magnetized), its collapsing field induces a small current in a separate sense wire threaded through the core. If a 0 was stored, there is no field in that direction to collapse and no current is induced in the sense wire. Figure 14-24c shows a 0 being stored in the core. Cores are threaded as shown in Figure 14-25 into an array called a *memory plane*. These planes are then stacked to make a 3-dimensional memory package. When a core is read, the data stored in it are lost unless it is immediately written back into the memory. However, data are not lost in the event of a power failure.

PUNCHED CARDS AND TAPE

The two most popular versions of paper mass memory are shown in Figure 14-26. Table 14-5 (see page 482) explains the codes used with the punch cards and tape.

MAGNETIC RECORDING

Magnetic recorders in several forms are used for digital memory (bulk storage). Inexpensive cassette recorders, expensive data tape

a. Current is applied; core stores as 1.
b. Core is magnetized; core remembers a 1.
c. Current is reversed; core reverses its magnetic state and stores a 0.

FIGURE 14-24. Magnetic cores.

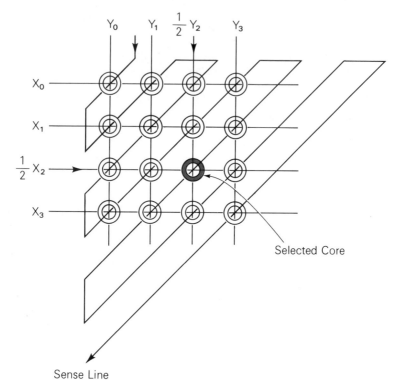

FIGURE 14-25. Core memory array.

machines, thin plastic magnetic-coated discs (floppy discs), precision magnetic discs, and drums are all common.

RANDOM-ACCESS MEMORY (RAM)

This is a memory that can be written into, read out of, and reused. The data stored in a RAM have definite addresses that can easily be located. The RAM memory is organized somewhat like city streets with each bit of data located at an intersection. Total RAM memory capacity is limited, and it must usually be made to forget previous information to make room for new information.

A RAM can be written into and read out of just as core memory can. Readout in semiconductor memories does not erase the data stored in the cell being read, but all data in the memory are lost when the power fails or is shut off. Figure 14-27 shows the arrangement of the memory cells in one type of 64-bit RAM, and the row and column

DIGITAL ELECTRONICS 481

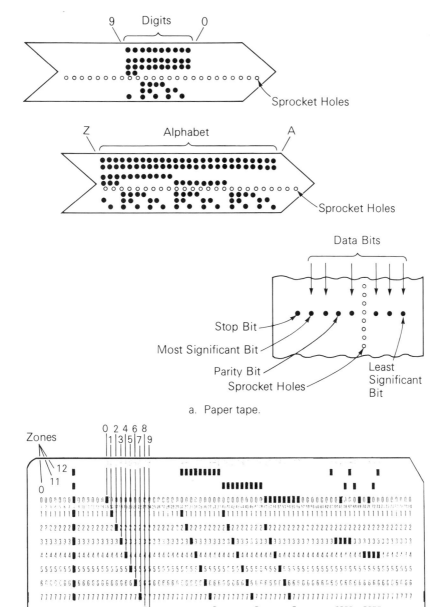

FIGURE 14-26. Punched paper tape and cards.

TABLE 14-5 PUNCH-CARD CODE

Numerical Row Only	Zone 12 Plus Numerical Row Below	Zone 11 Plus Numerical Row Below	Zone 0 Plus Numerical Row Below
0 = 0			
1 = 1	1 = A	1 = J	
2 = 2	2 = B	2 = K	2 = S
3 = 3	3 = C	3 = L	3 = T
4 = 4	4 = D	4 = M	4 = U
5 = 5	5 = E	5 = N	5 = V
6 = 6	6 = F	6 = O	6 = W
7 = 7	7 = G	7 = P	7 = X
8 = 8	8 = H	8 = Q	8 = Y
9 = 9	9 = I	9 = R	9 = Z

decoders and input- and output-buffer amplifiers. The decoder uses a 6-bit memory address to select any one of the 64 cells.

The number of cells can be expanded to 1024 with only 12 bits of address data required for the decoder to select any one of the 1024 cells. Many modern semiconductor memories are organized into groups of 8-bit words so that a 12-bit address can be decoded to select any one of 1024 8-bit words (a total of 8,192 bits). Larger memory chips use MOS or I^{2}L transistors and are available with up to 16,000 8-bit words on a chip, with larger memories soon to become available. Semiconductor memory is faster and cheaper than core, but is subject to data loss in case of power failure.

READ-ONLY MEMORY (ROM)

ROM memory is organized in the same random-access fashion as RAM, but it cannot be erased and used again. ROMs are preprogrammed units that determine the machine's basic behavior patterns and store special tables and other permanent data. ROM units that control the specific capabilities of the machine are called *personality modules*. However, these ROMs can be unplugged and replaced to alter the machine's basic "psychology." ROMs also serve a "rote" memory function much as we might memorize the multiplication

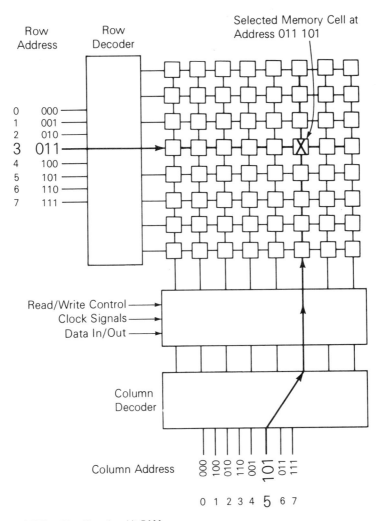

FIGURE 14-27. Sixty-four-bit RAM.

tables. ROMs can be preprogrammed with tables of often-needed information, special subroutines, or anything else that must be frequently used in the machine's normal operation. Computer instructions or data permanently stored in ROM are called *firmware* in computer jargon.

ROMs are made of transistors (either bipolar or MOS) that are permanently turned either off or on. Bipolar memories use fuse links that can be deliberately burned open. Some MOS devices are programmed during manufacture and are called *mask* programmed

ROMs. Programmable ROMs (PROMs) can be programmed by applying controlled overvoltage to appropriate cells. The EPROM (Erasable PROM) can be erased with ultraviolet light and reprogrammed in the same manner as the PROMs.

BUBBLE MEMORY

Bubble memories are becoming available as replacements for magnetic-tape, drum, and similar mechanical memories. Bell Telephone Company has even been testing bubble memories to record those all-too-common "this is a recording" messages. The bubble memory consists of 150,000 or more cells arranged in a line like a train on a track. Data are entered at one end, one bit at a time. A bit is shifted one cell down the line to make room for the next bit. Data come out the other end (when called for) one bit at a time in the same order in which the data were entered.

The memory consists of thousands of microscopic magnetic-domain groups (bubbles) in a magnetic film of ferrite or garnet. A magnetic field built into the memory packages causes the little magnetic islands (bubbles) to form. Another magnetic field under the control of the computer advances the bubbles down the line. The bubble at the first cell is magnetized to represent a 1 or 0. It is then moved down the line and the next bit is recorded. The output is fed back to the input and the data can be recirculated as many times as desired without loss, much like an endless tape. The memory can be erased and new data entered. A special detector, analogous to a tape-recorder playback head, extracts the data when the computer requires it.

CHARGE-COUPLED MEMORY (CCD)

The charge-coupled memory is intended for the same applications as the bubble memory. The CCD memory transfers the charge from one tiny capacitor to the next instead of moving a magnetic field down the line. These 2 memory schemes are running a neck-and-neck competition for the market at this writing.

Study Problems

1. What is a RAM memory?
2. Can a RAM be written into?
3. What is a ROM memory?
4. Can a ROM memory be written into?
5. For what purpose might a ROM memory be used?
6. What is a PROM memory?
7. Is a shift-register memory faster or slower than RAM or ROM memories?

8. What is the purpose of a decoder when used with a RAM or ROM memory?
9. Name one of the most basic memory cells.
10. List 3 mechanical memory mediums.

14-9 OTHER DIGITAL SUBSYSTEMS

The Bare Facts

1. Nearly all digital systems require that all operations be accurately timed.
2. A special oscillator called a *clock* is used to synchronize digital operations.
3. The subsystem that performs arithmetic and special logic functions is called the *arithmetic logic unit* (ALU).
4. The ALU is a programmable device that can be instructed to perform any one of a number of numerical and logic operations.
5. Programmable logic devices are devices capable of performing many different logic tasks.
6. Programming the devices involves applying +5 volts or 0 volts in specific combinations to special programming inputs.
7. Software is the paperware that contains step-by-step instructions for the computer, or data with which it is to work.

CLOCKS

Almost all digital systems require a clock to ensure that each operation takes place at exactly the right time. A clock is a circuit that generates a continuous train of logic pulses. Logic-gate oscillators and special timing packages are used for noncritical applications. For more critical situations, crystal-controlled logic-gate oscillators or special crystal-controlled clock-generator ICs are used. As in most situations involving a sequence of events, order and timing are important. One does not open the car door, close the car door, and then get into the car; the order is wrong and will not work out. It is also necessary to allow enough time to get arms and legs all the way into the car before closing the door. In digital systems the time required for an operation is very small, but that time still must be allowed.

PROGRAMMABLE LOGIC

Most logic packages have all of the gates interconnected internally to do some special job. A programmable device with 7 program terminals can become any of 128 completely different logic circuits. For the number of gates available in the package there are only 128 different circuits possible. Thus, by putting +5 volts and 0 volts in particular combinations on the 7 pins, any desired circuit out of the 128 possible can be had from a single package—a kind of universal logic package.

THE ARITHMETIC LOGIC UNIT (ALU)

The ALU is a complex programmable device capable of doing addition and subtraction, comparing numbers, and performing a number of other arithmetic operations. It can also accomplish several ordinary logic operations. It is a multifunction circuit. A group of binary digits represent instructions as to exactly which of the operations in its repertory the ALU is to perform. A decoder decodes the coded instruction and activates the proper circuits in the arithmetic logic unit for the desired operation. This is the number-processing part of the computer. Multiplication is performed as a string of additions, and division is performed as a string of subtractions.

SOFTWARE

Software is a set of step-by-step instructions to guide the computer's operation. Software is generally prepared by the programmer, starting with a flowchart of the desired sequence of computer events. The flowchart is used as an "outline" from which a detailed set of instructions (the program) is prepared. These instructions are keyboarded onto punched cards or tape and read into the computer's (RAM) memory. The computer then reads each instruction from its memory, in the proper sequence, and extracts the data previously loaded into RAM to work with. The computer can execute the program any number of times, extracting new data from memory each time.

FIRMWARE

Computer instructions permanently recorded in ROM memory are called *firmware*.

Study Problems

1. What is the purpose of the clock in digital systems?
2. What is the subsystem called that can perform binary arithmetic?
3. What is software?
4. What is firmware?
5. What do you think would be classified as hardware?

Chapter 15

Early Model Digital Computer

COMPUTERS

15-1 INTRODUCTION

In Chapter 14 we examined digital electronics at the circuit and building-block level. Here we will look at the systems level of digital electronics, specifically at the computer-system level. In the past few years the digital computer has revolutionized business methods and has had a great impact on the way industry operates.

The microprocessor is probably the most exciting computer development in recent years. The basic idea behind large-scale, fully programmable computers of the data-processing variety was to have a single basic machine that could be made to handle nearly any computing or data-processing task simply by providing a set of instructions called a *program*. The program and data could then be loaded into a memory, and the machine could take it from there at a fantastic rate of speed. Such a powerful machine could also be used for such things as traffic-signal control, electronic instrumentation, electronic scales, automobile systems, and industrial control, but until recently the cost has been prohibitive.

Now the microprocessor is available and its cost is dropping in the same way the price of pocket calculators fell not long after their introduction. The microprocessor is a slower, less powerful, but very inexpensive equivalent of the central-processing unit of the large-scale digital computer.

The microprocessor is the heart of a system called a *microcomputer*. An entire microcomputer can be put on a single chip, although until recently this practice has been limited to military applications. A large part of the computer is commonly put on a single LSI chip and called a *microprocessor*. The microprocessor unit (MPU) nearly always contains the ALU (arithmetic logic unit), minimum necessary working and control registers, and often some specifically committed ROM memory, along with counters and control logic.

The MPU is designed so that it has very little internal direction. It is virtually uncommitted to anything but the simplest operations common to any computer operation. Although the MPU is the central part and coordinator of the system, it is virtually a slave to an external set of instructions. Every action is dictated by external programming in the form of firmware or software. The firmware is a programmed ROM (read-only memory) that is a kind of job description for the machine. The ROMs are external to the MPU chip and can be programmed for the functions required of a general-purpose machine or of a dedicated (special-purpose) machine.

The following are some examples of dedicated machines:

Automotive ignition control	Automated gas pumps
Automotive brake control	Fast-food cash registers

Home-appliance controllers
Vending-machine control
Electronic scales
Specialty calculators
Adaptive control systems
Automotive analyzers
Medical instruments
Machine-tool control
Electronic games
Film-processing control

Communication-line controllers
Printer controllers
Traffic-light controllers
Photocopy-machine controllers
Point-of-sales inventory control
Electronic typesetting
Automatic drafting systems
Automatic elevator control
Chemical-process control

Once programmed, the firmware becomes permanent software that directs the activities of the processor and permits communication with keyboards and displays and other peripheral devices. In dedicated systems, firmware may constitute the entire program (stored in the ROM). A plug-in ROM assembly is often called a *personality* module. These personality modules can be unplugged and replaced with a module that can change the computer's task. For example, the machine could be altered to cope with new traffic patterns in a traffic-signal controller. The same microcomputer might be used in any of the dedicated systems listed above by providing the proper personality module for each system.

Microcomputers can also be used for traditional data processing such as payroll and inventory control in smaller companies where the larger machines are just too expensive.

Many industrial and medical instruments now have built-in microprocessors that permit a relatively untrained person to use them. This frees physicians and highly trained technicians for more vital tasks.

In aircraft and space-vehicle systems, microprocessors can be distributed throughout the control system to provide local computer control under the direction of a central master computer. A distributed system can provide a much faster response time than a large, central control computer.

15-2 THE BUS

The Bare Facts

1. The bus is a group of wires connected to all units in the system.

2. The bus is bidirectional.

3. Each unit on the bus is associated with a bus driver that controls the direction of data flow.

4. Data flow either into or out of each unit on the bus line.
5. The bus driver controls the direction of data flow for each unit. When the bus driver is not being instructed to move data into or out of a unit, it disconnects the unit from the bus to keep it from interfering with other units.
6. The bus drivers are controlled by a part of the control unit known as the *instruction decoder.*

In order to keep the number of interconnecting wires down to a practical number, the processor, memories, and so on are all connected to a common bus through a bus driver. Most of the bus drivers in the system are bidirectional, with an *off* condition to allow data to either leave or enter, or to disconnect the device according to instructions from the control unit.

The bus provides a communication path between any 2 selected devices at a time. The *control unit* functions as a traffic-signal control system. The bus is a one-lane street. If the (traffic) control unit says "data may now move from memory to the general register," that is all that can happen. Other operations must wait until the control unit clears the bus. As soon as the data complete the trip from memory to the general register, the control system permits some other pair of devices to use the bus. In human terms, the system would seem inefficient and slow. But some 4,000,000 trips per second can be managed without an error. This high speed makes up for the single-lane inefficiency of the bus. It is typical in computers to find such limited, simple, operations. Such simplicity may seem inefficient, but when millions of these simple operations can be completed each second, the final results are impressive.

Figure 15-1 illustrates how traffic is controlled on the bus. Bidirectional bus drivers are used where data must flow into or out of a device. The ROM cannot be written into, so a 1-direction bus driver is used. In order to connect a unit to the bus, 2 commands are required: The enable command to turn on the driver, and a clock command that tells the circuit when to turn on. An inactive bus driver is completely disconnected from the bus. The bus itself is simply a group of wires connected to each device in the system.

15-3 HOW THE MICROCOMPUTER WORKS

The Bare Facts

1. The arithmetic logic unit performs arithmetic and logic operations on binary numbers.

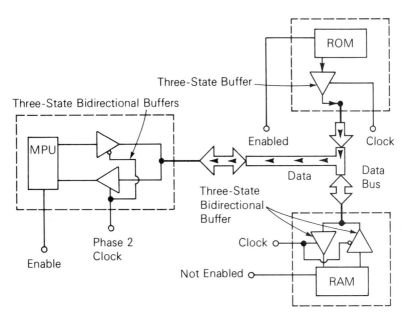

a. The processor reading the ROM.

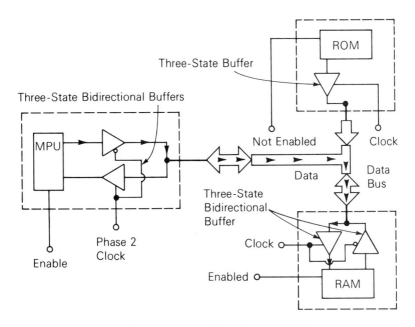

b. The processor writing into the RAM.

FIGURE 15-1. The bus.

2. The accumulator is a register composed of flip-flops. Data from the outside world generally pass through the accumulator.

3. Data and instructions from the outside world (or memory) enter the accumulator and are then transferred into specific memory locations (or to the outside) or to special registers.

4. The accumulator is closely associated with the arithmetic logic unit. It provides temporary storage for data for the arithmetic logic unit (ALU) and stores the results of ALU operations.

5. Data and instructions are usually loaded into memory before the computer begins its operation.

6. The computer then "fetches" instructions and data from memory at its own speed.

7. Instructions are transferred from memory to a group of flip-flops called the *instruction register*, where they are held until the instruction has been carried out.

8. The instruction decoder decodes the instruction, and connects the proper units to the bus with the proper direction of data flow.

9. The program counter produces a binary signal output that specifies the memory location of the next instruction or data required.

10. The program counter's binary signal is decoded by the memory address decoder.

11. The memory address decoder places data at the defined memory location at the *ready*.

12. The next time the computer fetches from the memory, the data are ready and waiting.

13. The general register is similar to the accumulator. It provides temporary data storage for the ALU, but it is also free for other special temporary storage needs.

14. Most of the circuits simply store data and instructions. Data and instructions are moved along the bus from unit to unit. The ALU operates on the data and places the results in the accumulator. The control circuitry then transfers the contents of the accumulator to other units in the computer or to the outside world, according to the prewritten program.

Figure 15-2 is the block diagram of a simplified microcomputer. Real machines will have some additional registers. A condition or

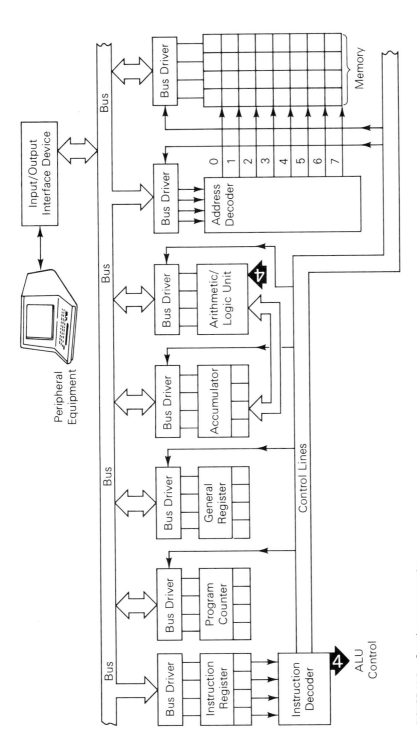

FIGURE 15-2. Sample microcomputer.

status register is used in nearly all machines to make certain decisions when an operation results in an overflow, a negative value, a zero result, and so forth. Other special registers are often used, depending on the design of the specific computer. Outside of some refinements, and a 4 bit limitation, the sample microcomputer in Figure 15-2 is reasonably typical. Clock circuits are omitted in the sample to keep it simple.

The following subsystems in Figure 15-2 would normally be on the integrated-circuit microprocessor chip:

1. Instruction register and bus driver
2. Instruction decoder and bus driver
3. Program counter and bus driver
4. General register and bus driver
5. Accumulator and bus driver
6. Arithmetic logic unit and bus driver.

There may be a number of integrated-circuit memory chips. Each memory chip will contain the memory cells, the memory address decoder, and a bus driver for the decoder and one for the memory itself.

THE MEMORY

The memory is loaded with both instructions and data before any computing is done. The computer will fetch data and instructions from memory as they are needed. Results from the various computing operations are stored in the memory until needed. The memory is usually a combination of RAM (random-access memory) and ROM (read-only memory). Data in the RAM are constantly changing as they are processed. The ROM contains certain permanent instructions—often-used numerical values such as log tables and so on. Memory size can be expanded by connecting additional memory units to the bus. The microcomputer in our illustration processes 4 binary digits (bits) at a time. Most contemporary microcomputers process 8 or 16 bits simultaneously. A typical 8 bit microcomputer can address 65,000 memory locations, each containing one 8 bit (1 byte) word.

THE ADDRESS DECODER (ON THE MEMORY CHIP)

The address decoder decodes a 4, 8, or 16 bit binary address (depending on the particular processor) and selects a particular word in memory to be read out of or written into. Most 8 bit microcomputers use 16 bit addresses, which allows the decoders to select any one of 64,000 individual word locations in memory. A word is 8 bits in an 8 bit machine, 16 bits in a 16 bit machine, and so on.

THE ARITHMETIC LOGIC UNIT

The ALU performs various arithmetic and logic operations under the command of the control unit. This is called the *arithmetic section* in some machines. The dark arrow (4) coming out of the instruction decoder connects to the dark arrow (4) entering the ALU. This group of lines provide programming voltages to the ALU that cause it to add, subtract, or perform one of its other functions.

THE ACCUMULATOR

The accumulator is a temporary memory (register) composed of 4, 8, or 16 flip-flops. Processors are classified according to the word length. In an 8 bit machine the memory word will consist of 8 bits; the accumulator and other registers will also be 8 bits wide. The accumulator works directly with the arithmetic logic unit. It always stores one of the 2 operands for the ALU and after each ALU operation it stores the results of that operation.

In most microprocessors the accumulator is the gateway to the outside world. Data from keyboards, displays, and so on pass through the accumulator on their way to memory or from the processor to the outside world.

THE GENERAL REGISTER

The general register is a one-word-long group of flip-flops. It is used to store data that will be acted upon. It is often used to store the second operand for the arithmetic logic unit. The general register (and the accumulator) are much faster than main memory, and the data stored in registers are immediately available for processing. Some machines use the accumulator and general register as interchangeable registers and designate them as Accumulator A and Accumulator B.

THE PROGRAM COUNTER

The program counter is set to some binary number. Each time it is advanced (incremented) it addresses the next consecutive memory location. The binary number stored in the program-counter flip-flops represents a definite memory location. The program counter tells the memory address decoder to select a particular memory location and set it up for reading the information stored there. The next step in the computer's operation will transfer the contents of that memory location along the bus to the appropriate register. The memory address decoder decodes the contents of the program counter and selects the proper location in memory. The program counter can be instructed to *set* to any memory location and start its count at that location. This allows the machine to jump from one group of memory locations to another. Much of the computer's flexibility depends on the ability to move around in the memory to find the required data and instructions.

THE INSTRUCTION REGISTER

Instructions located in the main memory are sent along the bus into the instruction-register flip-flops. The instructions are in a binary code called the *operation code* or *op-code*. Each machine has a set of instructions. An instruction tells the machine what to do next. The instruction tells the machine what units to connect to the bus, and governs transfers from memory to registers, addition operations, and so on.

THE INSTRUCTION DECODER

The instruction decoder decodes the binary instruction, and through a group of AND and OR gates, places an enable voltage on the units that are to be connected to the bus or produces the proper code to tell the arithmetic logic unit what function to perform.

THE CLOCK

The clock pulse is divided into 4 discrete segments (4 phases), each occurring at a different time. No device can be connected to the bus without simultaneous commands from both the instruction decoder and the clock.

PERIPHERAL EQUIPMENT

The peripheral equipment is connected through an interface device to the bus. The peripheral devices form a communication link between the computer and the human operator or between the computer and devices the computer is intended to control. Data from peripheral equipment are normally loaded into or fed out of the accumulator.

PERIPHERAL INTERFACE DEVICES

Peripheral interface devices serve several purposes. The computer itself operates at very low power levels, far too low to actuate relays, control a printer, or light indicator lamps. Interface devices often have a power-amplification function.

Peripheral devices may also operate in a code that is not standard binary. The peripheral interface device may be required to make the translation. If the peripheral devices are analog devices, the interface circuits must make digital-to-analog or analog-to-digital conversions.

Many peripheral devices are very slow in comparison to the computer. The interface device must often have some buffer memory. The memory can store input data from the peripheral devices at whatever rate it is received, and discharge its contents at computer-compatible speeds. Buffer memory is also needed to record data at high computer speeds and unload it to a typewriter or other machine at a speed that it can keep up with.

15-4   A COMPUTER PROCESSING EXAMPLE USING THE SAMPLE MICROCOMPUTER

Suppose we follow a sample microcomputer through all of its operations in the process of performing an addition.

The Problem:

   Add: $0111_2 + 0001_2$

By hand we get:
$$
\begin{array}{r}
111 \quad \leftarrow \text{Carry} \\
0111 \\
+\ 0001 \\
\hline
1000
\end{array}
$$

In our decimal system:

8	4	2	1	← Headings
0	1	1	1	← Binary
$(0 \times 8) + (1 \times 4) + (1 \times 2) + (1 \times 1)$				= 7 in base 10

8	4	2	1	
0	0	0	1	← Binary
$(0 \times 8) + (0 \times 4) + (0 \times 2) + (1 \times 1)$				= 1 in base 10

TABLE 15-1 SAMPLE MICROPROCESSOR INSTRUCTION SET

Instruction	Mnemonic	Operation Code
Designate memory location 0000 (0) for reading	Start	0000
Load accumulator with contents of the following memory location	LDA	0001
Load register with contents of next memory location	LDR	0010
Halt	HLT	0011
Add the contents of the register to the contents of the accumulator and place result in accumulator	Add	0100

$$\begin{array}{rl} & 111 \longleftarrow \text{Carry} \\ \text{So, we are adding:} & 0111 \longleftarrow (7 \text{ in base } 10) \\ & +\ 0001 \longleftarrow (1 \text{ in base } 10) \\ \hline & 1000 \ =\ 8 \text{ in base } 10 \end{array}$$

We need to know two things to start: how data and instructions must be entered into the memory, and the operation codes for each of the instructions we will need. The instruction codes are shown in Table 15-1.

The instructions must be loaded into the memory in the following order:

1. First instruction
2. First word of data
3. Second instruction
4. Second word of data
5. Third instruction, and so on.

Now, we must write a formal program and load it into the computer memory. The program is shown in Table 15-2.

Now that we have written the program, Figure 15-3 shows how it is stored in the computer's memory after the program has been loaded.

Suppose we follow the computer through its processing of the problem we have programmed. The step-by-step computer operation is:

1. Select memory location 0000 (0) for reading

COMPUTERS 499

TABLE 15-2 THE ADDITION PROGRAM

Step	Instruction	Instruction Op-Code	Data	Memory Location	Decimal
0	Load the general register	0010		0000	(0)
1	With the data in memory		0111	0001	(1)
2	Load the accumulator	0001		0010	(2)
3	With the data in memory		0001	0011	(3)
4	Add the contents of the register to the contents of the accumulator and store results in the accumulator	0100		0100	(4)

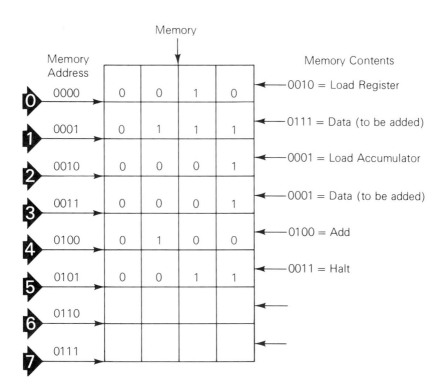

FIGURE 15-3. The contents of the memory before the computer starts.

2. Transfer contents of Memory Location 0000 into the instruction register.

> Instruction: 0010 load the register

3. Advance program counter (increment). Designate Memory Location 0001 (1) for reading.
4. Carry out the instruction in Step 2. Load the register with the contents of Memory Location 0001 (1).

> Data: Load 0111 into the register

5. a. Advance program counter (increment).
   b. Select Memory Location 0010 (2) for reading.
6. Transfer the contents of Memory Location 0010 (2) into the instruction register.

> The instruction is: 0001 = load accumulator with the contents in the next memory location.

7. a. Advance program counter (increment).
   b. Designate Memory Location 0011 (3) for reading.
8. Carry out the instruction in Step 6 and transfer the contents of Memory Location 0011 (3) into the accumulator.

> The data loaded into the accumulator consist of the number 0001.

9. a. Advance program counter (increment).
   b. Designate Memory Location 0100 (4) for reading.
10. Transfer the contents of Memory Location 0100 (4) into the instruction register.

    The instruction is: 0100 = add the contents of the register to the contents of the accumulator, and store the result in the accumulator.

11. Carry out the instruction in Step 10: Add the contents of the register to the contents of the accumulator.
12. Rest. The result of the addition in Step 11 is automatically stored in the accumulator.

Figure 15-4 shows each stage in the processing of the program. It is the programmer's responsibility to load the memory in proper order.

When the machine reads an instruction, that instruction not only tells the machine what action to take, but it also tells the machine

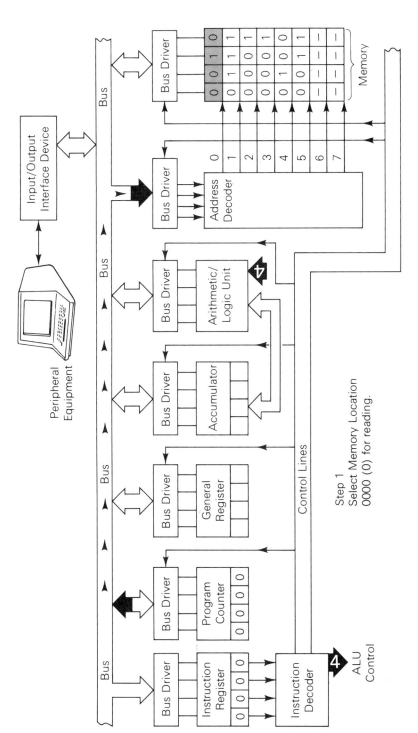

FIGURE 15-4. Stages in the processing of a program.

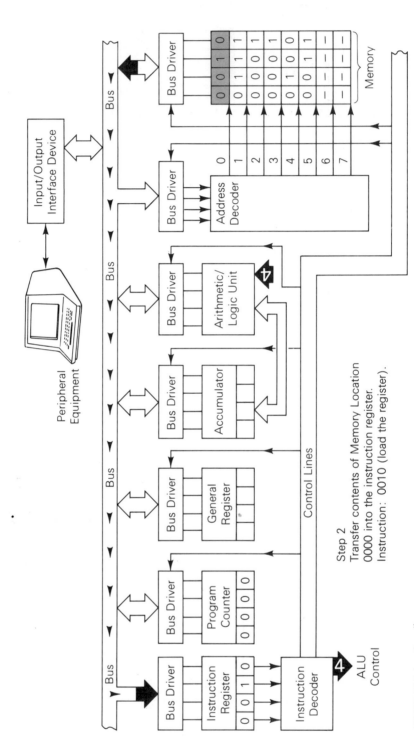

FIGURE 15-4. (continued)

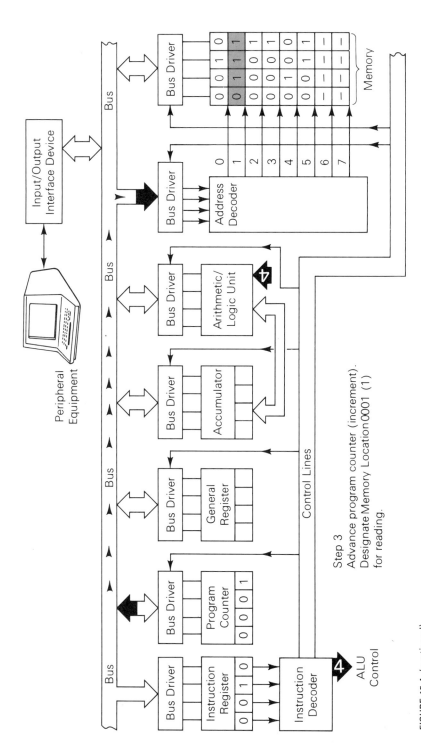

FIGURE 15-4. (continued)

504 CHAPTER 15

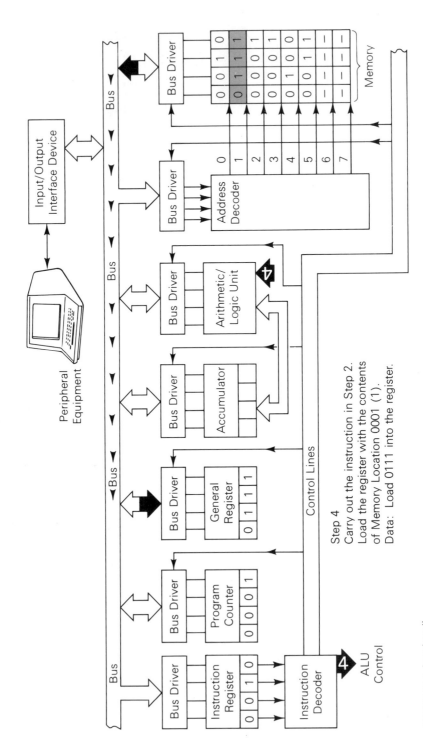

FIGURE 15-4. (continued)

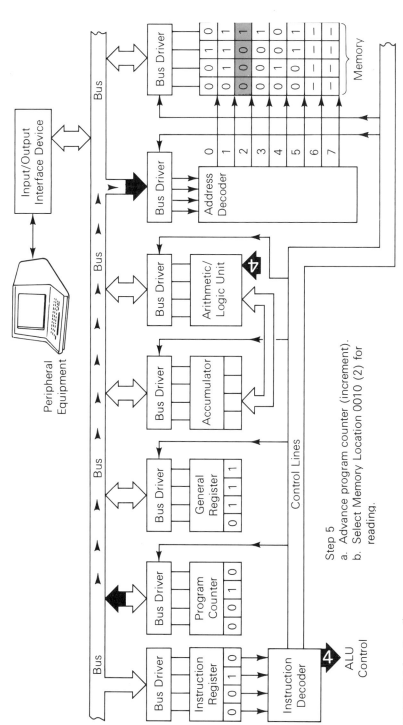

FIGURE 15-4. (continued)

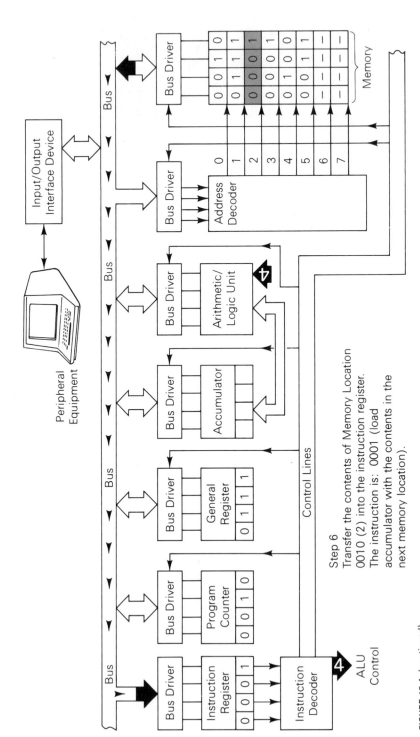

FIGURE 15-4. (continued)

COMPUTERS 507

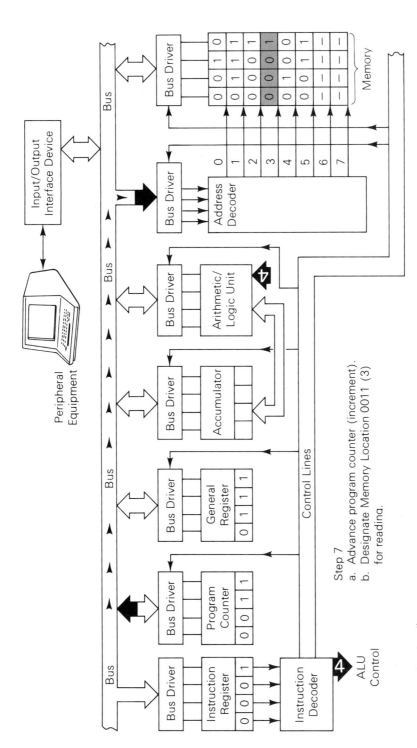

FIGURE 15-4. (continued)

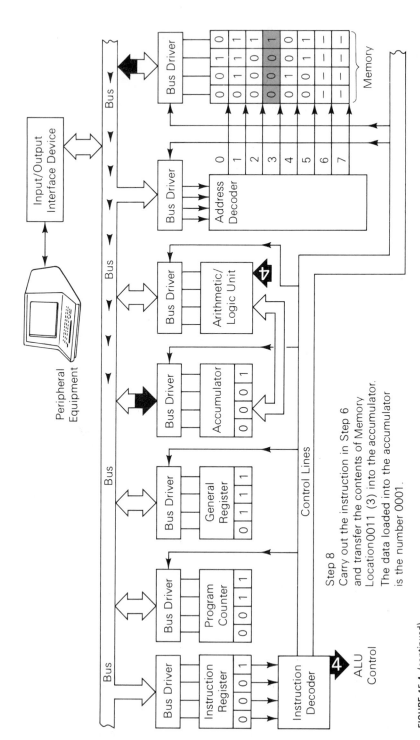

FIGURE 15-4. (continued)

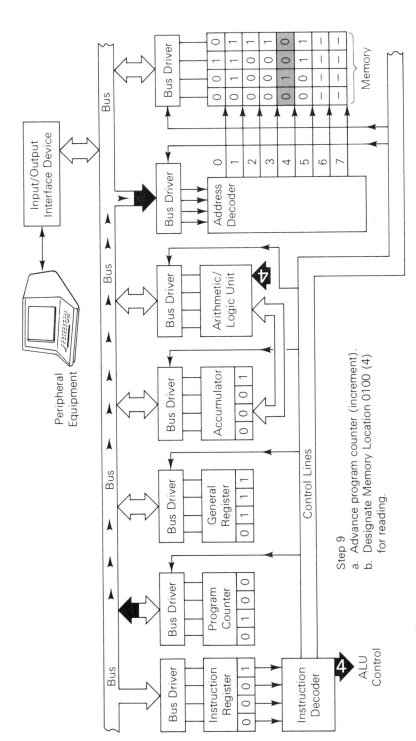

FIGURE 15-4. (continued)

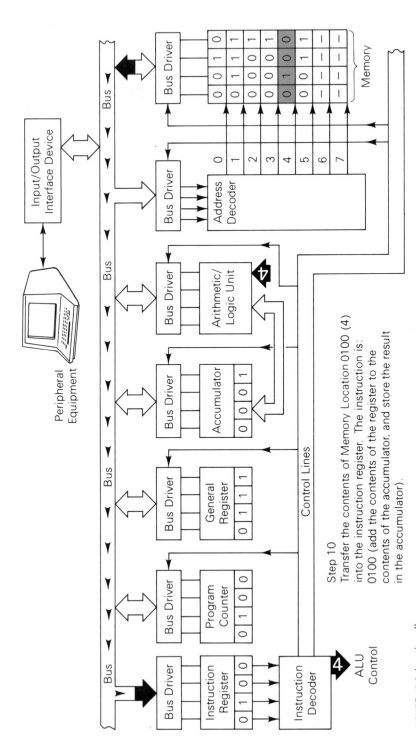

FIGURE 15-4. (continued)

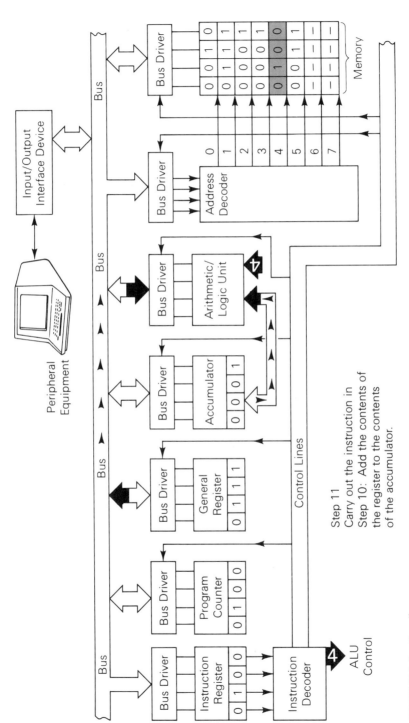

FIGURE 15-4. (continued)

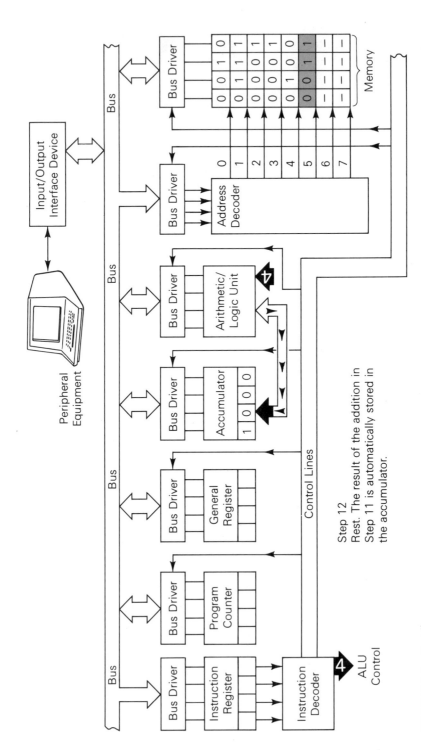

FIGURE 15-4. (continued)

that the next memory location in order will contain either the data or the memory address where the data can be found.

Definitions

Nybble: 4 bits
Byte: 8 bits

Most modern microprocessors use a word length 1 byte long. (Some have word lengths 2 bytes long).

Study Problems

1. What is the bus and what is its function?
2. What factor makes up for the bus's 1-way traffic limitation?
3. What is the function of the ALU?
4. When data are to be loaded into memory from a keyboard, what register stores the data temporarily before transferring them to the memory?
5. What is the function of the memory address decoder?
6. What is the purpose of the program counter?
7. What is the function of the general register?
8. What is the function of the instruction register?
9. What is the function of the instruction decoder?
10. When the adder performs an operation on 2 binary numbers, where is the result first stored?
11. What is the function of the bus drivers?
12. What are some of the functions that might be performed by an interface device?

15-5 PROGRAMMING LEVELS

The Bare Facts

1. A logic system designed for a specific job is known as a *hardwired* logic system. It must be redesigned for new applications.

2. In microcomputer systems, a nearly infinite number of different tasks can be accomplished by simply writing a new program.

3. A written program can be loaded into RAM. The written program is called *software*.

4. A program is a detailed set of instructions that defines a sequence of operations for the computer.

5. RAMs can be erased and a new program can be loaded in.

6. Programs that are intended to be more permanent can be recorded in ROM. The ROMs are plug-in units and can be exchanged for new ROMs with a new program. ROM programs are called *firmware*.

7. Machine language is the lowest-level programming language and consists of many binary 0's and 1's.

8. Machine-language programming is the most difficult of programming languages.

9. Machine language requires no extra memory and is the cheapest language when the machine must be programmed on a permanent basis.

10. When a machine must be reprogrammed often, higher-level languages become cheaper because of the fewer man-hours required to write the program.

11. The higher the program language level, the more nearly it resembles ordinary English.

12. A single statement in a high-level language results in several to many machine-language steps.

13. Ultimately, higher-level language statements must be converted into machine-language instructions.

14. The rules for translating from higher-level languages into machine language are stored in memory.

15. The memory space taken up by the translation rules subtracts from the total memory space available for the processing program.

16. There is always a tradeoff between language level and available memory space.

17. Machine-language-level programs utilize memory most efficiently. Higher-level languages utilize memory space less efficiently.

Programming is a broad term covering all methods of providing instructions for a computer or processor to follow. Some programs are hardwired and permanent, some are in the form of read-only memories (called *firmware*), and others are in the form of software, requiring no physical changes to change the program. A logic system designed for a specific job is hardware-programmed and cannot be adapted for other tasks without physical modification and possibly a complete redesign. A processor system using only ROM devices is firmware-programmed and can be transformed into a completely different system for an entirely different task by changing or reprogramming the ROMs. The most flexible of programs is software. The program is created on paper and transferred to the system by means of a keyboard, punched or magnetic tape, or punched cards. The machine stores the instructions in some kind of random-access memory and is dedicated to the performance of the special task directed by the instructions in the RAM. While a particular program is in the RAM, the machine is just as dedicated to a special task as a hardwired logic system. However, all that is necessary to rededicate it to an entirely new task is to erase the RAM and load in a new program. Once a library of programs has been built up, it takes but minutes to convert the machine from the performance of one task to another.

Many microprocessors are used in so-called *dedicated* systems rather than for general-purpose computing. For example, suppose a microcomputer is to be used to control elevators. The same microcomputer can be used to control the elevators in any building by simply programming a ROM. Before the advent of the microprocessor the entire control system had to be designed and hand-wired to fit the different requirements of each building.

MICROPROGRAMMING

Microprogramming is a technique that may pave the way for simpler communications with the microprocessor and perhaps make new processor-oriented languages a reality. Traditional internal computer control systems are composed of random logic, counters, flip-flops, and so on. Machines using microprogramming have a computer within a computer. The internal control computer has a processor and a memory: It is a complete computer on a small scale. The user is not aware of the internal computer and has no access to it. The control computer executes the user's programmed instructions by executing a sequence of its own preprogrammed microinstructions. By simply altering a ROM program, the machine becomes a completely different computer, emulating some specific machine or having a highly specialized instruction set. Microprogramming ROMs may also be available as an off-the-chip device allowing the microprocessor architecture to be altered from the outside.

MACHINE-LANGUAGE PROGRAMMING

This is the programming level we used in the previous addition example. Machine language consists of groups of binary digits that direct the actions of the logic circuits controlling the transfer and handling of logic-level data within the machine. Machine-language instructions consist of two basic parts: the operation to be performed, and the address or addresses of the data (operands) to be operated upon. Because machine language consists entirely of many 1's and 0's, it is difficult for direct human use. A hierarchy of languages has therefore been developed to interface human language and machine language. Each successively higher level of language more closely approaches standard English.

In machine language, all instructions must be expressed in binary or a binary code. An instruction set consisting of nothing but 1's and 0's is tedious to work with and prone to frequent errors. Actual numerical addresses must be spelled out for every instruction and piece of data. Editing or error correction in machine language is difficult because a single change usually requires that the entire program be rewritten.

The advantages of machine language are its direct access to even the simplest of operations, and reduced memory requirements. The directness of machine language permits direct, efficient programming.

ASSEMBLY LANGUAGE

One step up in the language hierarchy is assembly language. In this language a single instruction is machine-translated from a mnemonic (memory aid) code such as ADD MOV(e)CLA (clear and add) into the appropriate binary code groups. The human programmer can make statements in easily remembered mnemonic terms. A specially programmed ROM is often used to convert mnemonic codes into appropriate machine-language groups of 1's and 0's. Extra memory is required to store the assembly-language rules.

Assembly language is closely related to the machine language of a specific machine except that mnemonic symbols are used instead of 1's and 0's to specify operations and memory addresses. Both machine and assembly languages are machine-dependent; that is, their construction is dependent upon the organization of the hardware of a particular machine.

PROCEDURE- (OR PROBLEM-) ORIENTED LANGUAGES

Higher-level languages, such as FORTRAN (FORmula TRANSlation), a mathematically oriented language, COBOL (COmmon

Business/Oriented Language), and BASIC, are not machine-dependent. A translator (which requires considerable memory space) interfaces these standard languages to any given machine. With an appropriate translator these high-level languages can be used without regard to the kind of machine involved. They are called *procedure-oriented* languages because they are concerned strictly with the problem, not with internal machine operation.

A characteristic of these procedure-oriented languages is that a single statement results in a number of machine-language instructions. The housekeeping functions of loading data into memory, keeping track of memory locations for data and instructions, and retrieving the data when required are handled automatically.

Many high-level languages (FORTRAN, COBOL, BASIC, and others) contain a vocabulary of key or reserved words that cause the system to act, along with a set of optional words that can be used to make written statements more easily understood by humans. These optional words are ignored by the machine and are for human use only.

Programming languages are carefully constructed artificial languages. While we might debate about the use of commas in a term paper, there is no room for debate in programming languages. The syntax and punctuation must be correct or the computer will refuse to act on a statement and will call the programmer's attention to an error. At the machine-language level, an incorrect 1 or 0 entry would be executed by the machine and the final results would be in error. Because programming errors are inevitable, many mistakes are avoided by using high-level languages.

SELECTING A LANGUAGE LEVEL

Although higher-level languages allow for easier programming, they do take up memory space. In dedicated control systems, higher-level languages increase the size of the memory required and may well drive the cost of the hardware out of the competitive range. A small BASIC language program normally requires 8000 bytes or more of memory to store the language itself. When the program is to be placed in ROM for a dedicated control system, an in-house computer that operates in a higher-level language can be used to *burn* the program into the ROM in machine language. The programmer can work in a high-level language to produce machine-language ROMs that can be plugged into the dedicated controller. In this case it is not necessary for the dedicated microprocessor to carry (on board) the extra memory to allow the use of a higher-level programming language.

15-6 MINI AND LARGE COMPUTERS

The minicomputer is a direct descendant of the large-scale machines. Minicomputers are often as fast as the older full-scale machines, but provide less memory and a more limited capability. They have been used for the same kind of data-processing tasks as larger machines, as well as some dedicated control functions. Until recently, minicomputers used very little in the way of LSI chips because of the slower speed of LSI devices. TTL devices are the most common sort of IC found in minicomputers. Many large computers are 16 to 64 bit machines. Minicomputers are mostly 16 to 32 bit machines. The minicomputer has traditionally contained far more memory space than the microcomputer, enough to comfortably allow for the use of high-level languages, without problems. In microcomputers the limited available memory space must be used carefully. The minicomputer ordinarily contains enough memory space that one need not be so conservative in using it. As a result, minicomputers are often easier to write programs for than microcomputers.

There is a trend toward replacing the large machine with several minis. As microcomputer speeds increase, it is probable that they will replace the minicomputer in many applications.

It is doubtful that the traditional large machines will vanish entirely, but the concept of distributed, local computing power is gaining favor. A large time-shared central computer suffers from the disadvantages of limited memory space. Customers tend to increase their demands for computer time and memory space as they become more sophisticated in computer use. Although the big machines have a very large memory and are very fast, customers tend to push large central computers up to their limits in relatively few years. Large facilities must be expanded in increments of a million dollars or so. Small, local mini or micro systems can be expanded in increments of one or a few thousand dollars.

There has also been a security problem with large central units. Access to secret records has proved all too easy and computer crime has become a serious problem. Distributed small computing facilities make computer abuse more difficult.

Answers to Odd-Numbered Problems

Page 33

1. 5×10^2 **3.** 2×10^4 **5.** 3×10^{-2} **7.** 4×10^4 **9.** 6×10^{-7}

Page 35

1. 4.9×10^4 **3.** 6.4×10^7 **5.** 4.9×10^2 **7.** 4.33×10^7 **9.** 5.213×10^8

Page 36

1. 1.1×10^{-4} **3.** 4.6×10^{-3}

Page 49

1. Most important points are 1 and 3. The curve changes direction. Places on the graph where there is little or no change in slope or direction are much less important.

Page 55 (scientific notation)

1. 4.5×10^5 **3.** 3×10^3 **5.** 4.9×10^4 **7.** 4.9×10^7 **9.** 1×10^{-2}
11. 3.9×10^{-7} **13.** 1.43×10^{-3}

Page 56 (multiply)

1. 1×10^7 **3.** 12×10^{13} (1.2×10^{14}) **5.** 10×10^{-4} (1×10^{-3}) **7.** $4 \times 10^0 = 4$

Page 56 (divide)

1. 2×10^3 **3.** $.5 \times 10^{-2}$ **5.** 2×10^2 **7.** 2×10^{-22}

Page 56 (unknown)

1. V = 4 **3.** R = 2 **5.** R = 12 **7.** R = 9 **9.** V = 15

Page 94

1. There are approximately 100 elements, including "man-made" elements.
3. No. But they do combine. For example, oxygen is normally found as O_2.
5. Yes
7. The atom is composed of a central nucleus containing protons and (generally) neutrons, with one or more shells of electrons orbiting it.
9. Positive
11. There are the same number of orbiting electrons as protons in the nucleus.
13. Valence electrons

15. Valence electrons in a metal are more easily freed from the atom than those in an insulator.
17. True   19. True   21. a
23. See the case of the surprised brakeman.
25. Electromotive force (voltage)
27. The difference in the number of electrons concentrated in two areas in a conductor
29. Potential difference. The force that moves electrons through a conductor.
31. 1 volt
33. True   35. b

Page 113

1. 0.1 amp   3. 1 amp   5. 10 ohms   7. 2000 ohms

Page 122

1. 6 ohms   3. 100 ohms   5. 12 volt system   7. a. 1.83 amp   b. 0.916 amps

Page 132

1. 35.99 ohms   3. 27.79 ohms   5. 24.81 ohms

Page 140

1. 0.2 amp   3. Approximately ½ horsepower   5. 2.25 watts   7. 10 watts

Page 152

1. The group of domains broken off remains aligned.
3. A domain is a group of atoms with like electron spins that form minature magnets.
5. The domains are randomly oriented in unmagnetized iron.
7. There are no poles until a magnetic substance is placed in the field.

Page 157

1. Air, wood, vacuum, aluminum
3. Voltage = MMF; current = flux; resistance = reluctance
5. A gap between two low-reluctance poles

Page 164

1. The solenoid is a coil with a movable iron core.
3. A fixed magnetic field attracts a movable magnet mounted on a rotating shaft. A commutator reverses the current when north and south poles approach, keeping the motor rotating.
5. The field magnet
7. To provide an electrical connection to the rotating armature coil
9. Series   11. Pulsating

Page 172

1. By electro-magnetic induction

ANSWERS TO ODD-NUMBERED PROBLEMS/CHAPTER 7   521

3. To provide a starting phase shift
5. To disconnect the starting winding as soon as the motor reaches its normal speed

Page 176

1. A permanent magnet field motor
3. To increase a meter's current range
5. It produces very little torque.

Page 197

1. **a.** 38 volts   **b.** 75 volts   **c.** 96 volts

Page 206

1. 90°, current leading
3. A line whose direction shows the phase angle and whose length shows the magnitude of voltage, resistance, reactance, or impedance
5. 132.5 ohms

Page 219

1. 90°, voltage leading.   3. **a.** 5000 ohms   **b.** 269 ohms
5. The voltage is nearly 10V. The current is near zero. Only a very small fraction of a time constant has elapsed.
7. 17.76 H

Page 230

1. X_c is equal to X_1 at the resonant frequency.
3. **a.** Controlling the frequency of a transmitter signal   **b.** Tuning a radio to a desired station   **c.** Selecting the proper channel frequency in a TV set
5. Frequencies outside of the bandwidth will be rejected. Those within it will be accepted.
7. See answer no. 5 above.   **9.** Your problem!

Page 256

1. **a.** Torque step-down, velocity step-up
   b. The ratio of pedal diameter to wheel diameter is approximately 5.5. Therefore the input torque is about 5.5 times greater than the output torque.
3. The input power must be slightly greater than the output power.
5. Output current is reduced.
7. $$\frac{\text{primary resistance}}{\text{secondary resistance}} = \frac{\text{number of primary turns squared}}{\text{number of secondary turns squared}}$$
9. To isolate equipment and people from electrical shock from a AC power line
11. Yes

13. To prevent sound cancellation that could produce *holes* in the sound coverage
15. Current
17. Voltage-down, current-up
19. The sense of feel (heat)
21. The current flow must be periodically interrupted.
23. Work = force × distance
25. $\dfrac{\text{power in}}{\text{power out}} \times 100$ = efficiency (percent)
27. Voltage

Page 274

1. Voltage gain = output voltage/input voltage
3. Power gain = output power/input power
5. No
7. An analog amplifier can amplify a continuously varying signal. Digital amplifiers are on-off (only) devices.
9. The output signal goes high when the input signal goes high. Input and output signals are in phase.

Page 278

1. Positive feedback
3. It always alters it; it may increase or decrease it.
5. Your problem!

Page 279

1. Negative feedback subtracts from the input signal.
3. Always dynamic   5. Positive
7. Bias is constant. Feedback varies with the signal.

Page 296

1. An element with exactly 4 electrons in the valence shell. Semiconductors require outside energy to free electrons for conduction.
3. The temperature of the crystal determines its conductivity.
5. Arsenic, phosphorus, and antimony
7. A dangling bond is a silicon covalent bond adjacent to a tri-valent dopant atom where there is no electron with which to pair. The dangling bond is called a hole in semiconductor pariance.
9. Adding dopant atoms makes conduction in semiconductors less temperature dependent and more voltage dependent.
11. An N-type semiconductor is doped with donor dopant atoms.
13. The reverse-biased junction consists of two conductive layers separated by the insulating depletion zone. Two conductors separated by an insulator is a capacitor.
15. The covalent bond is a complex electron sharing phenomenon that increases the amount of energy required to free electrons.
17. Donor is a 5-valent atom that donates an extra electron to the semiconductor crystal.
19. See Figure 8-19.

ANSWERS TO ODD-NUMBERED PROBLEMS/CHAPTER 10   523

21. A zone on both sides of the junction that has been depleted of both holes and electrons

Page 309

1. The depletion zone is made wide or narrow to control the channel width.
3. Making the grid increasingly negative forces more electrons back toward the cathode, thus preventing them from reaching the plate.
5. The no-current-flow condition in a field effect transistor
7. Can you?

Page 324

1. Bipolar = enhancement mode
3. Beta = collector current ÷ base current
5. 1.a   2.c
9. A steady-state input current that sets the desired no-signal collector current
11. Bias is necessary if an AC signal is to be amplified.
13. (1) In a current mode circuit, the signal feedback lowers the voltage gain drastically. (2) In a voltage mode circuit the input impedance is lowered excessively by the signal feedback.

Page 336

1. See Figure 9-22.
3. When thermally produced electrons are accelerated by the external battery, they collide with atoms, forcing more than one electron to freedom. These free electrons also have collisions, multiplying the number of free electrons.
5. The current must be momentarily interrupted to turn the device off.
7. The triac can be triggered into conduction in both directions. It can control full-cycle AC.

Page 347

1. a. Typically 20 Hz to 20,000 Hz   b. Fairly narrow bandwidth   c. Typically from near zero Hz to more than 4 MHz.
3. Distortion is any signal information that appears in the amplifier output that was not part of the original input signal.
5. The sine wave   7. The harmonic content

Page 348

1. A signal containing an infinite number of frequencies
3. Sparking points, gas discharges, brush arcing in motors, or any random electron motion
5. Yes

Page 360

1. a. *approximately* 53°   b. *approximately* 63°
3. At low (audio for example) frequencies the inductor becomes very large and expensive.
5. For you to think about

524 ANSWERS TO ODD-NUMBERED PROBLEMS/CHAPTER 11

7. The original signal is noise. The tuned circuit filters the desired frequency out of the noise signal.
9. To provide the proper phase (0° or 180°) for oscillation. The extra winding may also have an impedance-changing function.

Page 362

1. The process of impressing audio, video, etc., information on a radio-frequency carrier
3. The power output of the transmitter varies. The output frequency remains constant.
5. Because the human eye varies in its sensitivity to different colors

Page 366

1. A cooler with high thermal conductivity and large surface area. The heat sink carries the heat away from transistors and power op-amps.
3. A transformer

Page 372

1. It provides a signal (455 kHz in the broadcast band) above the incoming signal frequency and mixes the two to get the intermediate frequency.
3. (1) incoming signal frequency (2) oscillator frequency (3) oscillator f + incoming f (4) oscillator f − incoming f
5. The AM signal contains two identical but opposite phase signals. The diode passes only one of them.
7. Principal differences (sometimes) are operating frequencies and the kind of detector required.
9. Neon signs, auto ignition, fluorescent lamps, etc.

Page 378

1. Most electronic devices require direct current. The most available power source is the AC power line.
3. Steps the power line voltage up or down as required.
5. To smooth out the pulsating wave from the rectifier

Page 394

1. In *spontaneous omission*, electrons are dropped to the ground state by random heat energy. In *stimulated emission*, electrons are returned to the ground state by feedback photons. Only stimulated emission produces coherent light.
3. Solid state, gas, and liquid
5. The principle is the same but the semiconductor laser is tuned to produce coherent light.
7. Coherent
9. To prevent the gas from accidentally being driven into avalanche
11. Dark current, glow discharge, avalanche

Page 398

1. See page 396.
3. Bulk photocells change resistance when light strikes them.

ANSWERS TO ODD-NUMBERED PROBLEMS/CHAPTER 14   525

5. Cadmium sulphide, lead sulphide, and cadmium selenide are the most common materials used in bulk cells.
7. They respond too slowly to variations in light brightness.

Page 405

1. Its frequency response is poor (narrow bandwidth).
3. The loudspeaker uses the same principle as the electric motor.

Page 411

1. 9 dB

Page 428

1. To make it easy for people anywhere in the house to hear sound from the stage.
3. 1 dB for trained ear
5. To prevent stray energy from getting into the amplifier and producing hum
7. Microphone design and placement, speaker design and placement. Electronic graphic equalizers.
9. Echo chambers, springs, tubing delay lines, electronic

Page 441

1. 60   **3.** To reduce flicker   **5.** 15750 Hz
7. Synchronizing pulses   **9.** Frequency modulation
11. Conductive film, bulk photoconductor
13. Red, blue, green   **15.** 342,000
17. To form one plate of a capacitor to filter the high voltage power supply ripple
19. Color

Page 446

1. To select the desired channel
3. To adjust for strong and weak signals from the antenna
5. To move the beam on the picture tube from top to bottom
7. To disable the color circuits when a black and white picture is being received. This prevents annoying color snow.
9. To decode the color signals into brightness values for each of the three colors

Page 453

1. a. 12 or fewer   **b.** 12 to 100   **c.** 100 to 10,000
3. See Figure 14-7.

Page 459

1. a. See Figure 14-5.   **b.** See Figure 14-4.   **c.** See Figure 14-6.   **d.** See Figure 14-7.
3. 1.a,c   2.b,d

Page 470

1. The truth table is a compact but complete specifications list.

3. The ordinary OR provides a high output for the 1,1 combination. The exclusive-OR provides a zero output for the 1,1 combination.
5. 16 **7.** Binary adder

Page 476

1. See Figure 14-16. **3.** Counting circuits
5. To convert the binary count to some other form; for example, decimal or 7-segment readout forms

Page 484

1. Random-access memory, also called read-write memory
3. Read-only memory
5. To store permanent data for immediate computer access. ROM is not erased when the power is shut off.
7. Slower **9.** The flip-flop

Page 486

1. To synchronize all of the circuits in the system
3. The written programs for a computer
5. Your problem this time

Page 513

1. The bus provides a bi-directional path for data or instructions to travel between computer subsystems.
3. The ALU performs arithmetic and certain logic operations.
5. The memory address decoder decodes binary addresses to select specific memory cells.
7. Primarily to hold one of the operands for future action by the ALU
9. The instruction decoder decodes binary instructions to connect the appropriate devices to the bus. It also tells the ALU what operation to perform.
11. The bus drivers control the direction of data flow to or from the bus. They also disconnect those devices that are not to be used at a specific time.

Index

Accumulator, 495
Acoustical feedback, 418
Acoustical transformers, 414
Acoustics, 418
Addition:
 binary, 469
 with scientific notation, 40
Address decoder, 494
Algebra, 51–53
Alternating current:
 with capacitors, 197
 effective value of, 89
 frequency of, 90
 graph of, 88, 91
 in inductive circuits, 215
 instantaneous value of, 91
 Ohm's Law for, 201
 peak value of, 90
 period of, 90
 RMS value of, 90
 and transformers, 241
Alternating current generators, 163
Alternating current motors, 167
ALU, 486, 495
Ammeter, 107
Ampere, Andre, 17
Ampere, 71
 turns, 159
Amplification, 261
Amplifier(s):
 analog relay, 269
 audio, 340
 audio (in TV), 444
 bandwidths of, 340
 bias concept, 271
 characteristics, 262
 classifications, 338
 complementary symmetry, 363, 365
 defined, 261
 differential, 325
 distortion, 342
 gain formulas, 284
 IC audio system, 364
 input resistance of, 269
 intermediate frequency (in TV), 443
 inverting, 269
 loading and gain, 268

Amplifiers (cont.):
 negative feedback, 277
 noninverting, 269
 operational, 281
 positive feedback, 278
 practical J-FET, 325
 radio frequency, 367
 relay, 264
 sound IF (in TV), 444
 specifications of, 264
 summary of characteristics of, 320
 switching, 326
 video (in TV), 443
Amplifier circuits, 321
Amplifier systems, 363
Amplitude modulation, 362
AM Receivers, 367
Analog meter, 173, 176
Analogs:
 acoustical transformers, 415–16
 amplifier, 261, 266, 275
 capacitor, 189, 192
 capacitor microphone, 402
 electrical-magnetic, 154
 feedback, 275
 hydraulic, 106
 inductive, 211
 magnetic circuit, 156
 resistor, valve, 103, 105
 resonant circuit, 225–26
 sky diver (resistance), 79
 the suprised brakeman (speed of electrical current), 81
 transformers, 232–33, 235, 237, 240, 244, 251, 253
 voltage (mechanical, acoustical), 86
 voltage source, pump, 107
AND gate, 455
Antenna coil, 367
Antennas, 255
Aperture mask, 435
Aquadag, 432
Arithmetic logic unit (ALU), 486, 495
Armature, 160
 (see also Electric motors and Relay)

Assembly language, 516
Atoms, 59
 acceptor, 290
 atomic numbers of, 64
 donor, 289
 electrical stability of, 62
 periodic table of, 64
 silicon, 287
 structure of, 61
 symbols for, 65
Audio equipment, 423
Audio power, 411
Automatic gain control (AGC) in TV, 444
Automobile ignition circuit, 242
Autotransformers, 252
Avalanche, in SCR's (thyristors), 329

Ballast inductor, 391
Bandpass, 228, 230
Bandwidth, 228, 230
 of amplifiers, 340
Bardeen, John, 21–22
Base ten, 27
BASIC, 517
Bass reflex, 417
Bell, Alexander Graham, 18
Beta (β), 315
Bias, 319
 in amplifiers, 271
 in diodes, 294
 in J-Fets, 319
 in thyristors (4-layer diodes and SCRs), 332
Binary addition, 469
Binary Coded Decimal (BCD), 467
Binary coded hexadecimal, 466
Binary Coded Octal (BCO), 467
Binary counter, 475
Binary number system, 464
Bipolar logic, 451
Bipolar transistor, 309, 312–13
Boolean algebra, 462
Brattain, Walter, 21–22
Bridge rectifier, 375
Brushes, motor, 162
Bubble memory, 484
Bulk photo cells, 398
Bus, computer, 489

Camera tube (TV), 434
Capacitance, in a PN junction, 296
Capacitive reactance, 200
Capacitors, 185
  and alternating current, 197
  analogs of, 189
  construction of, 187–88
  counter voltage in, 190
  dielectric constants and materials for, 187
  impedance in capacitive circuits, 203
  microphone, 402
  in parallel, 188
  phase shift in, 198
  picture tubes, 432
  reactance of, 200
  in series, 188
  start motors, 169
  time constants, 195
Carbon microphone, 400
Cathode, vacuum tube, 307
CCD memory, 484
Celsius-to-Fahrenheit conversion, 45
Centimeter, 42
Channel, in FETs, 299, 301, 304
Charging capacitors, 191
Charge-coupled memory, 484
Chemical compounds, 60
Chrominance, in color TV, 440
Circuit(s):
  automotive ignition, 242
  capacitive, 198–207
  complementary, symmetry, amplifier, 363, 365
  differential amplifier, 325
  digital binary adder, 469
  digital counters, 475
  digital decoder, 475
  digital multiplexer, 476
  digital shift registers, 475
  feedback oscillators, 356
  flip-flops, 471
  half-wave power supply, 374
  integrated, 450
  J-FET amplifier, 325
  magnetic, 153
  open, 125
  operational amplifier, 283
  parallel, 115, 123
  phase shift oscillator, 359
  power supply, 374
  practical transistor, 322
  regulator, 376
  relaxation oscillator, 352
  resonant, 222
  SCR lamp dimmer, 379
  series, 115

Circuits (cont.):
  series-parallel combined, 133
  short, 125
  switching amplifier, 378
  triac lamp dimmer, 381
Circuit breaker, 126
Clock, digital, 485
Clock, microprocessor, 496
COBOL, 517
Codes, binary, 466
Coherent light, 385
Coils (see Inductors)
Color killer (TV), 445
Color oscillator (TV), 445
Color synch circuit (TV), 445
Commutation, in SCRs, 379
Commutator, motor, 161
Complementary symmetry amplifier, 363, 365
Composite video signal (TV), 442
Computer programming, 513
Computers, 487
Condenser microphone (see Capacitor microphone)
Conductance, mutual, 301
Conduction, gas discharge, 390
Conductivity, in gases, liquids, metals, 67–68
Conduit, 182
Conservation of energy, 137
Conventional current, 71
Conversions, metric-english, 42, 44–45
Coulomb, 78
Counters, digital, 475
Covalent bond, 69, 287
Crystal lattice, 67
Crystal microphone, 401
Crystal oscillator, 355
Current:
  in capacitive circuit, 202
  conventional, 71
  electron, 70
  in series circuits, 116
  velocity analog of, 211
  water analogy of, 106
Current gain, 267
  in bipolar circuits, 321
  in transistors, 315
Cut-off in vacuum tube, 308
Cycles per second (see Hertz)

Dangling bond, 288
DC (direct current), 87
Decibel (dB), 408
  table of relationships, 410
  table of relative loudness, 410
Decoders, digital, 475
Deflection yoke, 432
"D" flip-flop, 471

De Forest, Lee, 21
Depletion mode, FETs, 301
Depletion zone, 291
Detector, video, 443
Diac, 381
Dielectric constants, 187
  of silicon PN junction, 296
Dielectric materials, 187
Differential amplifier, 325
Diffusospheres, 419
Digital clock, 485
Digital computers, 487
Digital counters, 475
Digital decoder, 475
Digital display, seven segment, 477
Digital logic, 449
Digital memory, 478
Digital shift registers, 475
Digital systems, hierarchy, 453
Diode:
  forward-biased, 294
  light-emitting, 393
  light-sensitive, 397
  PN junction, 291
  reverse-biased, 294
  varactor, 296
DIP, 459
Direct Current (DC), 87
  and transformers, 242
Direct current generator, 164
Direct current motors, 160
Discharge, in capacitors, 191
Disc recording, 420
Display, seven segment, 477
Distortion, 342
Division, with scientific notation, 40
Domains, magnetic, 146, 149
Donor atoms, 289
Doping in semiconductors, 290
Drain, in field-effect transistors, 299
Dual inline package (DIP), 459
Duty cycle, 329
Duty cycle control, 379
Dynamic microphone, 401

Echo methods, 425
Edison, Thomas A., 20
Efficiency, in transformers, 238
Electrical conduction, in gases, liquids, metals, 66–68
Electrical current, 70
Electrical hardware, 182
Electricity, defined, 70
  speed of, 81
Electric motors:
  armature of, 160
  brushes in, 162

INDEX 529

Electric motors (cont.):
  commutator of, 161
  DC, construction of, 164
  field coil of, 161
  fractional horsepower AC, 167
  series and shunt wound, 163
  shaded pole, 169
  slip in, 169
  squirrel cage, 167
  starting methods, 169
  theory of operation, 161
Electromagnets, 148
  with iron core, 151
  left-hand rule for, 149
  solenoid, 151
Electromotive force (EMF), 66
  analogs of, 86
Electron gun, 432
Electron-hole pair, 288
Electrons, excited, 386
Elements, 59
  atomic numbers of, 64
  electrical stability in, 62
  periodic table of, 64
  symbols for, 65
Emission, spontaneous, 385
Emission, stimulated, 385
Enclosure loudspeaker, 416
Energy, defined, 58
  conservation of, 237
English-to-metric conversions,
    42, 44, 45
Enhancement mode, FETs, 304
Equations, algebraic, 52
Exclusive-OR gate, 469
Exponents, 26, 27
  zero, 29

Fahrenheit-to-Celsius
    conversion, 45
Families, logic, 452
Faraday, Michael, 17
Feedback:
  acoustical, 418
  negative, 277
  positive, 278
Feedback network, 354
Feedback oscillators, 353
Feedback regulator, 376
Field coil, 161
Field, in TV, 438
Field, magnetic, 148
Figure of merit, 186
Filament, vacuum tube, 305
Filter, power supply, 374
Firmware, 486
Fleming, John A., 20
Flicker, in TV, 438
Flip-flop applications, 474

Flip-flops:
  clocked, 473
  type D, 471
  J-K, 472
  type T, 473
Fluorescent lamp, 391
Flux, magnetic, 153
FM, 362
FM receivers, 370
Folded horn, 417
Force, 58
Formulas, transformer, 245
FORTRAN, 516
Forward-biased junction, 294
Four-layer diode, 330
Fractions, 29, 31
Frame, in TV, 438
Franklin, Benjamin, 16
Frequency modulation, 362
Full-wave rectifier, 375
Fuse, 126

Gain, in amplifiers, 267
Galvani, Luigi, 17
Gas discharge tube, 389
Gas, inert (noble), 63
Gas laser, 393
Gate, in field-effect transistors,
    299, 302
Gates (see Logic gates)
General register, 495
Generators, 163
  alternating current, 165
  direct current, 165
  output waveforms, 165–66
  slip rings in, 165
Glow discharge tube, 389
Gram, 42
Graphs, 46
  linear, 48
  nonlinear, 48
  plotting, 48
  polar, 50
  rectangular, 50
Grid, in vacuum tube, 307

Half-wave rectifier, 374
Harmonics, 341
  and waveforms, 344
Heat and conduction in silicon,
    288
Heater, vacuum tube, 305
Heat sinks, 232, 254, 365
Heat transformers, 253
Henry, Joseph, 17
Hertz (Hz), defined, 90
Heterodyne, 346
Hexadecimal numbers, 465
High-voltage power supply (TV),
    445

Holding current in SCRs, 333
Hole, 288
Horizontal sweep oscillator (TV),
    445
Horn, folded, 417
Horns, muscial, 415
Horsepower, 137
House wiring, 181
Hue (in color TV), defined, 440
Hysteresis, in four-layer diode
    (thyristors), 333
  magnetic, 147

If amplifier, in TV, 443
Ignition, automotive, 242
Impedance, 203, 219
Impedance ratios, in
    transformers, 248
Incoherent light, 385
Incandescent lamp, 387
Interface devices,
    microprocessor, 496
Induction motors, 167
Inductive reactance, 215
  inertia analog of, 213
Inductors, 208
  and alternating current, 215
  analogs of, 211
  commercial types, 210
  and mass, 212
  time constants with, 214
  and vectors, 204
Inert gases, 63
Inertia, 213
Input resistance, in amplifiers,
    269
Instruction decoder, 496
Instruction register, 496
Integrated circuit
    (photomicrograph), 461
Integrated circuits, 450
Interfacing logic families, 452
Interlace, in TV, 439
Intermodulation distortion, 343
Inverting amplifiers, 269
Ions, 66
Isolation transformer, 252

J-FET, 299
  drain element, 299
  gate element, 299
  pinch-off in, 299
  source element, 299
  theory of operation, 299
Junction field-effect transistor
    (see J-FET)
Junction, PN, 291
Junction resistance, 315
Junction transistor, 309

530 INDEX

Kilogram, 42
Kiloliter, 42
Kilometer, 42

Lamp dimmer circuit, 379
Large-scale integration (LSI), 451
Laser, 383
　gas, 393
　ruby, 392
　semiconductor, 393
　theory of operation, 386
Lattice, crystal, 67
LDR, 398
Leakage current, transistor, 315
Leakage current, in thyristors, 329
Leakage flux, magnetic, 155
LED, 393
LED display, 477
Left-hand rule, 149
Leyden jar, 16
Light:
　coherent, 385
　incoherent, 385
　sources of, 383
　spectrum (graph), 384
Light-dependent resistors, 398
Light-emitting diode, 393
Light sources, 383
Linear graphs, 48
Liter, 42
Loads, in amplifiers, 268
Local oscillator, 369
Logic:
　bipolar, 451
　digital, defined, 449
　families of, 452
　family interfacing, 452
　MOS, 451
　programmable, 485
Logic families (TTL, P-MOS, C-MOS, I²L, ECL), 452
Logic gates:
　AND, 455
　exclusive-OR, 469
　inverter, 456
　NAND gate, 457
　OR gate, 456
　TTL, 457
Logic, levels in, 459
Loudspeakers, 402
Luminance, in color TV, 440

Machine language, 415
Magnetic:
　air gap, 154
　analogs, 154
　attraction, 147

Magnetic (cont.):
　circuit analogs, 156
　circuits, 153
　domains, 146, 149
　fields, 146
　flux, 153
　force, 153
　hysteresis, 147
　leakage flux, 155
　materials, 146, 154
　Ohm's Law, 155
　poles, 146
　reluctance, 153
　repulsion, 147
Magnetic machines, 158
　generators, 163
　electrical meters, 172
　electric motor, 160
　relay, 160
　solenoid, 159
Magnetism, 145
　and electrons, 145
Magnetomotive force (MMF), 153
Magnets, 143
　bar, 149
　electromagnets, 148
　shapes of, 155
Mass, 212
Medeleev, Dmitri, 63
Medium-scale integration (MSI), 451
Memory:
　bubble, 484
　charge-coupled, 484
　digital, 478
　magnetic core, 479
　mechanical, 479
　microprocessor, 494
　punched cards, 481
　punched paper tape, 481
　random access (RAM), 480
　Read only (ROM), 482
Metal oxide field-effect transistors (see MOS-FET)
Meters, electric (ammeter), 17
　moving coil, multimeters, ohmmeter, voltmeter, watt-hour, 174
Metric system, 41
　metric-English conversions, 42, 44, 45
　prefixes, 43
Microcomputer, 497
Microphone, 398
　carbon, 400
　crystal, 401
　dynamic, 401
　velocity (ribbon), 402

Microprocessor:
　accumulator, 495
　address decoder, 494
　arithmetic logic unit (ALU), 495
　clock, 496
　general register, 495
　how it works, 490
　instruction decoder, 496
　instruction register, 496
　memory, 494
　peripheral equipment, 496
　peripheral interface devices, 496
　program counter, 495
　subsystems, 494
Microprogramming, 515
Milligram, 42
Milliliter, 42
Millimeter, 42
Minicomputers, 518
Modulation, 361
　AM/FM, 362
Morse, Samuel F. B., 18
MOS-FET:
　depletion mode, 301
　enhancement mode, 304
　theory of operation, 301
MOS-LOGIC, 451
Motors (see Electric motors)
Multiemitter transistor, 459
Multimeter, 174
Multiplexer, digital, 476
Multiplication:
　of fractions, 38
　with scientific notation, 36
Mutual conductance (Gm), 300, 302, 310

NAND gate, 457
Negative feedback, 277
Neon lamp, 391
Neon lamp oscillator, 352
Noise, electrical, 347
　in receivers, 371
Noninverting amplifiers, 269
Nonlinear graphs, 48
N type silicon, 289
Number systems, 464

Octal numbers, 467
Octave, defined, 338
Oerstead, Hans, 14, 17
Ohm, defined, 79
Ohm, Georg S., 18
Ohmmeters, 174
Ohm's Law, 112, 118
　for alternating current, 201
　magnetic, 155

INDEX 531

Open circuit, 125
Operational amplifier, 281
  gain formulas for, 284
Optical spectrum, 384
OR gate, 456
Oscillator, sweep (in TV), 437
Oscillator-mixer, 369
Oscillators:
  crystal, 355
  feedback, 353
  four-layer diode, 352
  hydraulic relaxation, 351
  relaxation, 350
  requirements for oscillation, 353
  in TV, 438
Overtones, 341

Parallel circuits, 123
  (See also Ohm's Law)
Parallel resistance chart, 131
Parallel resistors, 126
Parallel-series circuits, 133
Peripheral equipment, 496
Peripheral interface devices, microprocessors, 496
Periodic table of the elements, 64
Phase angle, 207
Phase control, in SCRs and triacs, 379
Phase shift, in capacitors, 198
Phase shift oscillator, 359
Phasing in transformers, 251
Phasors, 204
Photo cell, 395
  photoemissive, 397
  resistive, 398
Photo diodes, 397
Photo transistors, 397
Picture tube, television, 431
Pinch-off, 299
Plate, vacuum tube, 307
PN diode conduction curve, 331
PN junction, 291
  capacitance, 294
  depletion zone, 291
  dielectric constant, 296
  diode, 291
  forward-biased, 294
  reverse-biased, 294
  transition zone, 291
PNPN devices (see Thyristors)
Polar coordinate graph, 50
Poles, magnetic, 146
Polycylinder, 419
Positive feedback, 278
Potential difference, 74
  analogs of, 86
  direct current, 87
  sources of, 77

Power:
  audio, 411
  calculations, 138
  commercial distribution of, 176
  consumed by common devices (table), 139
  electrical, defined, 137
  in the home, 179
  horsepower, electrical equivalent, 137
  related to work, 237
Power gain, 267
Powers of 10, 26
Power supplies, 372
  bridge, 375
  dual voltage, 378
  filtering in, 374
  full-wave, 375
  half-wave, 374
  waveforms in, 374–75
Preamplifier, 363
Prefixes, metric, 43
Prism, in color TV, 437
Program counter, 495
Programmable logic, 485
Programming:
  assembly language, 516
  computer, 513
  language level selection, 517
  machine language, 515
  microprogramming, 515
  procedure-oriented language, 516
Programming languages (BASIC, COBOL, FORTRAN), 517
P type silicon, 290
Punched cards, 481
Punched paper tape, 481
Push-pull, 363

Q, 223
Quartz crystal, 355

Radiation, coherent, 385
  incoherent, 385
Radio frequency amplifier, 367
Radio receivers, 367
Random access memory (RAM), 480
Ratios, in transformers, 241
Reactance, capacitive, 200
Reactance formulas, 218
Read only memory (ROM), 482
Reciprocal, 29
Recording, disc, 420
Recording heads, 421–22
Recording, magnetic tape, 422, 424

Recording, optical, 423
Records, disc, manufacture of a, 421
Rectangular coordinate graph, 50
Rectifiers (bridge, full-wave, half-wave), 374–75
Reflected load, 246
Regulators, feedback, 376
  zener diode, 376
Relay, 160
Relay amplifier, 264
Relaxation oscillators, 350
Reluctance, magnetic, 153
Resistance, 78
  friction analog, 106
  junction, 315
  ohm defined, 79
  water analog, 106
  in wire, 105
Resistors:
  color code for, 99
  commercial types, 99
  light sensitive, 398
  omega symbol (Ω), 103
  in parallel, 126
  parallel resistor chart, 131
  standard RETMA values, 104
  in series, 116
  symbols for, 99
  tolerance in, 101
  valve analog, 103, 105
  variable, 99
  wattage ratings, 98
Resonance, 222
Resonant circuit analogs, 225, 226
Reverberation systems, 425
Reverse-biased junction, 294
Reverse-bias leakage current (thyristors), 329
Ribbon microphone, 402
RMS, 90
ROM, 482
Ruby Laser, 392

Saturation, in color TV, 440
SCR, 326, 333, 379
  case styles, 334
  lamp dimmer, 379
Scanning, in TV, 437
Scientific notation, 26, 29
  with addition, 40
  with division, 40
  in fractions, 38
  with multiplication, 36
  standard form of, 33
  with subtraction, 40
Second harmonic distortion, 343

532 INDEX

Semiconductors, 285
  covalent bonds in, 287
  defined, 63
  heat effects in, 286
  junction, 291
  laser, 393
  silicon crystals, 287
Series circuits, 115, 118
Series, motors, 163
Series-parallel circuits, 133
Shadow mask, 435
Shaded pole motor, 169
Shielded cable, 424
Shift registers, 475
Shock hazard, in TV, 432
Shockley relationship, 315
Shockley, William, 21–22
Short circuit, 125
Shunt, motors, 163
Silicon:
  conduction in, 288
  conduction summary for, 292
  doping, 290
  electron-hole pairs in, 288
  N type, 289
  PN junction, 291
  P type, 290
Silicon controlled rectifier, 326, 333
  commutation of, 379
Silicon crystals, 287
Silicon solar cell, 395
Sine wave (curve), 90–91
Slip (motors), 169
Slip rings, 164–65
Small-scale integration (SSI), 451
Software, 486
Solar cell, 395
Solenoid, 151, 159
  field strength of, 159
Sound, in TV, 442
Source, in field-effect transistors, 299, 302
Speaker enclosures, 416
Speakers, 402
Spectrum diagram, musical instruments, 346
Spontaneous emission, 385
Squirrel cage motor, 167–68 171
Stability, in bipolar circuits, 324
Stability factor, 320
Standard form, 33
Stand-off ratio, in UJTs, 335
Stimulated emission, 385
Substrate, 301
Subtraction, with scientific notation, 40
Switching amplifiers, 326, 378

Synch circuit, color TV, 445
Synchronization, in TV, 440
Synch separator circuit (TV), 444

Tank circuit, 223
Television, 430
  camera tube for, 434
  color definitions, 440
  how images are formed, 430
  picture tube, 431–32
  primary colors in, 432
Television receiver, 441
  block diagram of, 443
Temperature
  (Celsius-Fahrenheit), 45
"T" flip-flop, 473
Thyristors, 328
  theory of operation, 329
  (See also SCR)
Time constants:
  graph of, 196
  resistor-capacitor, 195
  resistor-inductor, 214
Torque, 212
Torque transformers, 243
Transformers, 232
  acoustical, 414
  and alternating current, 241
  auto-, 254
  and direct current, 242
  efficiency of, 238
  failures in, 252
  formulas for, 245
  gear analog, 244
  heat, 253
  impedance ratios in, 248
  intermediate frequency, 369
  isolation, 252
  multiple winding, 244
  phasing of, 251
  reflected load in, 246
  step-up and step-down ratios, 240
  torque, 243
  what they do, 235
Transistor, J-FET, 299
  MOS-FET, 301
Transistor, bipolar, 309
Transistor, bipolar, specifications, 315
Transistor circuits,
  base bias, collector feedback, 322
  base bias, emitter feedback, 322
  emitter bias, emitter feedback, 323
Transistor, first one (photo), 22
Transistor, light sensitive, 397

Transistor-transistor-logic (TTL) 452
Transistors, reverse polarity types, 313
Transition zone, 291
Triac, 329, 334
  dimmer circuit, 381
Triode vacuum tube, 307
Truth tables, 455
TTL logic family, 452
TTL logic gates, 457
Turns per volt, 245

Unijunction transistors (UJT), 335
  in dimmer circuit, 379
Universal time constant graph, 196, 216

Vacuum phototube, 397
Vacuum tubes, 304
Valence electrons, 63
Van Musschenbroek, Pieter, 16
Varactor diode, 296
Vectors, 204
Velocity, current analog, 211
Velocity (ribbon) microphone, 402
Vertical sweep oscillator (TV), 444
Video amplifier, in TV, 443
Video detector, in TV, 443
Video signal, in TV, 442
Vidicon, 435
Volta, Alessandro, 17
Voltage, 72
  analogs of, 86
  sources of, 77
  water pressure analog of, 106
Voltage gain, 267
  in bipolar circuits, 323
Voltage regulators, 376
Voltmeters, 174

Watt-hour meter, 174
Waveforms:
  and harmonics, 344
  of common musical instruments, 346
Wire nut, 182
Wire, resistance in, 105
Work, defined, 58

Yoke, picture tube, 432

Zener diode, 376
Zener knee, 331
Zero crossing distortion, 342
Zero exponent, 29